Media Art and the Urban Environment

FUTURE CITY Volume 5

Future City Description

As of 2008, for the first time in human history, half of the world's population now live in cities. And with concerns about issues such as climate change, energy supply and environmental health receiving increasing political attention, interest in the sustainable development of our future cities has grown dramatically.

Yet despite a wealth of literature on green architecture, evidence-based design and sustainable planning, only a fraction of the current literature successfully integrates the necessary theory and practice from across the full range of relevant disciplines.

Springer's *Future City* series combines expertise from designers, and from natural and social scientists, to discuss the wide range of issues facing the architects, planners, developers and inhabitants of the world's future cities. Its aim is to encourage the integration of ecological theory into the aesthetic, social and practical realities of contemporary urban development.

More information about this series at http://www.springer.com/series/8178

Francis T. Marchese
Editor

Media Art and the Urban Environment

Engendering Public Engagement with Urban Ecology

Springer

Editor
Francis T. Marchese
Department of Computer Science
Pace University
New York, NY, USA

ISSN 1876-0899 ISSN 1876-0880 (electronic)
Future City
ISBN 978-3-319-15152-6 ISBN 978-3-319-15153-3 (eBook)
DOI 10.1007/978-3-319-15153-3

Library of Congress Control Number: 2015936191

Springer Cham Heidelberg New York Dordrecht London

Cover image credits: Keith Deverell (left) and Josh Hill Photography (right)

Printed on acid-free paper

Springer International Publishing AG Switzerland is part of Springer Science+Business Media (www.springer.com)

Preface

More than half the residents of the globe now live in cities, and must strive to flourish in densely crowded urban environments characterized by inequality, social conflict and enormous potential for growth and change. New media artists working in urban settings are stimulating awareness, altering our perceptions of, and relationships with, urban space though their digital inventions and interventions. These innovations can be used in the development of future practices for helping residents, not only cope with the daily challenges created by urban environment, but also flourish despite the inequality, spatial density, social conflict, and disorganization that are symptomatic of cities.

The purpose of this book is to formally appraise the ways new media artists engage urban ecology. It gathers together essays from artists, architects, urban planners, and critical theorists to consider these new modes of seeing, representing, and connecting within the urban setting. What makes this book unique is that it spotlights artists instead of technologists as innovators and agents of technological change. Because these individuals create socially motivated artwork, they possess a clear understanding of the societal issues and values they wish to address. And because these artists know how to exploit state-of-the-art technology, they can create artworks that transcend the technology's original purpose, thus expanding the language of environmental engagement.

For the purpose of this volume urban ecology may be defined in a two-fold way.[1] From the natural science perspective, urban ecology investigates the biological structures and environmental processes that support them in urban spaces. Here, urban ecology focuses on plant populations, animal communities, and environmental effects, including human, on them. A complementary definition takes an anthropocentric view where humans are at its center. The field of urban

[1]Endlicher, Wilfried, Marcel Langner, Markus Hesse, Harald A. Mieg, Ingo Kowarik, Patrick Hostert, Elmar Kulke et al. "Urban Ecology-Definitions and Concepts." *Shrinking Cities: Effects on Urban Ecology and Challenges on Urban Development* (2007): 1–13.

ecology then expands its emphasis to the improvement of living conditions for human residents as part of an augmented characterization of an ecosystem that includes theories from social and planning sciences.

This volume is composed of 14 chapters distributed across three general areas: urban ecology and its engagement, reimagining and transforming the city, and the view from the streets.

In the first chapter, "Toward an Ecological Urbanism: Public Engagement in Contemporary Art Practice," Maria Michails surveys a diversity of artists working within the urban environment, considering artworks and projects designed to inspire community awareness and stewardship of urban ecological systems. In Chap. 2, "Exploring Environmental Stewardship Through Data-Driven Practices," Tega Brain and Jodi Newcombe expand on Michails survey by exploring the strategies that artists and designers employ to engage urban audiences. In particular, they assess how data from smart cities may be used to create artworks that can recast residents' understanding of urban space. Grisha Colemana and Daragh Byrne follow with a discussion of how arts-driven processes can be used to bridge artistic practice and research in "Experiential Ecologies: A Transdisciplinary Framework for Embodiment and Simulacra," by considering how varied forms of ecological information can be combined to create foundations for exhibition, engagement and performance. These chapters are complemented by two chapters in which artists discuss their work. In Chap. 4, "Uncultivated: an Evolutionary Drama in the Urban Environment," Lynn Cazabon explores how her public art project documenting wild plants in urban landscapes with geo-referenced photographs and public displays increases awareness of overlooked plant life. In "Alone Together in the Dark: Horror Based Artworks and Fan Participation in Urban and Extra-Urban Space," Jillian McDonald recounts several projects in which the performances of horror fans in urban and rural stimulates new awareness of, and conversations about, the environment and society.

Urban information, its representation, perception, and use are important issues in the transformation of urban ecology. In Chap. 6, "Mobile Maps of Chamaleonic Cities: Urban Cartographies and Methodological Procedures and Experiences," by Pedro Marra and Carmen Aroztegui Massera, explore how temporality and spatiality are constructed in contemporary cities. By investigating the flow of capital, the staging of mega urban events such as Brazil's World Cup, and the use of artistic interventions, they propose methods for better understanding urban mobility and enabling greater residential participation in its evolution. Alice Arnold follows in Chap. 7 with "Electric Signs," a discussion of the background behind her documentary film of the same name. At its core, the chapter considers urban space, the transformation of the ecological dynamics of major international cities though corporate installation of large LED displays, and the socio-political ramifications of these transformations. The transformation of urban space is reflected upon as well in Chap. 8 by Annette Weintraub. "Overload/Absence: The Collapse of Space to Surface in Representations of Urban Space" considers the rapid diminishment of urban public space through gentrification and privatization, and its transformation through mediazation. By comparing hyperactive urban media zones with the numbingly bland façades of contemporary urban architecture, Weintraub elucidates the perceptual shift from urban space to surface, and explores how surfaces have been utilized to generate new representations of urban space. In "Design as Topology: U-City," Ulrik Ekman further explores the dynamic transformations of urban space by discussing the design of a busy traffic intersection in the South Korean u-city (ubiquitous city) Songdo as a topological problem. The emergence of u-cities or smart cities creates challenges and opportunities for an emergent, mobile, and globally connected citizenry, and the technological substrate that must be developed to support it. With a focus on the nature of a single traffic intersection, Ekman ferrets out the layers of design issues anticipated to support the dynamic flow of information and flow of people, focusing on a critical comparative discussion of a variety of ontological and epistemological approaches to design as topology, including cultural theory and technical studies. In Chap. 10, "The Emergent City: 2004–2012," the artist Stanza complements Ekman's presentation by discussing how his visual artworks are informed by critical analysis of city spaces. As an artist who designs urban experiments and gathers data through networks of sensors and video cameras, he captures the dynamic patterns of urban dwellers and re-imagines them as information visualizations.

The remaining four chapters consider street art and its ability to engage urban residents. Francis Marchese's chapter, "The Art of Urban Engagement," reviews how digital media artists have exploited pervasive technologies to explore urban ecology. By their interventions, they challenge a city's inhabitants to reconsider their daily urban experiences. Vaughn Whitney Garland discusses how new online community collaborations transform urban sites into spaces for new art in Chap. 12, "Our Place on That Wall: Community Online Art Projects." In particular, he focuses on the nature of online community art projects as works orchestrated by artists exercising the interconnected and participatory nature of the Internet. Brian A. Brown argues in "Digitized Street Art," that 'street art' is an ephemeral means of artistic expression dependent upon the whim of the urban environment for its display, and relies on digital technologies for its documentation and dissemination, thus transforming urban street artists into digital artists. Finally, the artist Malin Abrahamsson considers her temporary public art installation, "Solar Cycle 24: 15 Nightly Projections," which emphasizes place-making and digital aesthetics in the urban environment. Installed in a storefront, the vibrant, projected visualizations engaged New York City residents through a simulation of the aurora borealis, a phenomenon visible around the North and South Poles.

New York, NY, USA Francis T. Marchese

Acknowledgements

I am extremely grateful to have had the opportunity to work with the contributors to this volume throughout its preparation. Their insights, energy, and creativity have contributed to the creation of a book which I hope readers will find both stimulating and thoughtful. I would also like to thank the editors at Springer-Verlag, especially Simon Rees and Wayne Wheeler, for their support of this project.

Contents

Contributors

Malin Abrahamsson Independent Artist

Alice Arnold A2Studio.org, New York, NY, USA

Tega Brain Purchase College, State University of New York, New York, NY, USA

Brian A. Brown Department of Communication Media & Film, University of Windsor, Windsor, ON, Canada

Daragh Byrne School of Architecture, Carnegie Mellon University, Pittsburgh, PA, USA

Lynn Cazabon Visual Arts Department, University of Maryland Baltimore County, Baltimore, MD, USA

Grisha Coleman School of Arts, Media + Engineering, Arizona State University, Tempe, AZ, USA

Ulrik Ekman Department of Arts and Cultural Studies, University of Copenhagen, København S, Denmark

Vaughn Whitney Garland Independent Curator and Multimedia Artist, Richmond, VA, USA

Francis T. Marchese Department of Computer Science, Pace University, New York, NY, USA

Pedro Marra Centro de Convergência de Novas Mídias – UFMG, Fafich, Pampulha, Belo Horizonte, MG, Brazil

Universidade Federal Fluminense, Icaraí, Niterói, RJ, Brazil

Carmen Aroztegui Massera Centro de Convergência de Novas Mídias – UFMG, Universidade Federal de Minas Gerais, Belo Horizonte, Brazil

Jillian McDonald Art Department, Pace University, New York, NY, USA

Maria Michails Treia Studios, USA

Jodi Newcombe Carbon Arts, industry partner with the Centre for Art Society and Transformation, Royal Melbourne Institute of Technology, Melbourne, VIC, Australia

School of Design, Queensland University of Technology, Brisbane, QLD, Australia

Stanza Independent Artist, London, UK

Annette Weintraub Department of Art, City College of New York, New York, NY, USA

Chapter 1
Toward an Ecological Urbanism: Public Engagement in Contemporary Art Practice

Maria Michails

Abstract This survey essay presents the artworks of ecologically engaged artists working within the context of the urban environment and its relationship to broader bioregions. The author reflects on the interdependency of urban infrastructure on living systems and their services, focusing on artworks that align with this ethic and weaving them within three areas considered imperatives toward an ecological urbanism. The projects vary in their approaches to engage communities toward awareness and stewardship of the natural world through an understanding of ecological systems.

Introduction

> Imagination is an artist's greatest asset. It can produce bold visions of what a sustainable future might be like. People can be moved and aroused by powerful environments, innovative designs, and practical demonstrations of active engagement.[1]

I've been trying to imagine what it would be like to live in an ecological city. These two words, "ecological" and "city," seem rather contradictory. What constitutes an ecological city? Who sets the benchmarks? What models do we have to go by? I know what I would like to see and experience in such a city, but I sense there is more involved and much more at stake than wanting plenty of green space, public transportation, and bike lanes. Even these fundamental criteria are sorely lacking in many American cities.

The term "ecological urbanism" has become ubiquitous in design and architectural circles these days. Implementing its tenets, though, requires considerable upgrades to systemic infrastructure beyond green buildings and green spaces, although these are important aspects to reducing greenhouse gasses and are an encouraging start. Planning the ecological city of the twenty-first century takes into

[1] Victor Margolin, "Reflections on Art and Sustainability," 28.

M. Michails (✉)
Treia Studios, USA
e-mail: maria@treiastudios.net; http://www.treiastudios.net

F.T. Marchese (eds.), *Media Art and the Urban Environment*, Future City 5,
DOI 10.1007/978-3-319-15153-3_1

consideration more than the systemic urban needs of a city, as has been the directive in recent history. It also considers the city within a larger set of ecosystems within a region, a country, a continent, and the planet as a whole, as well as the diverse groups within that urban center. There are more of us now living in cities than in rural areas, and this is expected to grow as populations become displaced and mobilized or simply need to migrate toward urban centers for economic reasons, placing greater strain on cities with aging infrastructures.

The newest reports from the IPCC (Intergovernmental Panel on Climate Change) have now shifted focus from reporting on the physical science to informing about adaptation in the face of climate change.[2,3] Have we gone beyond the tipping point? Clearly, the strain that cities place on resources and their impacts on ecosystems put the responsibility squarely on the shoulders of policymakers and planners who have the power to put forth the changes needed. "Ecological design in the urban context faces a dual challenge of meeting ecological imperatives and negotiating meaningful expressions for the coexistence of urban infrastructure, human activities, and ecological processes."[4]

To effect widespread acceptance of this imperative, a shift in perception is needed and a recognition that a hybrid form of processes could propagate a move away from the conventional, essentially antagonistic, and binary notions of architecture versus landscape, infrastructure versus ecology, and urban versus rural. "[I]t allows for negotiations between different processes in the urban landscapes and encourages critical and nuanced expressions of the social, ecological and structural complexity of the contemporary metropolis."[5]

At the core of ecologically directed artworks is an ethic of social and environmental justice in its content and/or form, often drawing on systems theory and the complexities found within natural processes.[6] The works presented in this survey activate the interrelationship between social, economic, political, cultural, and ecological systems, employing varying approaches to public engagement, whether technological, physical, or both, setting the urban environment and its relationship to the larger bioregion.

According to poet and essayist, Gary Snyder, "a bioregion is an area defined by its natural boundaries, and is 'posited on the idea that the human community

[2]Intergovernmental Panel on Climate Change, Working Group II, Climate Change 2014: Impacts, Adaptation, and Vulnerability, report number 5, Spring, 2014. Retrieved March 10, 2014. http://www.ipcc.ch/report/ar5/wg2/

[3]Ibid. Working Group I, Climate Change 2013: The Physical Science Basis. September, 2013. Retrieved March 10, 2014. https://www.ipcc-wg1.unibe.ch/

[4]Jeffrey Hou, "Hybrid Landscapes," 1.

[5]Ibid., 6.

[6]Not all artists presented in this essay consider themselves as "eco-artists" or that their practice is exclusively "eco-art." I use the words "ecologically directed" rather than "eco-art" but borrow nuances of an evolving definition of the term, which falls under the umbrella of "environmental art." See the environmental art page on Wikipedia.org.

is only one of the communities on any given part of the planet, and that the other communities – plant life, animal life, mineral life – inside the landscape with its watershed division, its soil types, its annual rainfall, its temperature extremes, all of that constitutes a biome, an ecosystem, or, as they like to say, a natural nation.'"[7] In getting to know one's bioregion, whether this may be rural or urban, one becomes familiar with and gains an understanding of "the natural context within which we live," including where our resources (water, food, energy, materials) come from, where our waste goes, and "how best to live within our surroundings."[8]

Focusing on artworks that align with this ethic, the works have been grouped within three themed sections, although many projects easily overlap: *ecosystem services*, *managing living systems in urban infrastructures*, and *empowering a culture of sharing*. I am attempting to frame the projects within the contexts of what I consider to be imperatives toward an ecological urbanism that coexists within the broader ecological systems, or bioregion(s), that support it.

Ecosystem Services and the Urban Environment

> ...when human beings remove themselves from the natural environment, the biophilic learning rules are not replaced by modern versions equally well-adapted to artifact.[9]

The term "biophilia first appeared in 1979, meaning the innate human love of the natural world, a universal affinity toward all living beings. It was Edward O. Wilson, a Harvard biologist, who, in his book *Biophilia* (1984), proposed the existence of a genetic basis for this human tendency and suggested that there is an instinctive bond between human beings and other living systems or life forms. Initially, his concepts were not easily received, but nearly a decade later, an anthology of essays forming an academic volume titled *The Biophilia Hypothesis* appeared with a variety of disciplines represented. Although many of the essays sought to describe our relationship to the natural world through an agricultural lens, others focused on nature's influence on cultural development.[10]

In the context of an urban environment, the norm was to eradicate signs of the natural world and its chaotic systems and replace them with controlled environments and architectural objects reflective of that era's aesthetic desires. Biophilia was experienced on weekend getaways, like visiting a friendfrom one's childhood.

---

[7]Gary Snyder quoted in "Between Social Ecology and Deep Ecology: Gary Snyder's Ecological Philosophy" by Paul Messersmith-Glavin, 20.

[8]Paul Messersmith-Glavin, "Between Social Ecology and Deep Ecology: Gary Snyder's Ecological Philosophy," 20.

[9]Edward O. Wilson, "Biophilia and the Conservation Ethic," quoted by David Stairs in "Biophilia and Technophilia," 40.

[10]David Stairs, "Biophilia and Technophilia," 37.

Could this loss of "philia" (Greek word for kiss and friendship) of something that is "innate" explain our present relationship with the natural world and dissociation with vital ecosystems that support life?

Ecosystems are communities of interacting organisms and the physical environment in which they live and are made up of plants, animals, minerals, and people in a given area on the earth. The combination of living organisms and their tasks are part of a complex web of life that provides key services to sustain life on earth. For example, they purify the air and water, store carbon and generate oxygen, and stabilize our climate. Organisms decompose and detoxify our waste, while others help create the soils that grow our food, maintain the soil, and process nutrients. Thousands of small species, bees in particular, pollinate and fertilize plants, protect them from pests, and disperse seeds.[11] These services have recently entered the dialogue in respect to economic valuation but have yet to be meaningfully incorporated into the bottom line.[12]

Cities tax ecosystems heavily with their insatiable thirst, hunger, and energy needs, all of which are furnished far from the source of consumption, sometimes supplanting local needs. The burdens placed on these ecosystems are rarely seen, felt, or otherwise noticed. More than three and a half billion people, half the planet's population, currently live in cities with an expectation that 7–9.5 billion will be living in cities by 2050.[13] As cities continue their growth, the exploitation of resources becomes ever more critical. Despite the perception that urban centers are independent from ecosystems they derive benefit from, they exist within an interconnected web that includes rural lands and wilderness areas. Yet cities were often planned and built in isolation with an emphasis on technology and the built environment rather than a coexistence with the natural systems it replaced or the ones they rely on. Far from the eye are the mining sites requiring mountains to be lopped off and tailing ponds that seep into ground water in order to supply the demand for power, heating, and cooling.

Envisioning the future of urban centers requires a shift in the framework that shapes these environments. As Mohsen Mostafavi explains, ecological urbanism "utilizes a multiplicity of old and new methods, tools, and techniques in a cross-disciplinary and collaborative approach toward urbanism developed through the lens of ecology."[14] But what makes a city livable? Enjoyable? If we are to imagine a city that coexists within the larger ecosystem of a region, certainly incorporating the natural systems of that region makes sense. Part of this means making allotment for indigenous plants, wildlife, and waterways to flow through a city.

---

[11] Cagan H. Sekercioglu, "Ecosystem functions and services," 45.

[12] The relatively new subfield of environmental economics attempts to include the costs and benefits of alternative environmental policies to deal with air pollution, water quality, toxic substances, solid waste, and global warming. It differs from ecological economics, which emphasizes the economy as a subsystem of the ecosystem with its focus upon preserving natural capital (Wikipedia.org).

[13] http://www.who.int/gho/urban_health/situation_trends/urban_population_growth_text/en/

[14] Mohsen Mostafavi, "Why Ecological Urbanism? Why now?" 26.

For several years, I lived in Phoenix and felt a deep disconnect with the desert, even though it was the desert I sought when moving there. The plethora of nonnative trees, green lawns, and man-made lakes reminded me of the east coast I had left behind. Growth may have slowed with the economic downturn, but water consumption remains steady despite decreasing levels in supplies.[15] Could it be that our physiological requirement for water is expressed through a psychological desire? It seemed to me that nostalgia for the landscape that was left behind took precedence over or was a mechanism for adapting to the harsh desert reality.

Fostering a sense of caring for the desert and its native fauna is a key objective of desertArtLAB, a socially engaged art collective from Arizona. Cofounders, Matt Garcia and April Bojorquez, began this interdisciplinary initiative as a means to engage the predominately, but not exclusively, Hispanic community of Phoenix to explore connections between the native desert ecology, cultural history, and community. In *Mobile ECO- STUDIO* (2013) (Fig. 1.1), the artists use the medium of performance and land installation, distributing to community members native plants to be planted in one of the many dead, empty lots that dot the Phoenix landscape. The "performance plantings" are documented and locations mapped for future monitoring.

The project succeeded on several fronts: it brought diverse, intergenerational members of the community to participate in the plantings, individually and collaboratively; using social media and digital mapping techniques, each plant

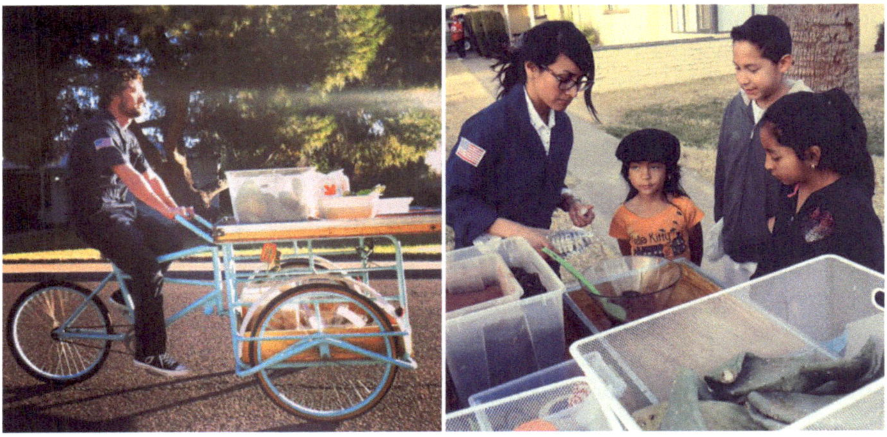

Fig. 1.1 desertArtLAB, *Mobile ECO-STUDIO* (2013)

[15]To track trends see USGS site http://water.usgs.gov/edu/wateruse-trends.html. Although this shows consumption from 1950 to 2005, it does not indicate supply levels. For this see http://tinyurl.com/mowtfbt. For water data both current and past and surface and ground, see USGS Water Data Discovery, http://water.usgs.gov/data/. For climate change impact on water resources, http://www.epa.gov/climatechange/impacts-adaptation/water.html

was then kept track of for future harvesting, therefore maintaining the public's interest. The prickly pear cactus has a several-thousand-year history in the Sonoran desert. The plant was once culturally important as well as ecologically significant, having been used as a source of food, medicine, intoxicant, water purifier, and mythology to the region's inhabitants. Reconnecting the community to the prickly pear's cultural significance cultivated an interest and a care for the larger goal that of replanting of open lots that were once thriving ecosystems. To date, over 300 cacti have been planted.[16]

The southwestern open landscape was very appealing to the 1960s and 1970s conceptual artists wanting to leave the city and make use of, what seemed to them, an infinite space compared to the confines of the gallery and its commodity system. While these early earthwork artists were bulldozing the desert or pouring tar downhill slopes with dump trucks, making monumental conceptual works in remote areas, Alan Sonfist chose to situate his work within the city rather than escape it.

"For almost 40 years, Sonfist has dedicated his work to linking city-dwellers and suburbanites to a nature that civilization has destroyed, with the hope that a greater appreciation of nature would encourage them to protect its future."[17] His *Time Landscape of New York* (1965–1977) sought to reverse the damage done by early settlement, returning the land back to its original state with indigenous plants and trees on a 45 × 200-foot city block in Greenwich Village in Manhattan. It continues to grow, with newer species finding their way into the park. Although it is based on scientific research and planning and infused with colonial history, Sonfist writes, "I am trying to provoke people to think about the environment, to focus on the poetry of the forest, not to present a science exhibit."[18] The work contributes to a cultural and natural history of New York City and functions as a space of repose for city dwellers.

With a similar objective as Sonfist, Canadian land artist, permaculture teacher and activist, Oliver Kellhammer, creates botanical interventions and public art projects that demonstrate nature's resilience in the face of degradation. His projects facilitate the process of environmental regeneration that continually evolve and change over time, thus engaging the botanical and sociopolitical underpinnings of the landscape. Kellhammer's projects also engage the communities in which they are situated.

In the early spring of 1991, Kellhammer began his first intervention, a guerilla style community garden on an unused city land slated for development as a "truck route" freeway through a low-income neighborhood in East Vancouver, BC. Kellhammer approached community residents, many of whom were immigrants he had met at a community garden group, to persuade them to get involved. Without permits or any permission, *Cottonwood Community Gardens* slowly went from a

[16]The artists' website is comprehensive with a well-produced video documentation of each project this collective does with the community, http://www.desertartlab.com/

[17]Sue Spaid, "Ecovention: Current Art to Transform Ecologies," 89.

[18]Jude Schwendenwien, Alan Sonfist quoted in "Breaking Ground: Alan Sonfist, et al." 41.

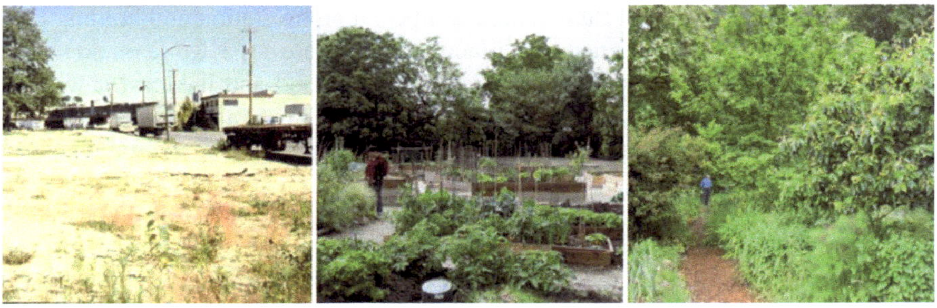

Fig. 1.2 Oliver Kellhammer, *Cottonwood Community Gardens* (1991, 2010)

desolate, 3-acre strip under the Vancouver Park Board and the City Engineering Department's joint charge to a thriving, biodiverse ecosystem of plants, shrubs, trees, and a pond, after a decade of continued community volunteering (Fig. 1.2).

Partially, to make up for the loss of plots at an existing community garden due to development, Kellhammer writes, "I figured that getting people to physically occupy the Malkin Avenue right-of way would be a useful impediment to the building of the freeway, which many of us in the neighborhood considered unsustainable and a threat to community well-being . . . A struggle soon ensued between some city staff, primarily from the Engineering Department, who wanted to evict us, and others, including several city counselors and Parks Board employees who saw what we were doing as a laudable example of grassroots community empowerment."[19] As the site continued to be cultivated, youth groups became involved, and grants were procured (after the city granted a lease) to enable the community to install a much needed irrigation system, garden shed with a greenhouse, and a pond to store rainwater that attracted wildlife. "A lot of the work I did during this initial pioneering period was to initiate and preside over an interactive, co-design process that would allow Cottonwood to evolve naturally, while meeting the needs of its diverse stakeholders – human and non-human."[20]

Urban-situated artworks concerned with ecological restoration and incorporate infrastructure, such as the *Time Landscape of New York* and *Cottonwood Community Gardens*, often do so with animal habitat and public engagement in mind. Patricia Johansons approach to integrating natural systems within an urban infrastructure is to design gardens that consider biodiversity and habitat in direct relationship to public engagement. Her water gardens, such as *Endangered Garden* (1987), incorporate the natural or original flood basins, dams, reservoirs, and/or drainage systems within the aesthetic design of the garden orpark. Her goal with *Endangered*

[19]Oliver Kellhammer quoted from his website: http://www.oliverk.org/art-projects/land-art/cottonwood-community-gardens

[20]Ibid.

Garden, a major work with Sunnydale Facilities, a pump station and holding tank for water sewage, was to create a "functional structure as a work of art and a productive landscape."[21] Her intent was to increase food and habitat for wildlife while maximizing public access to San Francisco Bay. Pathways and seating were planned to enhance the human-nature relationship, bringing them both closer together to interact.

The temporal aspect of these projects, both in the process of human "making" and natural processes, allows for a cultivation of a relationship between humans and other species and a closer understanding of the land and its systemic functioning. Executing these large-scale projects involves collaborations with practitioners outside of the arts and lengthy negotiation with city planners and bureaucrats. The planning process takes on its own form of time as it often takes years before the project can really begin. Bureaucratic time aside, these projects become fully realized after many years of cultivation and growth and perhaps never finish and may or may not take on the "aesthetic" intended.

These works sensitize city residents to the biodiversity of ecosystems within their local environment and foster a sense of care and well-being. Whether it is an urban forest, a neighborhood park, or a city wetland, we share these natural spaces with innumerable species and organisms. As part of the wider web of ecosystem functioning, these projects bridge our understanding for rural and/or wilderness far from our daily experiences because of their very presence. Migrating species, water flows, and wind patterns, these natural entities and phenomena may pass through our cities or environs without much notice, but they carry a thread from the larger web through our urban landscape, en route elsewhere.

Migratory life, though, is not always so easy. Building of large cities can interfere with migratory flight paths, by the lights at night and/or reflective glass windows during the day. In one North American city alone, it is estimated that between 1 and 9 million birds die each year from collisions with skyscrapers, the second leading cause of death after habitat loss.[22] Bird habitats, as with most nonhuman species, including forests, grasslands, farms, and wetlands, are severely on the decline due to development. To get a sense of the scale of such loss, since WWII, wetland habitat has been decreasing at a rate of about 250,000 acres a year in the USA alone.[23] Wetlands provide a complex series of ecosystem services offered by plant and animal species, many of which are listed as endangered or threatened. Urban backyards, urban forests, parks, and renewed wetlands have become a habitat for a diverse number of city wildlife, connecting urban areas with landscapes beyond city boundaries and entering our consciousness with regularity.

---

[21] Patricia Johanson, http://patriciajohanson.com/endangered-garden/

[22] http://www.flap.org/toronto-lights-out.php. Also see NYT article: http://www.nytimes.com/2012/10/28/world/americas/casualties-of-torontos-urban-skies.html?pagewanted=all&_r=0

[23] http://www.clemson.edu/extension/natural_resources/wildlife/publications/fs32_wetland_ecology.html

Although she is widely known for the Vietnam Memorial and other monuments, Maya Lin has had a prolific career in the field of land art and architecture. *What is Missing?* (2012) is a different kind of memorial and, perhaps, her "last monument."[24] An immense, internationally collaborative undertaking, *What is Missing?* gives voice to lost or close to extinction species, such as songbirds. The project was created in multiple formats that include a website, a large sculptural *Listening Cone*, 70 videos, and an installation, *Empty Room*, with projections of images of species illuminated beneath the floor that visitors can "catch." She describes the idea behind the project that of loss as a way to "wake [people] up" to compel them to take action.[25]

With similar concern, Beatriz da Costas installation, *A Memorial for the Still Living* (2010), and its accompanied mobile application, *Endangered Species FinderThe Endangered Species Finder* (2010) (Fig. 1.3), are a somber reflection on endangered species. For da Costa, the only modes of encounter remaining once a species becomes extinct are through imaging, description, sound, and taxidermy. Rather than focusing on already extinct species, da Costa puts the spotlight on the "still living" species that are listed as threatened and that still have a chance for survival if action is taken. The mobile application "facilitates encounters with other species within their 'natural' environments. [da Costa] believe[d] that experience and encounter, not just policy and regulations, are what ultimately change our behavior towards our environment. Through her encouragement of a 'go out and

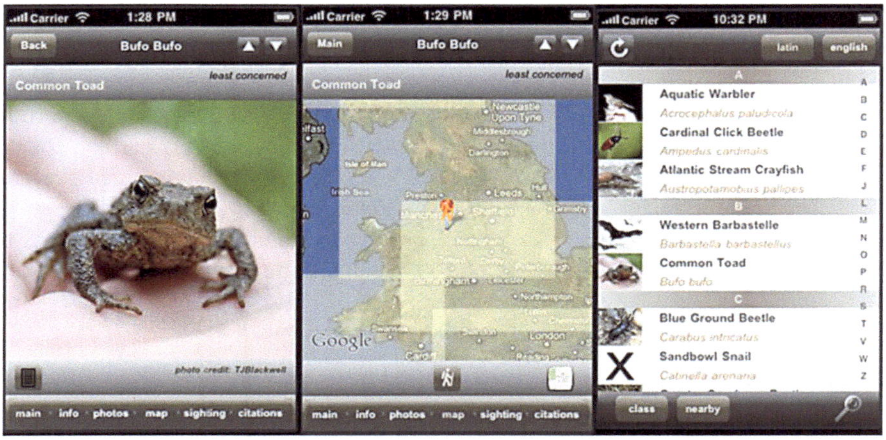

Fig. 1.3 Beatriz da Costa, the *Endangered Species Finder* (2010)

---

[24]Susan Platt, "The Planet According to Maya Lin: What is Missing? and Confluence Project," 144.

[25]Maya Lin interview, http://www.cornell.edu/video/artist-maya-lin

meet the species before it's too late' attitude, da Costa hope[d][26] to make a small contribution to the collective effort of examining our current relationships to non-human species."[27]

The proliferation and accessibility of new technologies has opened opportunities for artists to work with data to create visualizations in many forms. Where da Costa's mobile application enabled direct physical engagement, it also relied on a rich database that the user could add to. Through a different engagement, David Wicks interactive installation of *Drawing Water* (2010) uses real-time data to visualize water resource locations and consumption placed into a digitally constructed landscape, changing with the user's selection on a touch screen.

The diversion of water, pumped and siphoned from one location to serve another, generally from rural to urban centers, raises political and socioeconomic issues and implications of entitlement. When water is channeled far from where it falls, who should have the rights to this resource? When laws are passed prohibiting rain catchment so as to replenish ground water levels, does it inhibit progress toward sustainably harvested resources?[28] According to Wicks, "Although the paths are imagined, *Drawing Water* is based on real data and it reveals a clear truth about water resources and use. It is a parallel landscape constructed from our anthropogenic and natural hydrologic environment . . . [and] plays upon the 19th century theory that "rain follows the plow". As long as people plowed fields, they believed, water would come to them . . . Americans still live with an illusion of resource availability following need."[29]

The illusion of an inexhaustible supply of clean water comes into question when water supply becomes contaminated, affecting residents and species in riparian zones. Artists working with urban planners, water treatment management, and the public find effective means to "address the meaning, form, and function of the natural ecosystems and public spaces which attend our post-industrial properties and waterways,"[30] explains artist, Tim Collins, about his and, his partner, Reiko Gotos involvement in the *Nine Mile Run* (1997–2000). The project, initiated through Carnegie Mellon University's STUDIO for Creative Inquiry, was a collaboration between artists, scientists, and students. The 3-year project took a larger view of Pittsburgh's water system of three rivers, 52 streams, and riparian banks.

---

[26]Parenthesis added to indicate the artist's passing in December, 2012.

[27]The project was a commissioned by the Arts Catalyst in the UK. Both the exhibition and the mobile applications were developed to focus on species in the UK. http://www.artscatalyst.org/projects/detail/a_memorial_for_the_still_living/

[28]Several western US states make it illegal for residents to collect rainwater. Colorado reversed its laws in 2009 after a study found that it would not "rob water owners of their rights." See NY Times.

[29]David Wicks, "Record of Creative Work," 29–30.

[30]Tim Collins quoted in Glenn Harper's "Tim Collins and Reiko Goto: Art Has Everything to Do With It," 119.

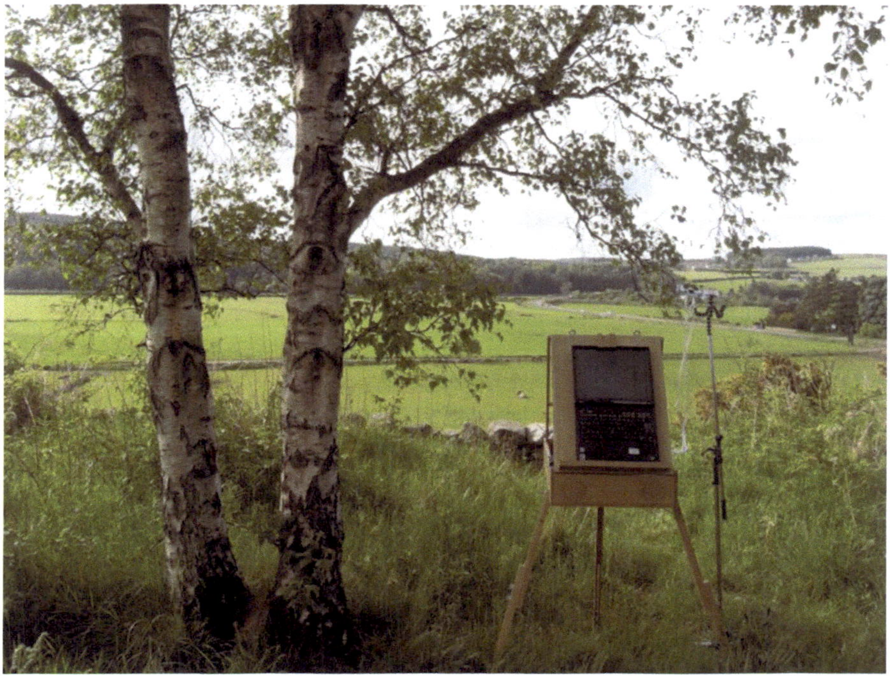

Fig. 1.4 Tim Collins and Reiko Goto, *Eden 3: The Breath of a Tree* (2006–ongoing)

After the pair moved to Scotland, Goto began to think about the challenges of cultivating an empathetic approach to the work. The series *Eden 3* is an ongoing project in collaboration with Collins that began with *Eden 3: The Breath of a Tree* (2006–ongoing) (Figs. 1.4 and 1.5). A device, in the form of a nineteenth-century-styled plein air painting easel, was developed "that gives a voice to trees, turning their reactions to changing climatic conditions (specifically the CO_2 content of the air) into sound."[31] The easel, with this data-collecting device, was placed in seven sites next to seven trees along an imaginary trail on the Aberdeen River and parks. The device produces scientific data while creating sounds that correspond to changes in levels of CO_2, essentially enabling the tree's breathing to be audible. Along with this tree monitoring, the artists created an indoor installation, including a greenhouse with plants, where the artists work in proximity to the monitored sounds of the trees. The project is expected to last 5 years and includes monitoring, research, documentation, and exhibition.

Eden 3 is more than a symbolic gesture of empathy for trees and, by extension, ecosystems and the vital services they provide to sustain life on this planet. The project's data contributes as a baseline for understanding a little more about the

[31] Glenn Harper, Ibid.

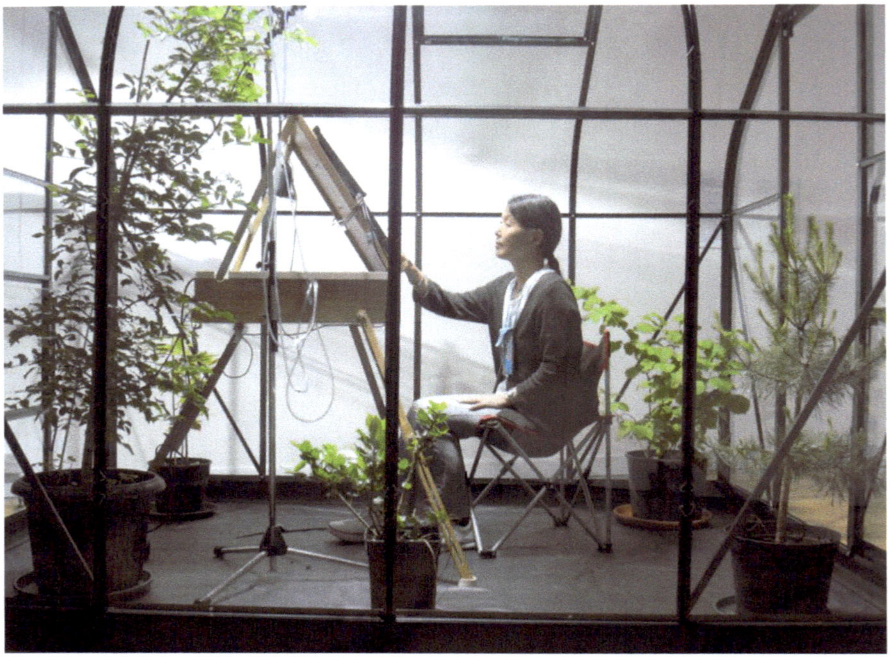

Fig. 1.5 Tim Collins and Reiko Goto, *Eden 3: The Breath of a Tree* (2006–ongoing)

process of photosynthesis in a form we can relate to (that of sound) and into visual graphs, giving scientists relevant information about pollution. Although Collins and Goto have stated *Eden 3* is, in a sense, a period of grieving, it also is hopeful because their objective is to seek a true empathic exchange with the tree itself. "[T]he human need [is] to hear the breath of people and living things we care for to assure ourselves of their well-being. This technology provides one way of listening to the breath of trees."[32]

Conceptually innovative projects such as *Eden 3* can offer a compelling engagement with the public; by using the familiarity of the newest technologies available, they encourage a biophilic appreciation of nature through immersion. The emphasis of such projects is on the need to view the fragility of the planet and its resources. Technological innovation is used as a means to an end. The collaborative processes of such projects broaden the outreach while relying on collective knowledge to inform the work and in turn inform the public. Such collaborative efforts can also lead to creative innovation, speculating on a positive future.[33]

[32]Reiko Goto quoted in Ibid.

[33]Mohsen Mostafavi, Ibid., 17.

Managing Living Systems in Urban Infrastructure

> Cities are some of the most profoundly altered ecosystems on the planet; within their boundaries are also found some of the most diverse ecological conditions. If there is a laboratory where ecological change can be viewed at close hand, it is the city.[34]

It was not that long ago that the city had the reputation of being a cancer on the planet and that spending time in the country in the tranquility of nature was necessary for the soul's spiritual renewal. The idea of cities as separate and detached from their broader life-support systems has its roots in the Chicago School of urban sociology of the 1920s. The modernist understanding of urban life was an essentialist reality separate from rural life and was reinforced by technological advancements in transportation.[35] The train enabled progress in urban industrialization and the movement of goods in and out of the city, but industry also generated increased pollution. Rapidly increasing industrialization brought migrants from rural and foreign land, requiring expanding infrastructures.

Although cities have made a remarkable improvement in services, much remains from the initial design and infrastructure that makes shifting to an ecological urbanism challenging. Many cities were not designed to have their living systems visible, disconnecting the urbanite from understanding where their wastewater and garbage go or where their electricity, water, and food sources are coming from. Few cities reveal the complexity of their transportation and communication systems and even fewer have easy accessibility to redesigning of these infrastructures. If the urban ecosystem is inapprehensible, how then can we understand intrinsically the broader ecosystems and supply lines that a city relies on for its survival? "Imagining an urbanism that is other than the status quo requires a new sensibility – one that has the capacity to incorporate and accommodate the inherent conflictual conditions between ecology and urbanism."[36]

Artists, even more so than designers, are well poised to rise to the challenge, having creative freedom to venture into contentious territory. They are joining urban ecologists and other professionals in attending to urban ecosystems and infrastructures by engaging policymakers, planners, and the general public. In his essay, *Art and Ecological Conscience* (1972), György Kepes states, "Environmental homeostasis on a global scale is now necessary to survival. Creative imagination, artistic sensibility, can be seen as one of our basic collective, self-regulating devices that can help us register and reject what is toxic in our lives."[37] Kepes was referencing what is known as a feedback loop. Natural processes have inherent self-regulating functions, as do our bodies. Most industrial processes, unfortunately, do

---

[34] James P. Collins, Ann Kinzig, et al., "A New Urban Ecology," 416.

[35] Stephan Barthel and Christian Isendahl, "Urban gardens, agriculture, and water management," 224.

[36] Mohsen Mostafavi, Ibid., 17.

[37] György Kepes, "Art and Ecological Conscience," 6.

not have this built-in mechanism; therefore, waste, for example, is not utilized but stored or released into the environment. In Kepes's opinion, this concept of a self-regulating mechanism was exemplified in the works of Pulsa, an interdisciplinary group that created projects in the late 1960s and 1970s, linking a variety of media with biological systems.

Pulsa's work imagined the utopic, where life consisted of harmony with natural systems, community, and creative activity. In addition to interactive sound and video installations, Pulsa set up an experimental social system that included communal living, self-sufficient agriculture, inflatable passive heating moveable architecture, and collaborative art and music projects. *Harmony Ranch* (1972) (Fig. 1.6) sat on several acres of partially cleared land in Oxford, Connecticut, not far from Yale University's School of Art and Architecture where several members taught.

Patrick Clancy, one of its members, writes in *The City as an Artwork* (1972), a contribution to Kepes's edited volume[38] that "the city *should* [italics mine] be recognized as separate from nature" because it "contains no facilities which enable

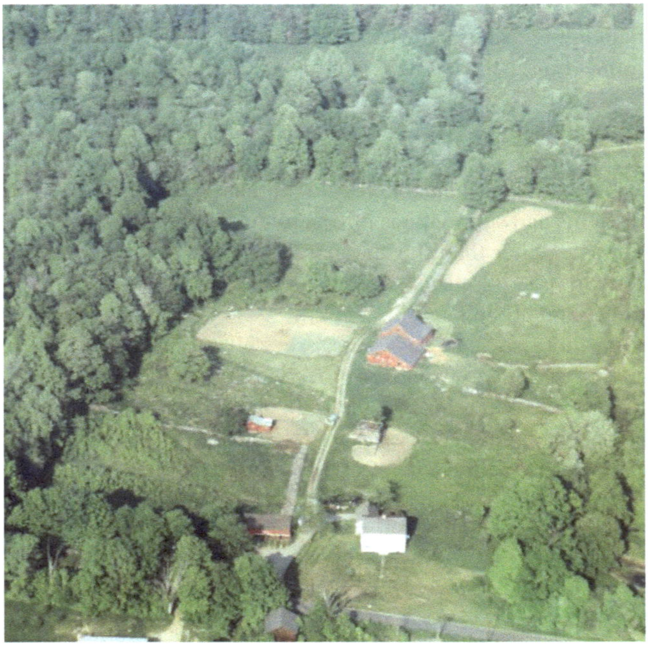

Fig. 1.6 Pulsa, *Harmony Ranch* (1972)

[38]Patrick Clancy (Pulsa Group), "The City as an Artwork," 210. This essay presents Pulsa's aspirations to transform the information systems which support man-made environments exemplified in the city into life-enhancing experiences comparable to those primordially enjoyed by humankind in nature.

one to perceive the city and to apprehend its beauty, nor are there unrestricted public experiments to redefine the urban environment [that] support the reality of man's existence within the city." This may be in part due to the fact that cities are designed to alienate its residents from the life-support systems of that environment. Clancy argues that such projects "should not be speculative,"[39] yet I would argue that the underpinnings of Pulsa's projects would be considered speculative today, to the extent that they may not have foreseen how quickly the advancing of climate change effects would have on urban centers or the rise in social unrest at the political apathy toward mitigation. The increase in social movements and the number of artists working toward speculative solutions reflect Pulsa's broader intentions in creating a just world.

Indeed, the speculative movement has gained a great deal of traction recently. Primarily concerned with current issues and trends and stimulated by a rapidly changing technologized culture, they postulate on a potential future. Speculative projects are borne out of a recognized need to change what is perceived as dysfunctional, signaling shifting social priorities. Such activities can fuel innovation and, although are potentially risky undertakings, which may mean doing it alone or with like-minded cohorts against the mainstream, have a wide effect on public awareness and action. Speculative eco-directed art imagines a possible future, whether dystopic or promising, sometimes presenting unusual solutions to impending or escalating crisis and other times raising red flags.

Mary Mattinglys projects seem to plan for survival in impending environmental or economic catastrophe, both of which are deeply intertwined. Her earlier project, *Waterpod* (2006–2010) (Fig. 1.7), a floating, eco-habitat, public art project built atop of a 100-foot barge floating in the Hudson River, complete with geodesic domes, gardens, chicken coops, outhouse, solar panels, and water filtration, speculates on rising sea levels and disappearing land masses. Unlike the city it floats around, its life-supporting systems are made visible and accessible.

Continuing this trajectory, her later series, *Flock House* (2011–2014) (Fig. 1.8), are mobile, self-sufficient living systems which she calls "living sculpture" – an experiment in urban sustainability in the face of displacement and mass migration. Drawing on the widely popular DIY movement, the concern here is not an aestheticized sculptural or architectural object, but rather a process in systemic, "real-time" improvement and constant redesign. This ability to adapt to evolving conditions relies on an interdependency of community involvement, resourcefulness (especially with repurposed materials), and an ethic of ecological and economic equitability. It is indicative of Buckminster Fuller's ideas, bringing to mind his theory of a comprehensive anticipatory design science.

In 1950, Fuller conceived of a course called comprehensive anticipatory Design Science (or simply design science) that he taught at MIT in 1956.[40] The course

[39]Ibid.

[40]Brief course outline available at The Buckminster Fuller Institute website, see link: http://bfi.org/design-science/primer/eight-strategies-comprehensive-anticipatory-design-science

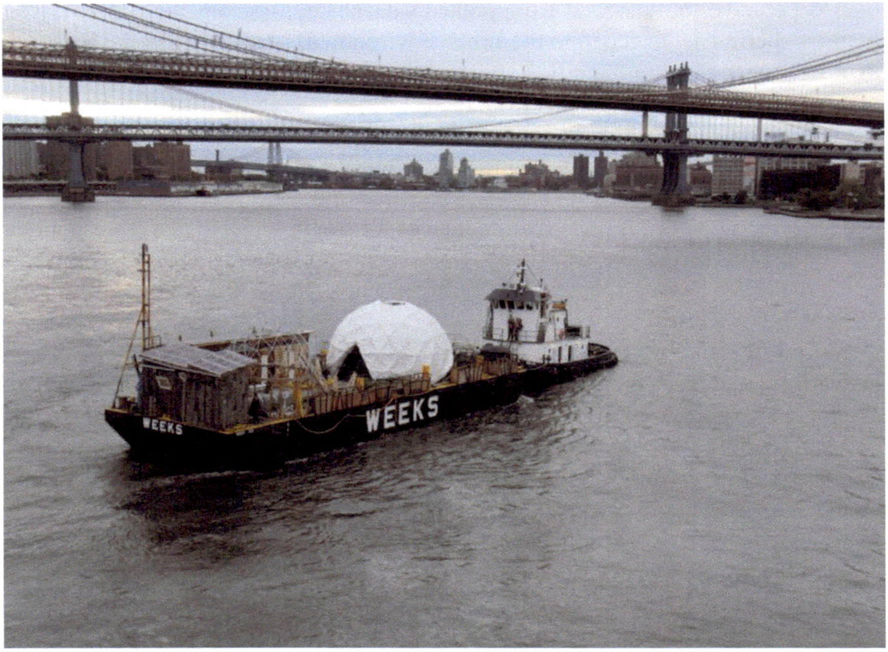

Fig. 1.7 Mary Mattingly, *Waterpod* (2009), a floating sculpture, ecosystem, and public space in New York City

outlined eight strategies that, if applied, would be an effective framework for change, particularly as we face the challenge of living more sustainably. Fuller's concept of Design Science reflected a more holistic approach to our built environment that was inclusive of the ecology it was placed in. In this regard, *Flock House* immerse themselves within the larger ecosystem of New York City and its five boroughs, migrating with ease from one location to another and adapting to that location's set of social, architectural, and ecological circumstances. These "alternative living models" reinforce Fuller's idea of providing simple, functional, and affordable habitats, available to anyone. Although partly fantastical, the habitats are conceptually flexible, accessible, and inclusive.

A resilient and flexible system is capable of adapting to real-time situations and events such as flooding, super storms, and drought. The term "real time" is often associated with computing, where data is processed and information is available immediately as feedback. This notion was brought into aesthetic dialectics in Jack Burnham's essay, *Real Time Systems* (1969), where he states, "real-time systems gather and process data from environments, in time to effect future events within those environments."[41] The services that ecological systems provide may

[41] Jack Burnham, "Real Time Systems".

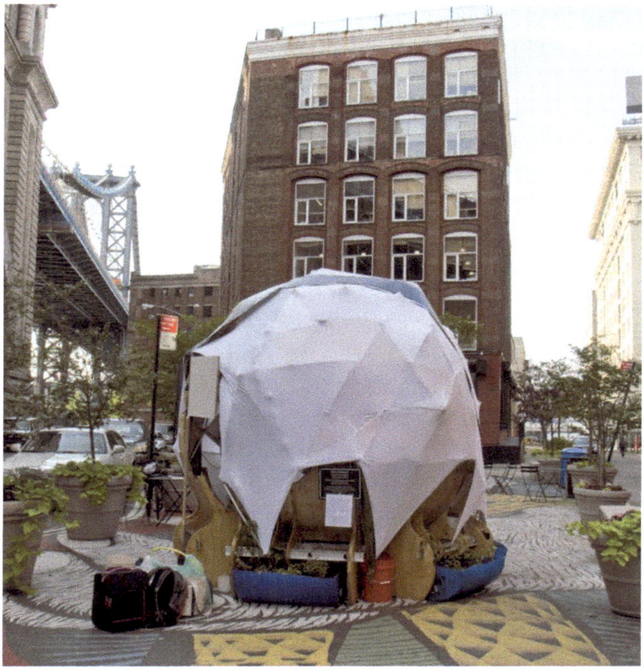

Fig. 1.8 Mary Mattingly, *Flock House* (2011–2014), modular environments inhabited in urban centers

not be immediately visible until their feedback mechanisms become overloaded, compromising their ability to self-correct and withstand and buffer extreme events. Information or "data" of imbalance is transferred through the biodiverse living matter (plants, amphibians, and insects) who act as harbingers to potential crisis if conditions are not reversed.

Hans Haacke, an early pioneer to utilize real-time concepts that paralleled computerized real-time systems, employed real-time feedback in works that addressed social and environmental problems that were not readily visible. In his 1972 work, *Rhinewater Purification Plant*, Haacke set up a filtration system inside the Museum Haus Lange in Krefeld, West Germany. Large glass vessels filled with contaminated water from the Krefeld sewage plant collected from the nearby Rhine River were processed through chemical treatment and then a charcoal and sand filtration "plant." The purification system cleaned the murky water so well that it could sustain goldfish in an artificial habitat created in a large acrylic basin. The excess water was then pumped outside to water the museum's garden. Haacke's setup was an early example of a reclaimed gray-water system and a potential solution for urban infrastructure.

In his critical essay, *The Politics of Sustainability: Art and Ecology* (2009), TJ Demos critiques Haacke's project in that it doesn't go far enough, because it is not interactive, failing to "involve the audience more directly within its feedback loop."[42] I would argue that although it does not involve the viewer's physical interaction, the work implicates both viewer and the institution in which it is exhibited in since both are complicit in the polluting of the river.[43] Furthermore, *Rhinewater Purification Plant* and its companion piece, *Krefeld Sewage Triptych* (1972), facilitate a transfer of complex scientific information in a form that is readily understandable and relatable. *Rhinewater Purification Plant* may not open itself to direct interaction; nonetheless, its success lies in the interrogation of the status quo that may instigate a change in behavior or at the very least raise awareness and questioning, what if we had to purify our own water?

Australian artist Tega Brains project, *Coin-Operated Wetland* (2011) (Fig. 1.9), could be viewed as an extension to Haacke's project. *Coin-Operated Wetland* is a large-scale interactive installation that fuses a washing machine and a biological wetland that processes the gray water from the washer, purifying and returning it for reuse. In this instance, *Coin-Operated Wetland* requires the audience's participation as an integral input. As she describes it, "for four dollars, the audience can wash their dirty clothes and be drawn into a complex set of relations with the many living and non-living actors of the system."[44] With the simple act of creating a pay-per-

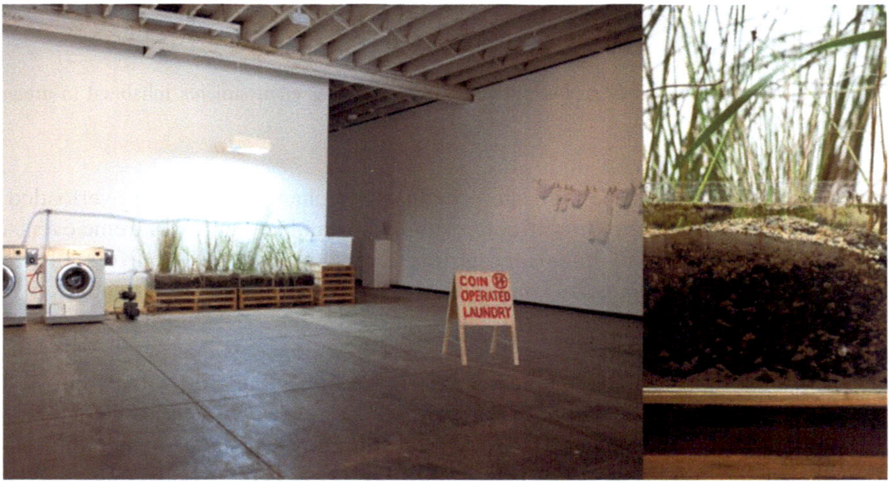

Fig. 1.9 Tega Brain, *Coin-operated Wetland* (2011)

[42]TJ Demos, "The Politics of Sustainability: Art and Ecology," 27.

[43]In a work shown concurrently, Krefeld Sewage Triptych (1972), Haacke presents conclusive data on the volume and types of industrial and household pollution levels found in the untreated sewage, tracing these back to the polluters, including that of the city of Krefeld, of which the Museum belongs to. Demos gives some credit to this institutional critique.

[44]Tega Brain, "The Politics and Poetics of Coexistence," 68.

user system, Brain connects a monetary value system with the ecosystem service value, bringing the ecological imperative into the realm of economics, a system that typically does not include externalities[45] within the bottom line.

Brain's project is an example of a closed-loop feedback system. The health of the plants and their ability to continue the service of providing purified water to continue washing are determined by a number of factors, including the number of loads the system can handle and the amount and type of soap used. *Coin-Operated Wetland* provides real-time feedback (the whiteness of the clothing) and a solution for the potential reconfiguration of our current domestic relationship with the environment. Restored urban wetlands could play a vital environmental role in improving water quality and conservation but also act as a key buffer to storm surges. Wetlands met with a devastating fate when urban centers began to form, having been filled in or paved over. As a river catchment service, they provide the necessary filtration from storm water runoff before ground debris and pollutants find their way quickly down impervious surfaces into the hydrological environment.

As artists continue to invent new models of social interaction with their artworks, they challenge and interrogate infrastructures that are ecologically illogical. Natalie Jeremijenko uses humor and playfulness to intervene in everyday city experiences. In *NoPark* (2008–ongoing) (Fig. 1.10), she converts (or returns) "no parking zones" such as fire hydrant spots to green cover using low growth mosses and grasses as storm water filtration systems, where 99 % of the time the spaces remain empty, barring the exception of the rare emergency. In this case, she redefines the "emergency" as being an environmental one. Jeremijenko calls these spaces "micro-engineered green spaces [that] prevent storm water run off, use foliage to stabilize the soil, and provide a durable low maintenance surface cover."[46] The "microparks" continue to provide space for fire trucks when needed, create additional green space for neighborhoods, and do a better job of capturing water than do green roofs because they can carry a thicker layer of soil than green roofs.

Conveying environmental data through real-time feedback in artworks can come in many forms, as those discussed above, that use an organic medium or method. Artists have also exploited digital forms to visualize their concepts, using a complex network of computers, satellites, or data sensors to expose ecological pollutants and their impact. Data visualizations are innovative and creative approaches for conveying real-time events that are, again, not readily visible and often purposely hidden due to inaccessible infrastructures.

One such inaccessible infrastructure is energy. In New York State nearly one-third (29.6 %) of all energy consumption is in residential homes, with commercial buildings slightly more (32.7 %), and transportation and industry sharing the

[45]"A side effect or consequence of an industrial or commercial activity that affects other parties without this being reflected in the cost of the goods or services involved, such as the pollination of surrounding crops by bees kept for honey," Google definition.

[46]http://www.environmentalhealthclinic.net/nopark

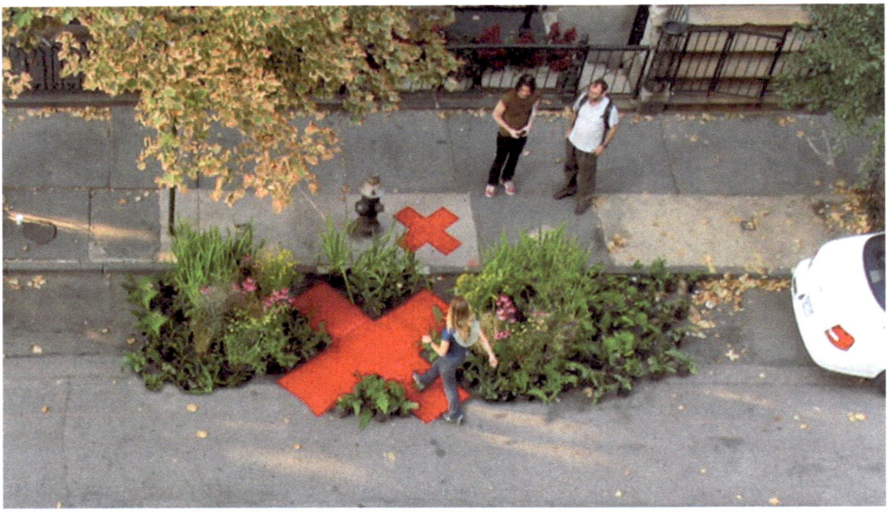

Fig. 1.10 Natalie Jeremijenko, *NoPark* (2008–ongoing), engineered microecosystem for emergency vehicle parking, such as in association with fire hydrants. The *NoPark* is designed to optimize infiltration of road-borne pollutants and foster urban biodiversity

remainder (28.2 % and 9.5 %, respectively).[47] If energy consumption data were available in real time on a daily basis, would it effect lower consumption rates? We have become so accustomed to having energy supplied on demand that when supply lines are cut during climatic events, for example, or when the grid is overloaded in the heat of summer, the effects can become dire. Would conservation efforts reduce these effects? How would an ecological city handle such events?

Commercial building developers have implemented complex systems that control HVAC (heating, ventilation, and air conditioning) in all new constructions, making them as efficient as possible. Although they are "smart" in their self-automation, the relationship with the user is opaque and noninteractive. Would knowing a building's carbon footprint trigger conservation by the occupants?

"Buildings breathe data. Our homes shelter technology that quietly counts in the background of our daily lives. We cook; we wash; we sleep. All the while, small electronic gadgets tally numbers that remain either inaccessible, or beyond our ability to interpret."[48] In *7000 oaks and counting* (2007), Tiffany Holmes exposed a building's energy consumption in the form of an interactive kiosk sitting in its lobby, as it counted the kilowatts per hour consumed by the building's occupants. The eco-visualization, as she calls it, is "composed of a sequence of animated clips

[47]Stats are as of 2011. Consult EIA website for various informative data sets. NYS data consulted May 21, 2014. http://www.eia.gov/state/?sid=ny#tabs-1

[48]Tiffany Holmes, "7000 Oaks and Counting (2007)," 20.

using a series [of spinning rings] of tree images that correspond to the carbon loads in the building." The project is a homage to Joseph Beuys planting of 7000 oak trees (*7000 Oaks – City Forestation Instead of City Administration*, 1982–1987) in Kassel, Germany, a pioneering social practice artwork that engaged the community in planting of the trees. Holmes calculates that it would take the planting of 5,600 trees to offset 800 pounds of carbon (or 533 kilowatt hours) into the atmosphere, an average amount usually reached by 5 pm on a winter day by a large building.[49]

The occupants of the building were invited to make a personal commitment to reduce their carbon footprint by filling out a web form. As each person inputs information, their name is incorporated into the animation, offsetting the carbon output of the building. "In reality, actions such as turning off lights and coffeepots and biking to work can do more in the long run for our climate than planting actual trees."[50] Trees act as carbon sinks, and mostly, every person has a positive perception of trees making it easier to picture and relate to than the electricity data. As the energy consumption of the building increases, the lush canopy begins to disappear, and in its place images of general appliances such as hairdryers, light bulbs, coffeepots, and light switches begin appearing, spinning furiously until the building's energy consumption lessens, usually after peak hours.

What if buildings could convey visually their hidden information about carbon load? In another one of her projects, *Kilowatt Hours* (2011), Tega Brain covertly installed an energy meter on the circuit board of a Sydney, Australia, building that the project was exhibited in. Employing DIY computing techniques, Brain wrote a software patch in the open-source processing language that uploaded the building's electricity consumption data to the Pachube and Twitter websites, making the information public on an hourly basis. She then used this data from the websites to inform a circular line drawing (resembling an architectural skyline) being projected on the floor in the exhibition space. As the electricity usage in the building increased, the circle's diameter would enlarge in a 360-degree rotation with a 1-h duration. Like Holmes, who makes a connection between the rings of a tree and small appliances to the associated carbon load, Brain tweets the electricity usage and equates it to the number of pieces of toast that it would produce, bridging the data into everyday experience.

The use of the tree is an effective object in relating information regarding its ecosystem service in countering pollution. Air pollution is one of the more pervasive and invisible environmental health hazards of cities and has been in the public consciousness for several decades. Using the air quality data collected at the local site and in real time, Andrea Pollis and Chuck Vargas *Particle Falls* (2010) (Fig. 1.11) visualizes particulate pollution focused on the smallest particle matter (PM 2.5) and outputs it in the form of a waterfall projected on the side of the building.[51]

---

[49]Ibid.

[50]Ibid.

[51]Andrea Polli and Chuck Varga interview about the project, http://vimeo.com/16336508

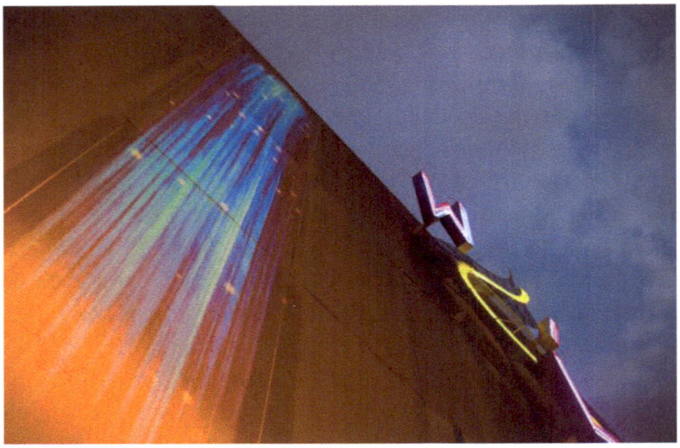

Fig. 1.11 Andrea Polli and Chuck Varga, *Particle Falls* (2013). Wilma Theater, Philadelphia, as part of Sensing Change exhibition, Chemical Heritage Foundation (Photo: Conrad Erb)

Particle Falls is a complex system using sensors that read the immediate particulate matter in the air surrounding the building and interprets it into an ever-changing laser light projection cascading on its surface. Passerby viewers can see the fluctuations as traffic increases and decreases in front of them. The project makes an immediate connection between traffic flow, air quality, and human health. The changing dot size and color correspond to the invisible particle size in the air. As part of the public works initiative, the city and the artists wanted the project to influence usage of the light rail system in San Jose, where it was initially installed.

Each of these works relies on access to urban architecture, linking its internal or external infrastructure to energy usage and local pollution conditions. As commercial building owners are recognizing the advantages that "smart building" technology offers in reducing costs and pollution associated with energy consumption, artworks such as these have greater resonance.

A recent article in Crain's New York Business touted New York City as a hotbed for green tech start-ups headed especially by young innovators who are focusing on data-intensive instead of capital-intensive technologies. The focus is less on hardware and more on software to better manage energy, water, and waste infrastructure.[52] New York City is not alone in this. Citizens around the world are organizing and mobilizing toward environmentally sound practices, despite the persistent lobbying by fossil fuel industries and their "business-as-usual" cohorts. With rising interest and institutional or governmental support in green tech, cities have become urban labs, perfect breeding grounds for innovative projects by tech-savvy artists and activists, becoming the necessary agents of change and using the already in place infrastructure to implement their ideas.

[52] Judith Messina, "Green tech powers up NYC companies," no pagination.

Michael Mandiberg is a New York City-based artist, programmer, designer, and educator who creates physical and digital objects as well as web-based projects. Mandiberg's conceptually "innovative work not only creates awareness in an art context but also anticipates and provides a model for similar applications in a larger social context."[53] His project, *The Real Costs* (2011) (Fig. 1.12), is a Firefox web browser plug-in that inserts CO_2 emissions data on air, bus, train, and car travel into e-commerce websites (such as Orbitz.com), giving real-time feedback on the impact of the travel along with the price. The user, seeing what their carbon footprint would be, can then make an informed decision. Using an existing online "infrastructure" (the browser interface and the website), Mandiberg's intervention can potentially alter future events through accessing and sharing of real-time data, empowering the user with important information and enabling him/her to act in a responsible way.

With a similar objective, my project, *S.OIL* (2012) (Fig. 1.13), entrusts users in operating a railway handcar to generate the electricity needed to power an electronically controlled irrigation system watering experimental grain crops and small video monitors. The project was inspired by the research being done at The Land Institute. Cofounded by Wes Jackson, The Land Institute explores the potential for a perennial food system through the manual hybridization (breeding rather than genetically altering) of perennial native grass seeds with that of annual grains. *S.OIL* links industrial agriculture to fossil fuel dependence and topsoil erosion through an immersive physical engagement.

S.OIL sets up contrasting agricultural systems: one focused on a biotechnological system reliant on fossil fuels and the other a perennial system using traditional methods of hybridization of grain crops with native grasses. The video playing on the three small monitors is a compilation of images of corn germination in labs, grown in row crops in the Midwest, and its ultimate processing. But the monitors get slowly taken over by the perennial plants growing in the three stainless planters, suggesting that an alternative to GMO (Genetically Modified Organism) -dominated agriculture is on the horizon. GMO corn has become the dominating cash crop and is produced and used as a raw material in a plethora of industrialized processes from sweeteners to automobile fuel.

The project focuses on the big picture concerning agriculture and its relationship to oil. With the rising popularity of supporting local farms as a means to provide for cities, the emphasis is on growing vegetables and fruits. But, can grains be grown on small local farms, and can they potentially disrupt industrialized agricultural production and distribution? The prospect of having perennial grains that provide large yields, contribute lower emissions (the plants' deep root systems sequester CO_2 and nitrogen, needing no synthetic fertilizers), and help to build topsoil for a sustained natural system agriculture may not be so far off in the future.

Laura Allcorn calls herself a designer, storyteller, and maker whose works reimagine how the world could be. *The Human Pollination Project* (2009) brings

[53]Edward A. Shanken, "Investigatory Art: Institutional Critique, Real-Time Systems, and Network Culture" (lecture).

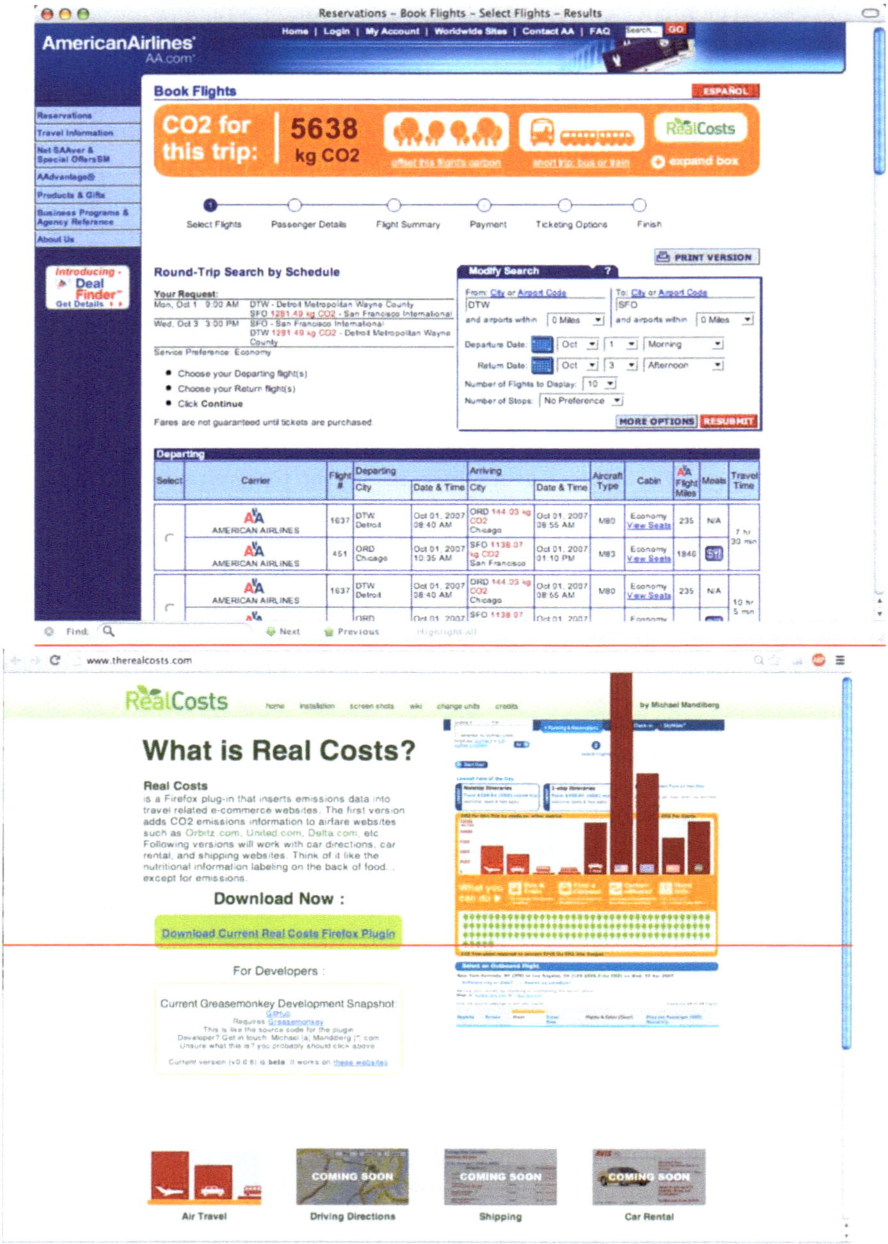

Fig. 1.12 Michael Mandiberg, *The Real Costs* (2007). therealcosts.com screenshot of American Airlines website with plug-in, CC BY-SA Michael Mandiberg 2007

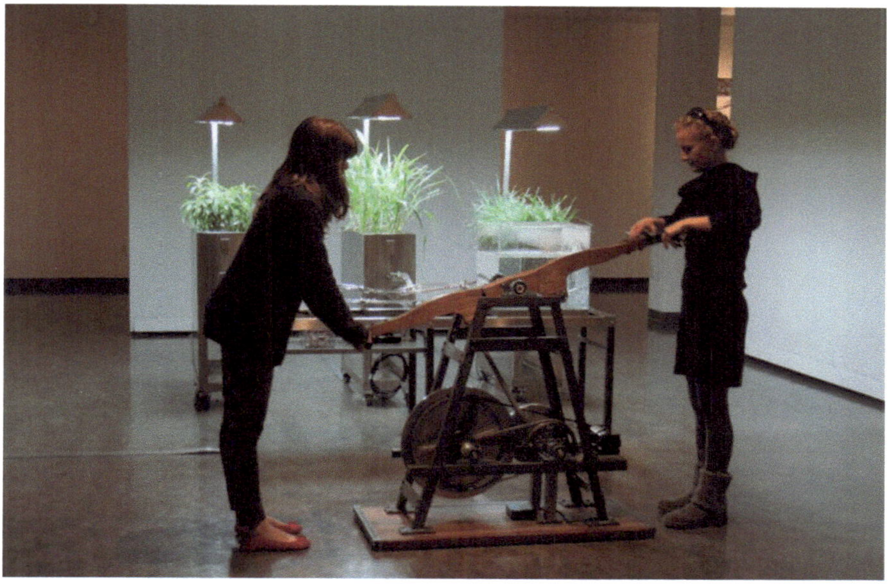

Fig. 1.13 Maria Michails, *S.OIL* (2012), a human-powered installation combining electronics, mechanical sculpture, video, and living plants

attention to our reliance on the services of bees. The project speaks of the future of our species. What if humans had to resume the job of the industrious honeybee? What tools would we need? The artist created a small-metal tool kit consisting of five objects hanging on a metal tool belt, delicately designed such as the tool belts worn at the hip by Victorian women trying to upkeep the house. Allcorn says it's meant to be a bit tongue-in-cheek because people do not want to do this – it's a laborious and tedious work. What would happen to our food supply, one-third of which is pollinated by honeybees, if their colonies collapsed? Humans have pollinated by hand for centuries, but to think about the scale and vastness of doing this ourselves puts things into perspective.[54]

The question of "What if... ?" is at the core of UK duo Anthony Dunne and Fiona Rabys series of design proposals, *Between Reality and the Impossible* (2010). What if we did run out of food because the population continued to grow as predicted? For this project, Dunne & Raby looked to evolutionary processes and molecular technologies to build DIY devices inspired by the digestive systems of other mammals, birds, fish, and insects. On the heels of synthetic biology, they "create "microbial stomach bacteria"... to maximize the nutritional value of the urban environment, making-up for any shortcomings in the commercially available

[54]See video interview of Allcorn at: https://dublin.sciencegallery.com/humanplus/human-pollination-project/

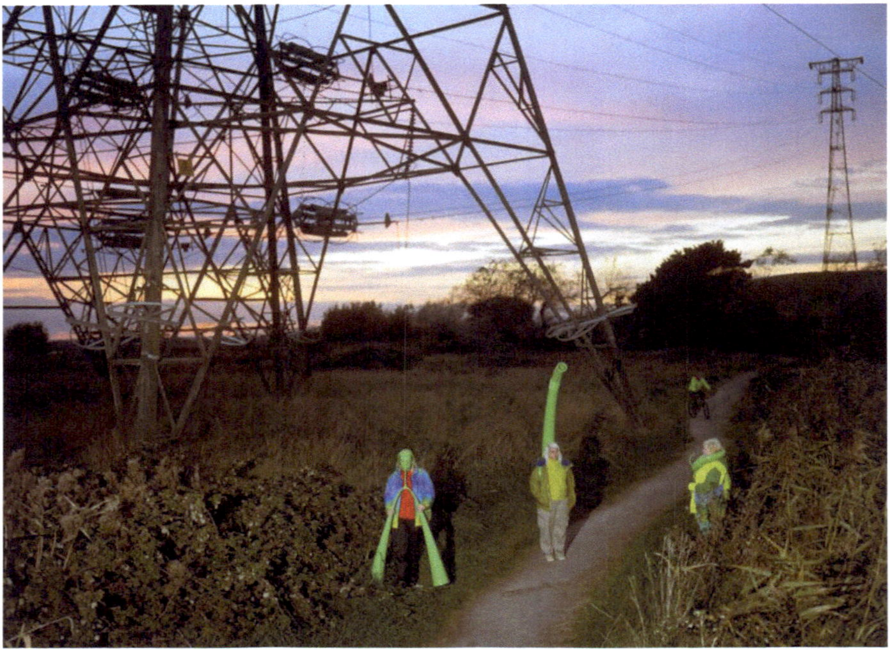

Fig. 1.14 Dunne & Raby, *Designs for an Overpopulated Planet: Foragers* (2009), exhibited at Between Reality and the Impossible (2010), commissioned by Constance Rubini for the 2010 St Etienne Design Biennale (Photo: Jason Evans)

but increasingly limited diet."[55] One of the scenarios, *Foragers* (Fig. 1.14), responds to the problem of food shortage, corporate control over our food system and supply lines, and the role of scientific and technological knowledge. It builds upon the already existing culture that exists on the fringe, including guerilla gardeners, garage biologists, hackers, and makers. As the new urban foragers, they become models of a speculative future.

The dominant discourse today celebrates the city as the future and the timing being ripe for innovation. The city offers smaller living quarters, closer neighborhoods, community cohesion, and cultural diversity. Public transportation means less driving and community gardens decrease economic disparity. Understanding the vital relationship between rural and urban ecosystems, both built and natural, within the context of a wider, even global perspective, will prove the greatest asset toward any innovation. It requires progressive thinking in planning and bold decision-making at the policy level, both of which are slow to be implemented, in order to integrate urban living systems with the region's ecological systems. When

[55]Artists' website, consulted on May 22, 2014. http://www.dunneandraby.co.uk/content/projects/510/0

included in decision-making, artists can bring fresh ideas and creative approaches to solutions. Artists, particularly those working with cross-disciplinary professionals and the public, are accustomed to navigating the necessary domains to realize projects.

Empowering a Culture of Sharing

Openness, in short is more than a commercial and cultural issue. It's a matter of survival.[56]

In the last decade, the explosion of Internet tools such as social media platforms and crowdsourcing sites has enabled social organizations, activists, and artists to reach mass populations as a means of getting their ideas, information, and projects out there and off the ground. Open-source technologies and a culture of sharing contribute to an openness of "institutional" or "expert" knowledge and can have a mobilizing effect on communities, particularly with vulnerable and/or underrepresented groups. These means have been especially useful in empowering individuals and communities in the face of environmental crisis, and artists' projects have been instrumental in many instances.

The ideals of open culture are at the core of the DIY movement, inspiring a return to making, modifying, or repairing without the aid of an expert. Do-it-yourself has been associated with consumer culture since the early 1900s.[57] Citizen science also has a long history. In the mid-1800s, William Whewell of England, who was studying ocean tides, enlisted the help of thousands of coastal communities on both sides of the Atlantic to measure and mark tide levels every 15 min, day and night for 2 weeks, compiling over a million data points.[58]

Where DIY is associated with the arts and crafts movement, Citizen Science is widely being used by the scientific community as a means to gather data farther afield than would be possible with a single researcher. The two often overlap, with projects being intrinsically collaborative in nature and often incredibly innovative. These methodologies can create a platform for critical engagement with regard to the politics behind scientific and technological information and access. When these techniques are used toward public engagement for environmental monitoring, protection, and awareness, information can have a powerful effect: mobilizing communities to take action, standing up to corporate manipulation, and swaying policy decision-making in their best interest.

The self-sufficiency that is inherent in doing it yourself has been especially relevant to issues concerning the food stream, as evidenced by a widespread

[56]John Thackara, "Into The Open," online: http://opendesignnow.org

[57]http://en.wikipedia.org/wiki/Do_it_yourself

[58]Caren Cooper interview with Diane Toomey, "How Rise of Citizen Science is Democratizing Research," online at: http://e360.yale.edu/content/print.msp?id=2733

discontent with its control by multinationals. Cities are massive consumers of food that travel great distances, even around the globe. The push for locally produced food is seeing the revival of small farms close to the city. "I'm not arguing for local food because it tastes better or because it's better for you. I'm arguing that we have no choice. [In] a world with less oil, we need the kind of small, mixed farms that can provide their own fertilizer and build their own soil," says small farmer, Pete Johnson, from Craftsbury, VT. Johnson has helped pioneer year-round farming by building solar greenhouses that move on tracks so that he can cover and uncover the crops when needed. His farm is supported by local community members (CSA as it is widely known in cities) who buy "shares" in advance of the growing season. The community invests in the farm, and everyone shares the risk with the farmer, a model that subverts commodity trading of food and the large-scale production and distribution industry along with it.[59]

Food growing can be one of the most defiant acts a civilian can participate in. To grow your own food to barter with your neighbors is a dissent against the capitalist system. In September 2012, after being silent for the entire growing season, the city of Toronto ordered its parks department employees to destroy the free community garden planted by Occupy Gardens in the People's Peas Garden at Queens Park (near the city hall). The city did so without warning, dumping the rare heirloom plant species just before harvest was to commence during a festival planned by the community. In a time of food crisis, this act sends a volatile message to the people: the establishment will not be countered. Community members fought back and took to social media, spreading the word to media outlets and quickly organizing to continue the festival despite the setback.[60]

When Agnes Denes planted her monumental public art project, *Wheatfield – A Confrontation* (1982), she did so as a comment on "human values and misplaced priorities."[61] The two-acre field of golden wheat replaced "a rubble-strewn landfill near Wall Street and the World Trade Center in lower Manhattan (now the site of Battery Park City and the World Financial Center)."[62] Denes cleared the lot of debris and brought in truckloads of soil. With the help of assistants and volunteers, an irrigation system was put in place, and the field was maintained over a 4-month period until harvesting could begin. The paradox of the wheat field amidst the backdrop of the world's most powerful symbol of capitalism is a clear defiance of commodity trading of wheat, the price of which is determined at the end of a stock trader's pencil, completely removed from the cycle of growth and the reality of global food shortages.

More than just a symbolic gesture, Denes' two-acre crop yielded nearly 1,000 pounds of grain (approximately 16 bushels), less than a quarter of the space's full-growing potential, and equal to approximately 672 loaves of white bread or 960

[59]Pete Johnson, quoted in "Breaking the Growth Habit" by Bill McKibben (63).

[60]http://toronto.mediacoop.ca/newsrelease/13169

[61]http://www.greenmuseum.org/content/artist_index/artist_id-63.html

[62]http://en.wikipedia.org/wiki/Agnes_Denes

loaves of whole wheat bread (based on one pound of flour per loaf, the grains going much further if left whole).[63] What if small-scale grain crops could be grown with perennial seeds (such as those developed by The Land Institute) that would reduce the labor involved and sustain the soil? How many acres are needed to feed a city the size of New York City? Can wheat and other grains grow in variable locations rather than centralized on big monocrops in the Midwest? Would it disrupt the market system if wheat could be grown and owned locally, outside of the market? When it comes to staple foods such as grains, is a DIY approach possible? Is the idea of small-scale farming, such as community gardens and local family farms, merely symbolic and hopeful, or can such methods really feed a large metropolis?

To answer such questions, one needs to go beyond metrics and consider the value of ideas that catalyze social transformation via projects, such as Denes'. Aesthetics play a key role in injecting into the collective conscience the state of our ecological life-support systems and, in this case, the overall problem of agriculture and the unfair distribution and waste of food. More than 30 years later, *Wheatfield* continues to have a significant cultural value long after it is gone from the site. *Victory Gardens* (2007–ongoing) (Fig. 1.15) by Amy Franceschini and the collective she founded, Futurefarmers, continues the discourse left off from *Wheatfield*, quietly challenging the large-scale farming model by creating and supporting urban agriculture planted on empty land, be it public or private.

Victory Gardens garners its name from the WWI and WWII initiatives by the US government that encouraged and supported community gardens as a means of feeding the US population so that farm production could be reserved for the war efforts. San Francisco was known to have the most vibrant system of civic gardens, the original garden planted at Golden Gate Park in 1943. In 2007, Futurefarmers (Fig. 1.16) began the project with a prototype of a rainwater harvest and documentation of items to be used in a trial garden. The "bikebarrow" and "pogoshovel," two of the many social sculpture objects that would follow, were part of the aesthetic tactic, what Franceschini calls "propaganda in sculptural form," aesthetic objects intended to raise people's interest in the broader project. Promotional posters, inspired by the art of the original Victory Gardens, were also disseminated, and a starter kit, from seeds to instructions, was created for gardeners interested in learning how to garden (Fig. 1.17).[64]

The collective approached city officials to support gardens and now has over 15 gardens that receive city funds. They were also successful in getting approval to create a permanent demonstration garden at the original site of Victory Gardens and have, since 2009, a large garden at Golden Gate Park where they also hold workshops and demos. The momentum achieved in a mere few years to convert open spaces to small urban gardens, thereby creating a wave of alternative forms of urban

[63] A bushel of wheat weighs 60 lbs. Gene Logsdon in Small-Scale Grain Raising conservatively estimates 40 bu/acre which converts to 42 lbs of white flour and 60 lbs of whole wheat flour per bushel.

[64] Paul Schmelzer, "Practical propaganda," Art21 blog, March 1, 2009.

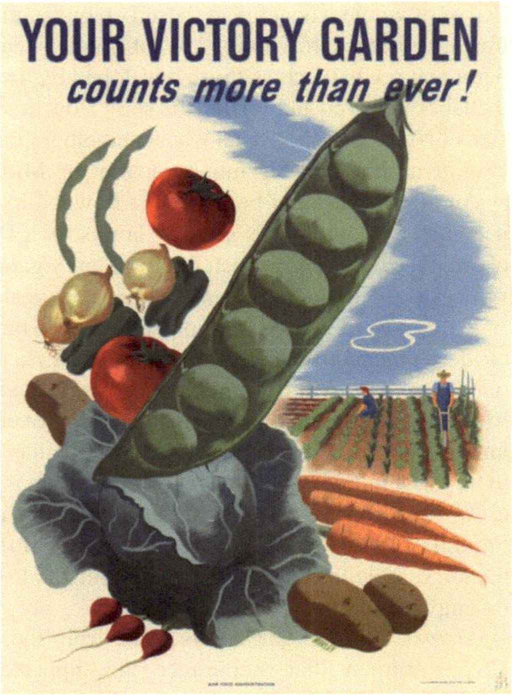

Fig. 1.15 US government. *Victory Gardens* poster (1945). Your victory garden counts more than ever! 1945. CC-PD 1.0 (Morley, Hubert. Your victory garden counts more than ever! [Washington, DC]. UNT Digital Library. http://digital.library.unt.edu/ark:/67531/metadc544/. Accessed 2 Oct 2014)

agriculture using ecologically compatible practices and reducing the production chain of large-scale farming, is a testament to the power that communities have when empowered with knowledge, tools, and a motivating factor. As a means of mapping and social networking, a website was set up for participants to register the location of their garden, public or private open space for a potential garden, or to list what excess food they may have to share. *GardenRegistry.org* acts as an online cooperative with the goal to "quantify the total amount of potential food production zones within the city in order to consider how to support, connect and cultivate these spaces."[65]

Several of the artists discussed here create and use objects as a means to stimulate transformations in society. They affect engagement through an aesthetic prerogative with the ultimate goal being the participatory aspect of the work. Joseph Beuys termed this as "social sculpture." Beuys believed that by applying creative thinking in all disciplines and working collaboratively, changes could come about in the environment, be it the political system, the economy, or a community, and this need for change was urgent.

Marjetica Potrč navigates between art, design, architecture, and social science, employing a "case-study" approach to her creative research, made visible in a

[65]http://gardenregistry.org/

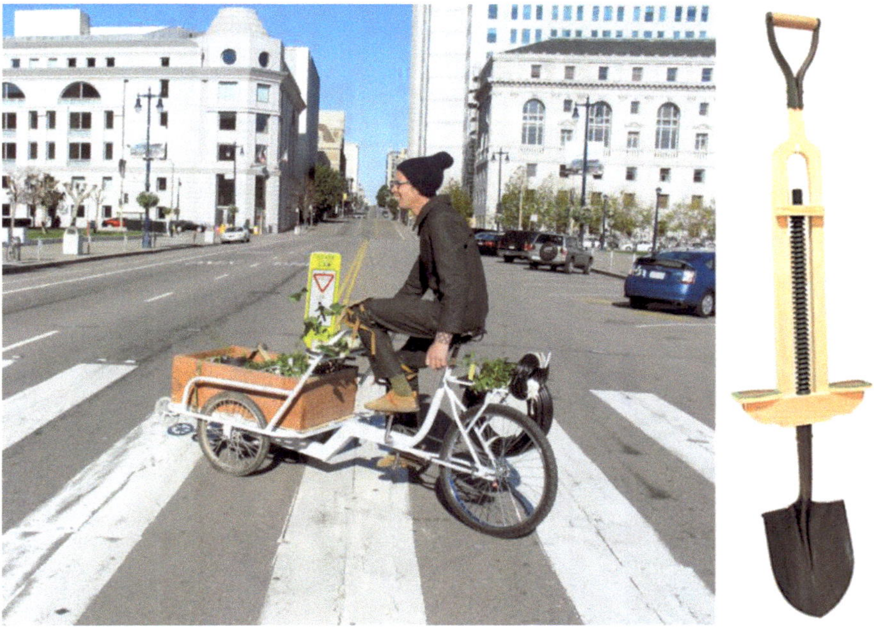

Fig. 1.16 Futurefarmers, *Victory Gardens*, and *pogoshovel* (2007–ongoing)

gallery context using two-dimensional drawings and sculptural installation. Her work comes out of an interest in the ad hoc architectural structures "exist[ing] around the edges and shadows of global metropolises" that she refers to as "informal cities" because of their impromptu formation.[66] She observes how these residents live under such conditions and how they solve infrastructural problems without municipal support, without permits, and sometimes outside the law. After spending 6 months in Barrio La Vega, an informal city on the outskirts of Caracas, Venezuela, she worked with the local residents to develop the project *Dry Toilet* (2003), an ecologically safe, waterless toilet in a district in Caracas that has no access to municipal water.

Her participatory design projects seek to provide aides to improve living conditions of impoverished populations who are especially vulnerable to climate change effects. In her ongoing *Power Tools Series*, Potrč re-presents small-scale, sustainably designed, commercial objects (such as solar-powered flashlights and clockwork cell phones) geared for residents from developing countries, "urban explorers" of the informal cities, as well as rural areas. One such object in the series, the *Hippo Water Roller for Our Rural Times* (2005), a large rolling drum with a handle that enables easier transport of water but also prevents the person from stepping on land mines

[66]Stephanie Smith, "Marjetica Potrč," Beyond Green, 108.

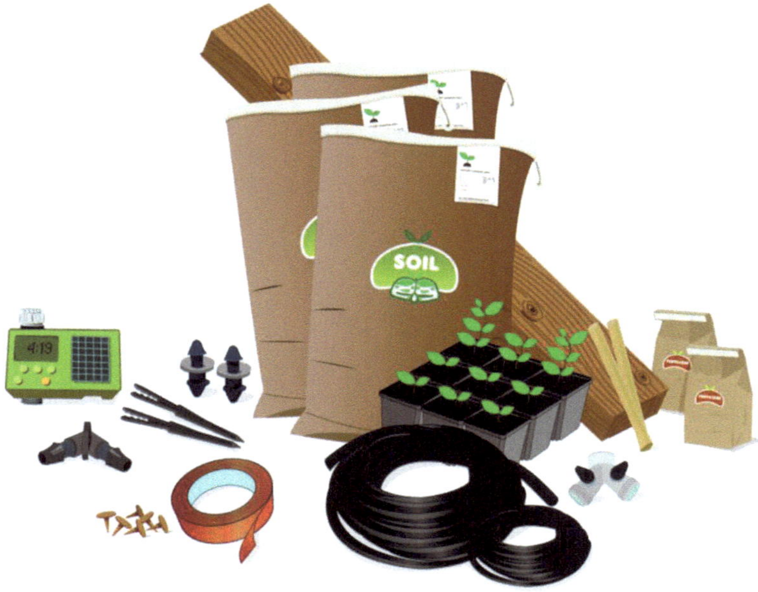

Fig. 1.17 Futurefarmers, *Victory Garden Kit* (2007–ongoing)

in the process,[67] makes a clear connection between geopolitical issues and the harsh social and natural environments endured everyday by these residents.

As Potrč embeds herself in cities of the developing world (some of the world's fastest growing), Nils Norman considers ad hoc spaces within the industrialized cities of Europe and North America. Drawing on a history of utopian experiments and thinking, such as New Mexico's Earthships and Brook Farm in Massachusetts, he creates alternative solutions to the homogenized consistency of urban regeneration planning. Using artists' books, digital drawings, models, and murals, he

[67] "Tests were conducted near Pretoria, South Africa, using Hippo Water Rollers to establish what the impact might be on a person should the roller trigger an anti-personnel land mine. A hippo Water Roller filled with 90 liters (20 gallons) of water was pulled over a land mine that had been planted in front of a soft cardboard model mounted on a steel frame. The shock wave and incredible heat (3,000°C) generated by the blast were absorbed so effectively by the water that not even a yellow flame was noticeable. Very little damage was evident on the cardboard model. In all three tests, indications were that no hospitalization would be required. Some bruising and lacerations may occur caused by bits of plastic from the roller." Source: http://www.rexresearch.com/hippo/hippo.htm. Also, "The Roller has been used in test cases as an anti-personnel demining device, whereby it is rolled along the ground to absorb the blast of landmines when filled with water." See: http://www.consultancyafrica.com/index.php?option=com_content&view=article&id=997:the-hippo-water-roller-technology-for-improved-access-to-water-&catid=90:optimistic-africa&Itemid=295

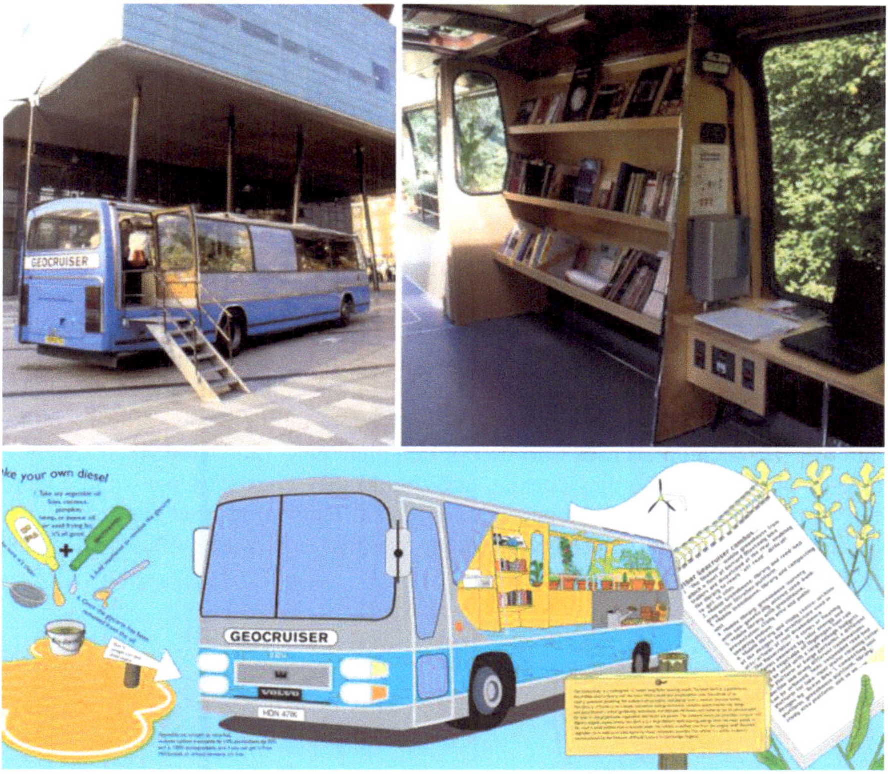

Fig. 1.18 Nils Norman, *Geocruiser* (2001–2004)

creates fantastical potentials for utopian urban spaces as a form of satire and irony to critique the business-as-usual capitalism.

In an interview, Norman responds to a question about his interest in adventure playgrounds and makeshift architecture as the "ideas that revolved around the concept of 'Non-Plan' planning." He imagines what it would be like "to experiment with the idea of taking a city area and removing all planning regulations, enabling local people to design and build whatever they wanted....radical models of alternative public space...."[68]

This is exemplified in one of his earlier projects, *Geocruiser* (2001–2004) (Fig. 1.18), a mobile public sculpture made of a refurbished old bus with a biodiesel-converted engine, solar panels, a greenhouse, and, at the front, a community library. The bus is an "information center devoted to city gentrification, experimental city

[68]Interviewed by Stephanie Smith, Beyond Green, exhibition catalogue, 26.

design, radical gardening, sustainable design, alternative energy, and utopias."[69] He considers the bus as a framing device and mobile propaganda machine. A public space where the urbanite can go for free to view the content, use the space (it has a solar-powered laptop and copier), and learn about the history of US utopian experiments in agriculture, economies, and communal living.[70] Just as Pulsa's *Harmony Ranch* aimed for a similar objective, but outside the city, Norman is more interested in situating his utopia within the city, giving equal importance to ecological issues within the context of public space and urban regeneration.

The ethic behind the non-plan, open hardware or software, crowdsourcing, do-it-yourself, street science, citizen science, and so forth are part of the people's common desire for an open society of sharing and equality. It goes without saying that the rise of this ethic and contemporary movement has been facilitated by new technologies. But technological advancements are a capitalistic enterprise mired in patents, competition, and exclusivity, completely the opposite of open culture thinking. Each camp accuses the other of being undemocratic. How, then, can new models of economic entities coexist, and should they coexist? Open-source ideas can empower amateur inventors, yielding radical innovations that can lead to the creation of technology-focused social enterprises that employ a new business model,

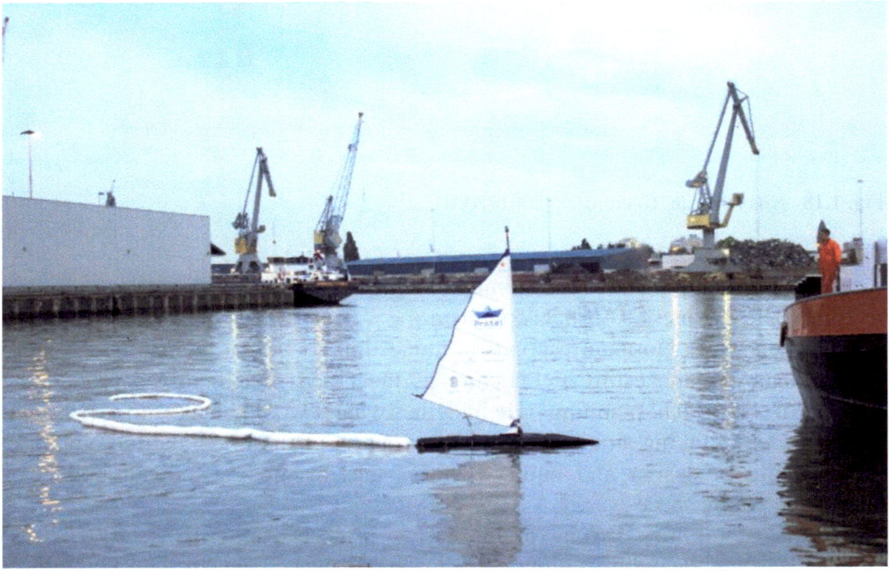

Fig. 1.19 Cesar Harada and Gabriella Levine, *Protei* (2011–ongoing)

[69]Nils Norman quoted in "Utopia Now: The Art of Nils Norman" by Jennifer Allen.

[70]Nils Norman quoted in "The Politics of Sustainability: Art and Ecology" by TJ Demos, 28.

placing human and environmental well-being over profit. These models place a monetary value on the future of social benefits that are accrued from their activities.

Cesar Harada and Gabriella Levine worked with an internationally assembled interdisciplinary team on the highly inventive project *Protei* (2011–ongoing) (Fig. 1.19), a revolutionary shape-shifting sailing robot that monitors and cleans ocean pollution. With combined backgrounds in music, glassblowing, sculpture, new media, performance, animation and film, interaction design, cancer research, and architecture, this duo's passion for DIY and open-source technology drives the newly formed social enterprise, *Scoutbots*, the home of *Protei*.

In 2011, Harada raised $30,000 on a kickstarter.com campaign to fund the prototyping of *Protei*, after quitting a lucrative job at MIT and moving to New Orleans to assist in the oil spill cleanup. Observing the techniques being used, most of which were inefficient and toxic to the ocean and fishermen's health, he designed the initial plans to a small, unmanned sailboat trailing a long material that could capture more oil than the fishing boats. Levine came on board to help prototype the project, implementing a mimetic, snake-like motion to the prototype. They made iteration after iteration with a young team of technologists with varied backgrounds in a rented warehouse in Rotterdam. They worked quickly to design, build, test, and then start again, until a well-functioning prototype could be presented for further funding.

From the technologies that they develop to their business model, Harada and Levine firmly believe that social change can happen only if a shift is made that opens the way for inclusiveness, collaboration, and sharing of tools and techniques in order to solve many of our environmental, social, and economic problems. They believe that profit can be made from sharing rather than patenting inventions. As such, *Protei* became an international network of shared technology, with contributors testing their own version of the project and then sharing it with the community. In his TED talk, Harada speaks about their philosophy of placing the environment and people first, technology and profits as secondary priorities to the enterprise.[71] *Protei* is being marketed as a tool available to companies, environmental activist groups, and smaller, simpler versions for educational programs.

Artists who utilize a DIY approach to their technological practice, sometimes without prior professional training, often bridge into the Citizen Science model. By working directly with the public or enabling those to collaborate through an online platform, the collection and sharing of information, for example, may result into an object, such as *Protei*, or a nonmaterial form, such as a website as a platform for open dialogue and engagement, expanding the concept of social sculpture.

Exploiting the Internet, despite it being increasingly controlled, artists with an environmental (or other social) agenda can reach a large global audience by using conceptually creative and innovative interactive sites that are not merely listings of eco-information but make meaningful connections for the users. Where *Protei* aims its focus on ocean toxins and waste, the web project *Superfund365* (2007)

[71] www.cesarharada.com

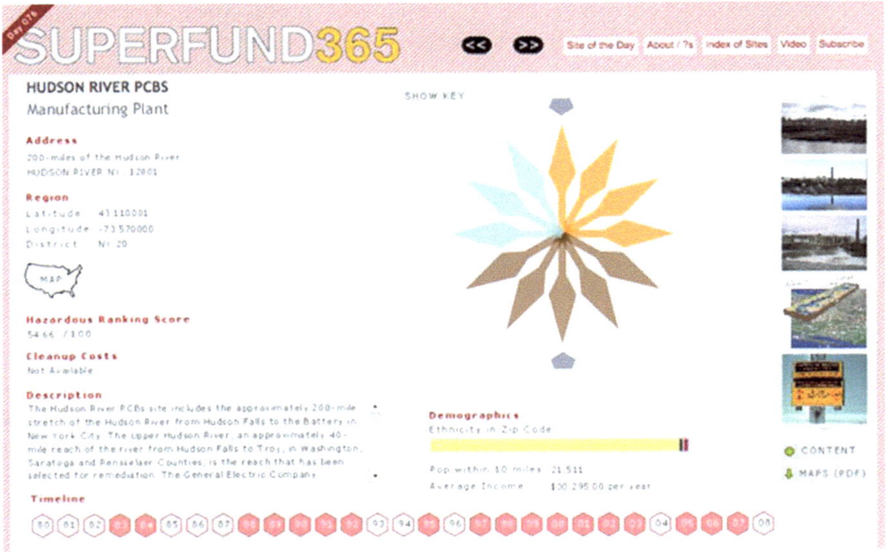

Fig. 1.20 Brooke Singer, *Superfund 365* (2007)

(Fig. 1.20) directs our attention to toxic waste sites on land. Artist Brooke Singer conceived of the interactive website as a means to visualize the enormous volume of EPA (United States Environmental Protection Agency) information from the government's National Priorities List and the Center for Public Integrity's list of the most dangerous Superfund sites.

After extensive research, Singer and her assistants narrowed the over 1,500 sites to 365, choosing some of the most toxic areas with more conclusive and complete data sets. She then plotted the sites on a map and set out to visit one each day, starting in New York City and ending, 12 months later, in Pearl Harbor, documenting each site. The website design is an attractive, clean interface that is simple and direct, based on a wheel with compass pointers that pop up information about that day's Superfund site. When scrolling over the pointers, information about the contaminants at the site come up as well as who the polluter was. Links to the Agency for Toxic Substances and Disease Registry (ATSDR) are also included, furnishing even more information. Users are encouraged to visit the Superfund sites nearest them to document, upload, and caption their images on the site's upload page.[72] As an extensively detailed resource, *Superfund365* stimulates a dialogue (literally available in the interface) and makes visible, through a ranking score of the polluters (such as GE, Dow, etc.) and the toxic sites that 1 in 4 Americans live near. *Superfund365* personalizes the data by adding demographic information into graphical visual, listing the number of populations at risk and making the data more meaningful.

[72]http://turbulence.org/Works/superfund/about.html

Three weeks after September 11, 2001, a cloud of dust from the rubble of the World Trade Center was still visible in the air in lower Manhattan. City officials were insistent that the air quality, although a nagging discomfort, was not a significant health problem. The EPA and the city's Health Department had repeatedly tested the air, and results showed only a few samples of heavy metals, asbestos, and other pollutants exceeded health safety levels. But the residents who lived in the area didn't buy it – they mistrusted the finding and trusted their own experiences, many of whom were complaining of rashes, sore throats, burning itchy eyes, and other pulmonary and overall ailments.[73]

What agency do people have when the science doesn't meet their reality? For months the city continued to insist that there were no serious health hazards to the lingering dust cloud as more people became sick, broadening the divide between science and public trust. What if tools were available to those residents that could accurately monitor the pollution in their immediate area enabling them to see the numbers? Would it have eased their concerns if the numbers match that of the experts' findings? Should numbers trump embodied experience? What obligations do experts have to incorporate community knowledge into their analysis?

The coproduction of knowledge between expert and amateur seems readily acceptable when the research led by a scientist has the assistance of public volunteers, enabling very large spatial and temporal scales that produce large data sets not normally possible with the scientist alone or even with a team.[74] One might conclude that it's the expert that primarily benefits from this collaboration. According to Cornell ornithologist, Caren Cooper, certainly the citizen scientist has enabled her research to ask "amazing questions that transcend single study sites." But she also observes "that these field experiences and this collaborative relationship between members of the public and scientists actually provide meaningful and really transformative experiences for individuals and for communities [because] it can change people's perspectives . . . it can also empower communities and individuals, because the whole point of science is to make reliable knowledge, and that's a powerful thing."[75]

When artists or artist groups engage in DIY and citizen science, but without the scientist as the principle investigator, they too are empowered, as agents of change, inventing and deploying tools that can in turn empower the public to bring about political action by the information they make attainable. The collective Preemptive Media – a group of interdisciplinary artists Beatriz da Costa, Brooke Singer, and Jamie Schulte – built a device that, had the lower Manhattan residents had access to, could have possibly influenced the experts' analysis and policymakers' decisions about what levels are acceptable.

---

[73] Jason Corburn, "Street Science," 1.

[74] Diane Toomey, "How Rise of Citizen Science Is Democratizing Research," online interview with Caren Cooper.

[75] Ibid.

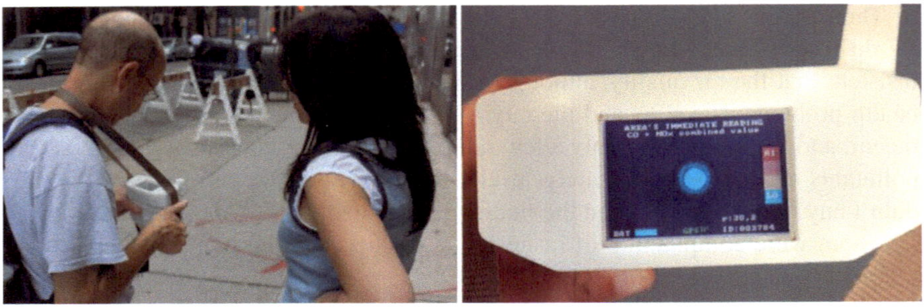

Fig. 1.21 Preemptive Media, *AIR* (2006)

Area's Immediate Reading or *AIR* (2006) (Fig. 1.21) is an ongoing experimental project that puts a portable device developed by the group into the hands of the public to monitor pollution levels in their local urban surroundings. The project was initially carried out in New York City and later deployed in Riverside and San Francisco, California, and Belo Horizonte, Brazil.[76] The device enables participants to self-identify the quality of the air wherever they may find themselves, tracing levels of nitrogen oxide, carbon monoxide and ground-level ozone, and chemical compounds associated with by-products of carbon combustion – or smog.[77] The difference between a personal device monitoring air quality and the monitoring of large-scale air quality index is that this monitoring is produced from a wide geographic distance, whereas the personal device monitors the exposure in a smaller area and with as much frequency as the user intends.

In his in-depth analysis of *AIR*, Michael Dieter writes, "As a techno-social experiment, the *AIR* project... function[s] as a subsidiary to government-based information-gathering through the *ethos* of open-source technology, peer production and activist-based politics."[78] The project not only investigates urban pollutants, it is a catalyst in the coproduction of knowledge by enabling collaboration between expert and amateur. By experimenting with already developed technology, the group took a DIY approach by modifying it in order to democratize the technology and the information, placing both into the hands of the general public.

Collaborative team Cary Peppermint and Leila Nadir formed EcoArtTech in 2005, merging her background in literature, critical theory, and environmental thought with his in solo performance, digital spaces, and technology. Their projects explore the complex relationship between ideas of nature (especially wilderness), technology, and culture. Never content to accept technology merely as a tool, by expanding upon their use, they imagine an environment where the natural and the technologized are symbiotically intertwined. Merging biological systems,

[76]http://www.pm-air.net/events.php

[77]Michael Dieter, "Processes, Issues, AIR: Toward Reticular Politics," 1.

[78]Ibid., 2.

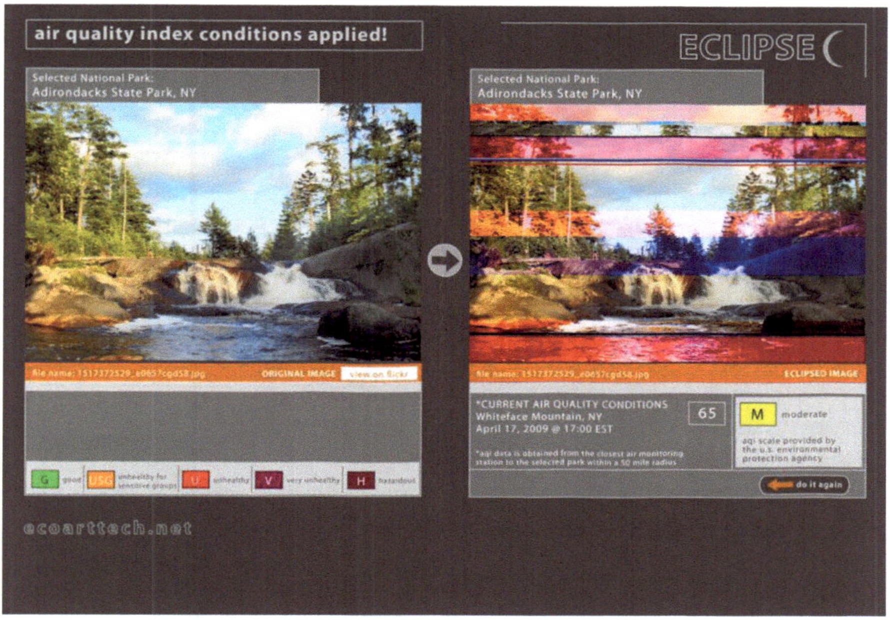

Fig. 1.22 EcoArtTech, Eclipse (2009)

primitive technologies, traditional narratives about nature, theories of modernity, and new media technologies, their projects might take the form of an architectural intervention, a solar-powered rover with a networking interface, a website linking city pollution to nearby national parks, or a mobile application enticing us to explore the "wilderness" in our urban landscape.[79]

The interactive web-based project, *Eclipse* (2009–2010) (Fig. 1.22), is an attempt to link the city with the country, demonstrating how urban activity, such as air pollution, affects ecosystems beyond its boundaries. The interface of *Eclipse* searches Flickr for idealized images of nature scenes based on the user's selected national or state park. Sets of algorithms affect the color, saturation, and contrast in the foreground, corrupting the pristine landscape according to real-time air quality index values of the nearest city. The higher the value (signifying greater pollution), the more the park image becomes distorted. This digital disruption of "wilderness" calls attention to the connection between the "dirty" city and "unspoiled" nature, visualizing the invisible damage accrued on natural systems.[80]

In a series of interconnected recent works, the artists recontextualize the "outdoor explorer" as one who takes on the urban wilderness adventure of dis-

---

[79] http://ecoarttech.org/

[80] Amy Lipton and Patricia Watts, "Public Art Ecology: From Restoration to Social Intervention," 52.

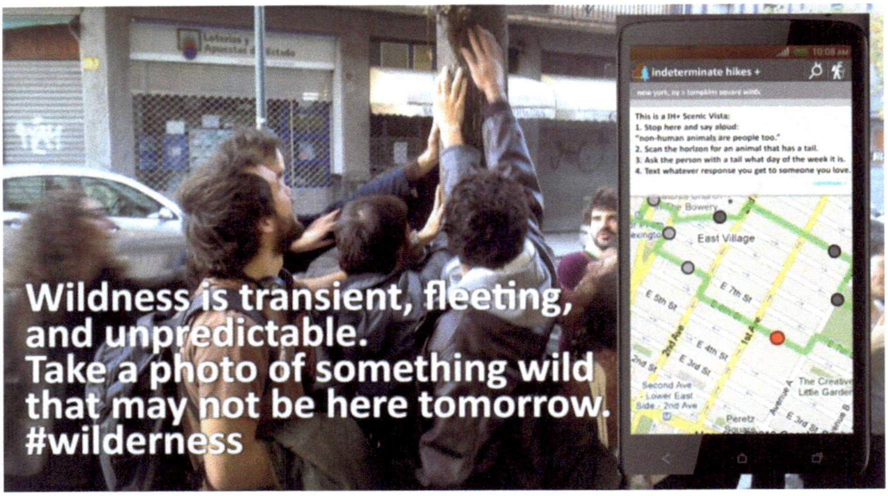

Fig. 1.23 EcoArtTech, *Indeterminate Hikes+* (2012)

covery. In the words of Peppermint and Nadir, "Our projects weave into one another, like an ecological web. *Basecamp.exe* intertwines with our *Indeterminate Hikes+* iPhone/Android hiking app, *Fir Tree Combinators*, and *Wilderness Collider*. Together, they create an interconnected network of new media, participatory eco-artworks, all of which seek to foster study and observation of the diverse ecological landscape of twenty-first century life."[81]

Indeterminate Hikes+ (2012) (Fig. 1.23) takes users on walking tours of the city, engaging them in "renewing awareness of biological, cultural, and media ecologies."[82] By reappropriating smartphones that are normally used for rapid communication, the artists subvert this by creating an application that does precisely the opposite and what would be expected of a remote wilderness experience – slowing down and taking notice of one's environment. *Indeterminate Hikes+* can be experienced in two ways: as a self-guided tour or as part of the *basecamp.exe* (2013–2014) installation with an EcoArtTech guide. Either way, the "hikers" can tweet or text their snapshots which are then collected by the generative web art project *Wilderness Collider* (2013) and projected into the indoor space where *basecamp.exe*, a workshop installation, is situated.

Wilderness expedition base camps provide supplies for backcountry recreation or exploration. *Basecamp.exe* (Fig. 1.24) mimics this by "creat[ing] a campsite for the modern world, comprised of digital, natural, and industrial materials collected by

[81] http://ecoarttech.org/

[82] Ibid.

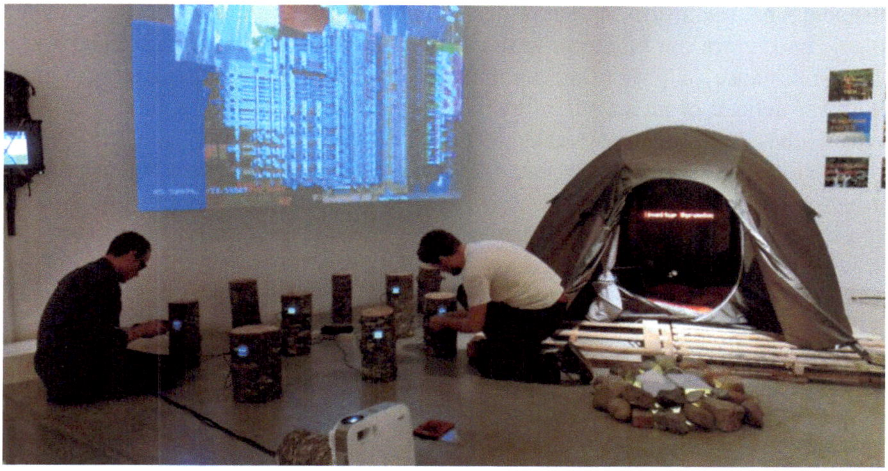

Fig. 1.24 EcoArtTech, *basecamp.exe* (2013–2014)

hikers exploring their immediate environment."[83] The projects imagine the "ecological city" where humans, human-built systems, and the natural world coexist. The artists' exploitation of new technologies combined with objects synonymous with wilderness adventuring obliterates the expected distance between traditional thinking about "wilderness" or "nature," the city, and technology. They point our attention to the city's ecosystems and the unaccustomed perception of urban "wildness." Taken further, the works speculate on the potential for the coexistence of the city with the wider bioregion.

Ecologically directed artworks may challenge our assumptions about nature and wilderness, such as EcoArtTech does directly, but they do not go so far as to eliminate the idea of nature, nor do they reiterate traditional objectifications or align with any romantic notions about nature. Science has enabled us to view the natural environment with greater understanding of the interconnectedness of all living things as part of a whole. We understand to a large extent how the earth's systems function and how they interact, sustaining life along with it. Fostering stewardship is the first step to a biophilic relationship with nature and its services, and artworks can promote this through an aesthetic engagement.

Artworks, such as Sonfist's *Time Landscape*, Kellhammer's *Cottonwood Community Gardens*, and Johanson's *Endangered Garden*, look to permanently affect the natural history of an urban space by remediating the land to a productive natural landscape, benefiting urban infrastructure while providing the opportunity for the return of a biophilic relationship. desertArtLAB engages a community through participatory remediation, in the process highlighting a cultural history in relation to

[83]Ibid.

the area's natural history. These projects, though, have a function in mind; they are poetic and return our attention to the subjectivity of our relationship with nature. The importance of promoting a subjective "reconnection" to the natural world, however defined or mediated, at the minimum, raises awareness and, at their most effectiveness, can empower a change in perspective and behavior.

When artists' projects intentionally incorporate a systems approach that includes urban infrastructure, by default, they bring bioregional systems into the picture. In particular are projects that focus on food systems, water, energy production, or wildlife, since these flow from elsewhere into and through the city. For example, Brain's *Coin-Operated Wetland* connects us directly to our water consumption and the services provided by a wetland. Her project is on the heels of Haacke's *Rhinewater Purification Plant* that raises critical awareness of the source of pollution in the Rhine River and the responsibility of the municipal sewage plant that dumps sewage in it. Others approach whole systems and infrastructure in a speculative mode, such as Mattingly's *Waterpod* and *Flock House*. Any notion of a "passive deferral of the public's responsibility to 'scientific expertise and governmental authority'"[84] is ameliorated through a physical, emotional, and intellectual public engagement.

I used to have, what some would call, a "romanticized" view of nature where my understanding of it was derived from the physical senses and processed through an embodied awareness. My knowledge of how nature worked was limited in terms of factual information but was open to perceiving its function through a reverential lens. I've since learned that this view is problematic for its objectification of nature, divorcing it from the social, political, and technological processes.[85] Although there is some truth to this, I was not so naïve as to think my consumptive behavior was completely removed from the degradation I was witnessing in "wilderness" areas subsumed by industrial activity.

Ecological writer, Timothy Morton, suggests that we must relinquish the idea of nature in order to have a properly ecological view. In other words, the data-filled, rational view of the natural world is what may help save it.[86] I would only partially agree with Morton on this general point. When addressing ecological issues through factual knowledge and confronting one's responsibility of collusion in the excessive consumption of natural resources, waste, and pollution, the influence of aesthetic valuing of nature needs to be part of this subjectivity.[87] After all, it is a subjectivity that awakens one to their actions in their everyday life and its connection to environmental degradation. This process of "subjectification [is] not rooted in

[84] TJ Demos, "The Politics of Sustainability," 18.

[85] TJ Demos, "The Politics of Sustainability," 20.

[86] Timothy Morton, "Ecology without Nature: Rethinking Environmental Aesthetics".

[87] I realize I am being unfair as I do not give full treatment of Timothy Morton's text, *Ecology Without Nature*, but rather interpret the overarching view he presents.

science [as we have seen with the lower Manhattan residents] but instead embrace[s] a new "ethico-aesthetic" paradigm as their primary source of inspiration."[88]

The discerning, even appraising, of the beauty of nature is inescapable to human perception and, as evidenced by the cultural significance placed on the protection of national parks, monuments and preserves, is valuable. One's sense of awe of nature is deeply affecting. Would dismantling this notion increase protection of the natural world? Would expanding this notion toward our immediate cities and bioregion promote a conservation ethic in our everyday consumption of natural resources? Would it promote a broad biophilic relationship if there were an understanding of ecosystem functioning?

Enabling a biophilic relationship through an aesthetic, as well as an understanding of nature's systemic function, can perhaps shift the culture-nature dynamic. Artists can effect this through a reimagining of the urban territory in relationship to the broader bioregion. It is here that ecologically directed artworks have the potential to make a significant impact – in the perceptions and actions of a connected, engaged, and mobilized citizenry.

Artists' Websites

Thank you to all the artists who provided images. For more information about the artworks cited here, please consult the following websites:

desertArtLAB http://www.desertartlab.com/
Alan Sonfist http://www.alansonfist.com/
Oliver Kellhammer http://www.oliverk.org/
Patricia Johanson http://patriciajohanson.com/
Maya Lin http://www.mayalin.com/
Beatriz da Costa http://bdacosta.net/
David Wicks http://sansumbrella.com/
Tim Collins and Reiko Goto http://collinsandgoto.com/
Mary Mattingly http://www.marymattingly.com/
Hans Haacke http://www.paulacoopergallery.com/artists/HH
Tega Brain http://www.tegabrain.com/
Natalie Jeremijenko http://nataliejeremijenko.com/
Tiffany Holmes http://tiffanyholmes.prosite.com/
Andrea Polli and Chuck Varga http://www.andreapolli.com/
Michael Mandiberg http://www.mandiberg.com/

[88]Mohsen Mostafavi, "Why Ecological Urbanism? Why now?" Discussing Felix Guattari's ethico-aesthetic paradigm, in which Guattari argues that for there to be a radical change to the ecological crisis we face, "a relational and holistic approach to our understanding of ecological issues" must be achieved, and that emphasis is placed on the "interrelations between individual responsibility and group actions," 22.

Maria Michails http://treiastudios.net/
Laura Allcorn http://www.lauraallcorn.com/
Anthony Dunne and Fiona Raby http://www.dunneandraby.co.uk
Agnes Denes http://www.agnesdenesstudio.com/
Amy Franceschini and Futurefarmers http://www.futurefarmers.com/
Marjetica Potrč http://www.potrc.org/
Nils Norman http://www.dismalgarden.com/
Cesar Harada and Gabriella Levine http://protei.org/
Brooke Singer http://www.bsing.net/
Preemptive Media (Brooke Singer, Beatriz da Costa, Jamie Schulte) http://www.preemptivemedia.net/
EcoArtTech (Leila Nadir and Cary Peppermint) http://ecoarttech.org/

References

Allen J (2002) Utopia now: the art of Nils Norman. ArtForum. Retrieved May 15, 2014. http://artforum.com/index.php?pn=interview&id=2281

Barker TS (2012) Information and atmospheres: exploring the relationship between the natural environment and information aesthetics. M/C J 15(3). Retrieved May 27, 2014. http://journal.media-culture.org.au/index.php/mcjournal/article/viewArticle/482

Barthel S, Isendahl C (2012) Urban gardens, agriculture, and water management: sources of resilience for long-term food security in cities. Ecol Econ 86(2013):224–234. Retrieved December 16, 2013. http://www.elseview.com/locate/ecolecon

Burnham J (1969) Real time systems. Artforum VIII(1):49–55. Quoted in Shanken EA (2012) Investigatory art: real-time systems and network culture. NECSUS. Eur J Media Stud 1(2): 77–89. Amsterdam University Press. Retrieved May 10, 2014. http://www.necsus-ejms.org/investigatory-art-real-time-systems-and-network-culture/

Brain T (2012) Chapter 4: Coin-operated wetlands and Chapter 3: Explicating environments, kilowatt hours. The politics and poetics of coexistence: experiments at the intersection of art and environmental engineering. MA thesis, Queensland University of Technology

Clancy P (Pulsa Group) (1972) The city as an artwork. In: Kepes G (ed) Arts of the environment. George Braziller, New York, pp 208–221. Retrieved May 10, 2014. https://www.sflan.net/details/TheCityAsAnArtwork

Collins JP, Kinziig A, Grimm NB et al (2000) A new urban ecology. Am Sci 88:416–425

Corburn J (2005) Street science: community knowledge and environmental health justice. MIT Press, Cambridge, MA

da Costa B (2008) Reaching the limit: when art becomes science. In: da Costa B, Phillips K (eds) Tactical biopolitics – art, activism, and technoscience. MIT Press, Cambridge, MA, pp 365–385

Demos TJ (2009) The politics of sustainability: art and ecology. In: Radical nature: art and architecture for a changing planet 1969–2009. Barbican Art Gallery, London, pp 17–30. Retrieved March 21, 2014. http://www.ucl.ac.uk/arthistory/about_us/academic_staff/dr_tj_demos/Demos-Politics_of_Sustainability.pdf

Demos TJ (2012) Art after nature. ArtForum, pp 190–197. Retrieved May 20, 2014. http://www.ucl.ac.uk/art-history/about_us/academic_staff/dr_tj_demos/Demos-Art_After_Nature.pdf

Dieter M (2009) Processes, issues, AIR: toward reticular politics. In: Australian humanities review, vol 46. ANU E Press. Retrieved May 27, 2014. http://www.australianhumanitiesreview.org/archive/Issue-May-2009/dieter.htm

Harper G (2011) Tim Collins and Reiko Goto: art has everything to do with it. In: Moyer T, Harper G (eds) The new earthwork: art action agency. ISC Press, Hamilton, pp 118–120

Holmes T (2007a) 7000 oaks and counting (2009). In: Speculative data and the creative imaginary – shared visions between art and technology. Exhibition catalogue. Rotunda Gallery, National Academy of Science, Washington, DC, pp 1–44

Holmes T (2007b) Eco-visualization: combining art and technology to reduce energy consumption. Updated version of paper originally published for C&C'07, Washington, DC, pp 1–13. Retrieved April 16, 2014. http://www.academia.edu/2996779/Eco-visualization_Combining_art_and_technology_to_reduce_energy_consumption

Hou J (2006) Hybrid landscapes: toward an inclusive ecological urbanism on Seattle's Central Waterfront. In: Cheng R, Tripeny PJ (eds) Getting real: design ethos now. ACSA 94th annual conference, March 30th to April 3rd, 2006. Salt Lake City, Utah, pp 245–250

Intergovernmental Panel on Climate Change, Working Group I (2013) Climate change 2013: the physical science basis. September, 2013. Retrieved March 10, 2014. https://www.ipcc-wg1.unibe.ch/

Intergovernmental Panel on Climate Change, Working Group II (2014) Climate change 2014: impacts, adaptation, and vulnerability. Report number 5, Spring, 2014. Retrieved March 10, 2014. http://www.ipcc.ch/report/ar5/wg2/

Kepes G (1972) Art and ecological consciousness. In: Kepes G (ed) Arts of the environment. George Braziller, New York, pp 1–12

Lipton A, Watts P (2011) Public art ecology: from restoration to social intervention. In: Moyer T, Harper G (eds) The new earthwork: art action agency. ISC Press, Hamilton, pp 52–61

Margolin V (2005) Reflections of art and sustainability. Beyond green: toward a sustainable art. Exhibition catalogue, pp 20–29. SMART Museum of Art, University of Chicago and Independent Curators International, New York

McKee Y (2008) The public sensoriums of Pulsa: cybernetic abstraction and the biopolitics of urban survival. Art J 67(3):46–67

McKibben B (2010) Breaking the growth habit. Scientific American, April 2010, pp 61–65

Messersmith-Gavin P (2011) Between social ecology and deep ecology: Gary Snyder's ecological philosophy. The Anarchist Library, pp 4–26. Retrieved June 4, 2014. http://theanarchistlibrary.org/library/paul-messersmith-glavin-between-social-ecology-and-deep-ecology-gary-snyder-s-ecological-philos

Messina J (2014) Green tech powers up NYC companies. Firms find data-intensive ways to end inefficiencies wherever they lurk. Crain's New York Business, April 22, 2014. Retrieved May 8, 2014. http://www.crainsnewyork.com/article/20140422/TECHNOLOGY/304209994/green-tech-powers-up-nyc-companies#

Morton T (2009) Ecology without nature: rethinking environmental aesthetics. Harvard University Press, Cambridge, MA

Mostafavi M (2010) Why ecological urbanism? Why now? In: Mostafavi M (ed) Ecological urbanism. Harvard University Graduate School of Design, Lars Muller Publishing, Baden, pp 12–51

Platt S (2011) The planet according to Maya Lin: what is missing? And confluence project. In: Moyer T, Harper G (eds) The new earthwork: art action agency. ISC Press, Hamilton, pp 144–149

Pulsa (1972) The city as an artwork. In: Kepes G (ed) Arts of the environment. George Braziller, New York, pp 208–221

Schwendenwien J (2011) Breaking ground: Alan Sonfist, Helen Mayer and Newton Harrison, Mel Chin, Michele Oka Doner, Jody Pinto, and Aviva Rahmani (1991). In: Moyer T, Harper G (eds) The new earthwork: art action agency. ISC Press, Hamilton, pp 41–46

Science Gallery (2011) Human+: the future of our species. Exhibition catalogue. Trinity College, Dublin

Schmelzer P (2009) Practical propaganda: Amy Franceschini reinvents the Victory Garden. Art21 Blog, March 1, 2009. http://blog.art21.org/2009/03/01/practical-propaganda-amy-franceschini-reinvents-the-victory-garden/#.U4kj6BnI8lb. Accessed 30 May 2014

Sekercioglu CH (2011) Ecosystem function and services. In: Sodhi NS, Ehrlich PR (eds) Conservation biology for all. Oxford University Press, New York, pp 45–72. Retrieved January 27, 2014. http://www.conbio.org/images/content_publications/Chapter3.pdf

Shanken EA (2012) Investigatory art: real-time systems and network culture. NECSUS. Eur J Media Stud 1(2):77–89. Amsterdam University Press. Retrieved May 10, 2014. http://www.necsus-ejms.org/investigatory-art-real-time-systems-and-network-culture/

Shanken EA (2013) Investigatory art: institutional critique, real-time systems, and network culture. Lecture, University of Memphis, April 2, 2013. Retrieved May 10, 2014. http://artexetra.files.wordpress.com/2009/02/shanken_invest_art_um_sm.pdf

Smith S (2005) Beyond green: toward a sustainable art. Exhibition catalogue. SMART Museum of Art, University of Chicago and Independent Curators International, New York

Spaid S (2002) Urban forests. In: Ecoventions: current art to transform ecologies. The Contemporary Arts Center, pp 89–96. greenmuseum.org/ecoartspace.org

Stairs D (1997) Biophilia and technophilia: examining the nature/culture split in design theory. Des Issue 13(3):37–44. MIT Press

Thackara J (2009) Metric or aesthetics? In: Green platform/Art ecology sustainability. Exhibition catalogue. strozzina centro di cultura contemporanea a palazzo strozzi, Italy. Retrieved May 28, 2014. http://www.strozzina.org/greenplatform/e_thackara.php#content

Thackara J (2011) Into the open. In: van Abel B, Klaassen R, Evers L, Troxler P (eds) Online design now: why design cannot remain exclusive. Available online June, 2012. http://opendesignnow.org/index.php/article/into-the-open-john-thackara/. Accessed 1 June 2014

Toomey D (2014) How rise of citizen science is democratizing research. Interview with Caren Cooper. Yale Environment 360. January 28, 2014. Online at: http://e360.yale.edu/content/print.msp?id=2733. Accessed 25 Apr 2014

Wallen R (2012) Ecological art: a call for visionary intervention in a time of crisis. Leonardo 45(3):234–242

Wicks D (2011) Record of creative work. MFA thesis in Design|Media Arts, University of California, Los Angeles. June 10, 2011. Retrieved March 10, 2014. http://sansumbrella.com

Chapter 2
Exploring Environmental Stewardship Through Data-Driven Practices

Tega Brain and Jodi Newcombe

Abstract What potential do artists working with environmental data in public space have for producing new forms of engagement with local environmental conditions? Operating on the edge of heavy bureaucracy, these types of data-driven artistic experiments probe the politics of environmental metrics and explore methods of engaging audiences with issues of environmental health. This discussion considers a small collection of cases studies representative of this growing field of practice. These are works by Natalie Jeremijenko and The Living, Tega Brain and Keith Deverell. The case studies considered are examples of strategic design, works that soften, reveal and potentially shift existing regulations and bureaucratic norms. In doing so they open up new possibilities and questions as to what the smart city is and how it might be realised.

Introduction

Public artworks that engage and perform environmental data in urban space construct and disseminate representations of local environmental conditions that usually escape quotidian perception. They materialise practices that seek a public engagement with environmental informational layers and material urban conditions. Operating on the edge of heavy bureaucracy, these experiments not only probe the politics of environmental metrics and explore methods of engaging audiences with issues of environmental health, but their productions also reveal the complexities of governance and cultural norms.

The projects and practices discussed in this paper have been gathered from Jodi Newcombe's experiences as director and founder of the Australian curatorial

T. Brain (✉)
Purchase College, State University of New York, New York, NY, USA
e-mail: tega.brain@gmail.com

J. Newcombe
Carbon Arts, industry Partner with the Centre for Art Society and Transformation,
Royal Melbourne Institute of Technology, Melbourne, VIC, Australia

School of Design, Queensland University of Technology, Brisbane, QLD, Australia
e-mail: jodi@carbonarts.org

F.T. Marchese (eds.), *Media Art and the Urban Environment*, Future City 5,
DOI 10.1007/978-3-319-15153-3_2

organisation, Carbon Arts, and from Tega Brain's experience as an artist often working in public space. Carbon Arts actively commissions, produces and curates creative work that responds to complex environmental challenges. It is an initiative that facilitates forms of engagement between the disciplines of science, engineering, arts and policy in order to bring new perspectives and means of public participation to environmental issues. Tega Brain, who has also worked closely with Carbon Arts, creates work exploring environmental infrastructures and the interpretation of environmental conditions.

This chapter discusses a series of data-driven public artworks that engage water quality data, aquatic biodiversity and electricity consumption data. These are *Amphibious Architecture*, a collaboration between Natalie Jeremijenko and architectural studio The Living, *Kilowatt Hours* by Tega Brain and *Building Run* by Keith Deverell with Carbon Arts. All three projects have been realised as temporary installations in public space but are yet to be installed permanently. Thinking through the successes and challenges of producing these works to date enables a productive and reflexive discussion of the existing state of play of the environmental performance and management of the contemporary smart city, as well as the criteria upon which successful urban design is measured. They offer small yet important opportunities to reflect on the current feasibility of realising experimental data works, why there is motivation to do so and what this reveals of the dominant framing narratives relating to environmental data in the urban space.

Our discussion of each project explores the tactics employed by the artist or designer and the audience engagement these produce. Each project is also considered as an example of strategic design, a term used by Dan Hill (2012) to describe projects that operate in multiple dimensions, often shifting and shaping policy frameworks, culture and governance in the process of their production. Hill refers to soft structures of policy, culture and governance as dark matter which includes existing organisations, culture, rules and regulation that constrain shifts in the inhabitation and operation of the city. He observes: '[Dark matter] gives a name to something otherwise amorphous, nebulous yet fundamental' (Hill 2012). This discussion considers how each artwork illuminates and potentially shifts the dark matter surrounding the environmental conditions and systems that they engage. It is structured in two parts: firstly, through considering the process of the work's implementation and, secondly, by exploring how the work reconfigures public space and engages audiences with otherwise peripheral systems.

Strategic Design and Adversarial Design

In *Dark Matter and Trojan Horses* (2012), Dan Hill focuses on reshaping cultures of public decision making at institutional and individual levels, thereby enabling systemic change through a practice he calls strategic design. Strategic design is described by a trajectory between the 'matter' (the designed object) and the 'meta' (the context) and a 'recalibration' between the two. If this trajectory is a football field, then the 'plays' available to the strategic designer map the ways in which matter can be used to shape meta, or how design objects can be used as strategic vehicles.

Hill describes several possible 'plays', including the MacGuffin, the Trojan Horse, the Platform and the Layer, which are described here as sequences that can be enacted to embed long-term strategic change into the institutions of urban governance. Whilst Hill focuses on the design and the built form, in particular an expanded concept of the building itself, these 'plays' could equally apply to works of socially engaged art that are driven by a strategic agenda.

The MacGuffin is a term borrowed from a film, which according to Hill is 'a plot element that drives the plot, but is somewhat inconsequential in the end'. In this way, an artefact, such as a building with a vision to be carbon neutral, can drive a change in a policy apparatus, such as an entire building code. The building in the end is a 'mere detail', but one that has the gravitational pull to affect a wider set of strategies. The MacGuffin is the motivation for strategic change.

The Trojan Horse is an artefact that possesses a host of 'hidden agendas'. Rather than presenting a one-off manifestation, the Trojan Horse offers many platoons, capable of strategically addressing the wider culture, pointing to replicable solutions through demonstration. This is facilitated by a multidisciplinarity that produces many strategies. Thus, an artwork acting like a Trojan Horse can contain the seeds of multiple strategic outcomes.

Hill also uses the metaphor of the layer, borrowed from Brand (1995) in conceptualising design projects as consisting of a series of layers moving at different speeds, 'shearing and slipping', a kind of informational exchange of learning and adaptation. For example, when designing digital platforms for a public service, faster layers (e.g. the user interface) can inform and be supported by slower layers (e.g. those regarding the delivery of core government services), thereby offering different opportunities for experimentation. Moving between layers is likened to moving between matter and meta, and structuring design in this way allows platforms such as this to have long life as layers can be specifically addressed and updated.

DiSalvo (2012) offers the concept of *adversarial design* to describe the area of cultural production that straddles contemporary art and design and that expresses or enables a particular political vision. Unlike strategic design, central to adversarial design is agonism, a political theory that approaches democracy as ongoing dissensus, where a contestation of power is intrinsic to its nature (Mouffe 2000). The work of adversarial design is to facilitate greater participation in these spaces of contestation in order to contribute to thriving democracies. DiSalvo's particular focus is designed artefacts and systems that make use of the qualities of computation as a medium. As the dominant technologies of the day, computation and how it both designs us and potentially enables new modes of political expression are a fundamental question for political design (DiSalvo 2012).

In categorising artefacts and systems into three groups – informational design, social robots and ubiquitous computing – DiSalvo reveals the different tactics of adversarial design that are afforded by each one. Moving from left to right, these strategies progress from a position of revealing or representing political conditions (revealing hegemony) to one of enacting and challenging these dominant conditions (reconfiguring the remainder), to then offering and facilitating alternatives (articulating collectives).

Dreaming of Smart Cities [Smart Citizens[1]]

The combination of the growing urban populations, environmental pressures and pervasive digital networks, connecting, measuring and monitoring these populations and the places they inhabit, continues to inspire the utopian idea that these technologies will adequately address environmental and civic challenges. The rhetoric of the smart city embodies the idea that the thickening informational layer of so many urban spaces will inevitably lead to more effective governance, a strengthened social fabric and improved environmental health (Haque 2012). This techno-optimism underlies the premise for numerous corporations such as IBM, Cisco and Siemens to be developing and selling smart city solutions, solutions that tend to reinforce top-down methods of governance with an emphasis on efficiency and optimisation as the cornerstones for improving urban life. Yet, as many critics (Greenfield and Kim 2013; Hollands 2008; Townsend 2014) have suggested, the rhetoric of these players hides a questionable political ideology and a certain naïveté about how thriving cities actually function and how environmental stewardship and pathways to positive environmental change are actually forged.

Projects like *Amphibious Architecture*, *Building Run* and *Kilowatt Hours* provide entry points to alternative perspectives to this debate. They emphasise the performance and presentation of environmental data in an attempt to shift the relationship between the observer and the system being monitored. By using spectacular, poetic or humorous strategies, they are interfaces that attempt to redistribute attention and produce empathetic relationships with environmental conditions. Again, to draw on Hill (2012), they can be thought of as strategically orthogonal strategies designed to deform existing regulations and institutional cultures.

[1]In reaction to this corporate vision of a techno-utopia with its authoritarian overtones has emerged an opposing vision – that of the smart citizen (Haque 2012; Hill 2012; Hemment and Townsend 2013). The smart citizen resists being seen as part of the city government's optimisation problem, being measured and managed, e.g. through behaviour change programmes, and is actively engaged through a form of participatory, self-organising governance. In the tradition of Jane Jacobs, the smart citizen manifesto calls for a diversity of responses; it embraces the city for its autocatalytic nature (de la Peña 2013), for its inherent 'messiness' and the messy nature of data, too (Haque 2012); and it calls for the engineering of serendipity in the face of a homogenising, neoliberal commercialisation of space and puts faith in the bottom-up processes of management.

Case Studies

The artworks discussed here have been realised in both Australia and the USA in the period from 2009 to 2013. They engage various environmental data streams, from water quality to electricity consumption patterns.

Amphibious Architecture

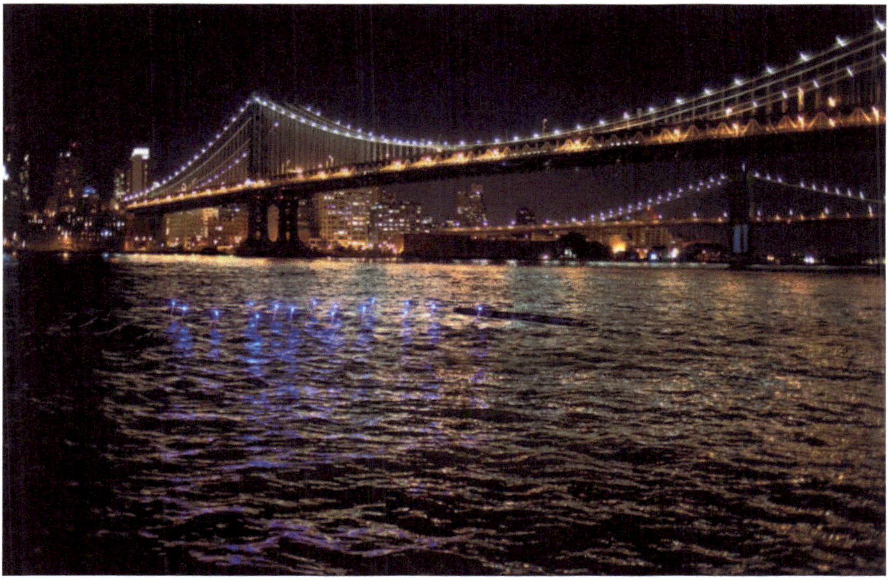

Amphibious Architecture by Natalie Jeremijenko and The Living

Amphibious Architecture is a temporary floating, interactive light display that was installed in two New York City rivers and commissioned for the *Toward the Sentient City* exhibition in 2009. The work is a collaboration between architects David Benjamin and Soo-in Yang (The Living) and artist Natalie Jeremijenko (The Environmental Health Clinic and Lab). Described by Benjamin and Yang (2011) as a horizontal, dynamic envelope akin to a building envelope or skin membrane, these networked arrays of sensors and LED lights mounted on grids of 8–10 buoys provide a portal through which to interpret the health of the surrounding urban river environments.

Dissolved oxygen sensors are submerged in the river, and they display an indication of water quality via the colour of a LED light that varies from red (poor) to blue (good). Ultrasonic fish finders indicate the presence of fish by activating other LED lights to flash on and off when movement is detected. Both sets of sensor data are relayed to a remote monitoring site and then displayed on a website. Furthermore, the public is invited to interact with the work through an SMS interface, entering into text conversations with fish and other charismatic fauna, including a beaver. This interaction is intended to deepen the public's empathy via a heightened awareness of both the existence and well-being of non-human species in an environment which is predominantly conditioned by humans (Jeremijenko, 2011, interview conducted by Jodi Newcombe).

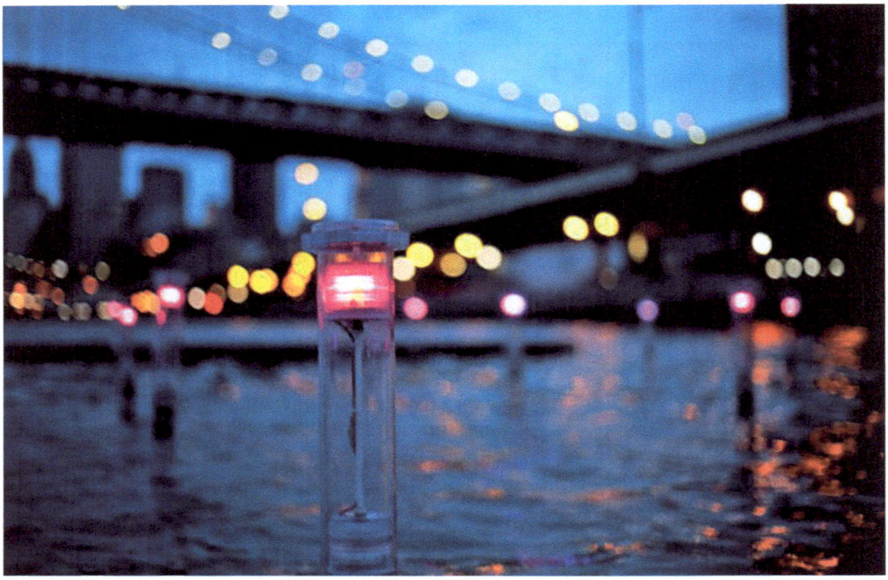

Amphibious Architecture detail

Although NYC water quality data is made available by the city's public agencies, *Amphibious Architecture* displays the real-time data as situated public spectacle. The work reshapes public space by offering new forms of environmental legibility aiming to shift audience interactions with the river. It is an open-ended experiment intended to be broadly educative and to inspire generate remediative action over time.

Amphibious Architecture was installed for 8 weeks; however, the interest and conversation generated by its public exposure have led to a permanent commission being undertaken at a new site along New York's foreshore, at a site called Eco-Pier 35. It is to be installed along with another bio-sensing artwork by Jeremijenko called *Mussel Choir* which harnesses the water filtration activities of live mussels (i.e. the bivalves), measuring their activity to trigger a sound piece in an adjacent park.

Benjamin and Yang (2011) describe *Amphibious Architecture* as a sort of horizontal dynamic envelope, an idea that resonates with Hill's notion of layers. The work can be thought of both as a physical informational layer and as a metaphorical one, operating like an outer layer of government whereby society can interface with issues of water quality management and potentially engage in a contestation of environmental health. Such experimental, censored skins, whether on the surface of buildings or waterways, shear against the slower-moving layers of government policy and regulation, potentially influencing in turn how governance is conducted. Furthermore, as a material layer, the work offers a framework for additional interventions such as media campaigns, performances, workshops and other artist or non-artist initiatives. It has the potential to operate as a platform to support other ways that informational interventions might deliver a new level of civic engagement.

Kilowatt Hours

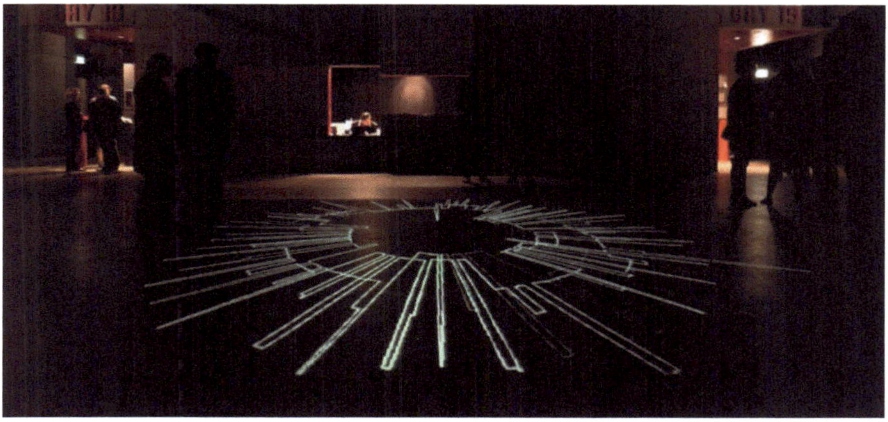

Kilowatt Hours by Tega Brain (Image credit, Alex Wisser)

Kilowatt Hours (2011) is an installation that visualises electricity consumption data from an energy metre installed on the circuit board of its exhibition space. Custom software publishes this data online every 6 s to the data-sharing websites Pachube[2] and Twitter. The work also consists of a large-scale floor projection of an illuminated visual generated from the measured data. This element of the piece references the Anthony McCall piece, *Line Describing a Cone*, where an ellipse is drawn over the course of an hour producing the form of a cone in a smoke-filled space. However, in *Kilowatt Hours*, the diameter of this ellipse changes in response to the magnitude of the electricity being consumed. As the electricity use increases, the circle diameter becomes larger, and the work thereby presents the viewer with an image showing fluctuations in energy use over time and back through the past

[2]Pachube has since been rebranded as Xively; see their website at: https://xively.com

hours. Just as McCall's work extends the medium of film to form, moving it beyond that of a flat representation, *Kilowatt Hours* uses software to extend form to become an index, a real-time metric of conditions in the surrounding building.

This work was installed at CarriageWorks, a public performance and exhibition space in Sydney, and required access to the electricity consumption data of this space. At the time, the space was neither measuring real-time electricity data nor making any of their operational data public. In light of this, an important part of the *Kilowatt Hours* piece was the installation of a hacked electricity metre onto the circuit board of the building's electricity infrastructure. The curators, anxious that the senior management would be unsupportive of publishing this data stream, requested that the metre be installed unsanctioned and that this activity be kept quiet. As per these instructions, the metre was installed unbeknownst to management, yet the electricity data was published online for the duration of the exhibition.

Building Run

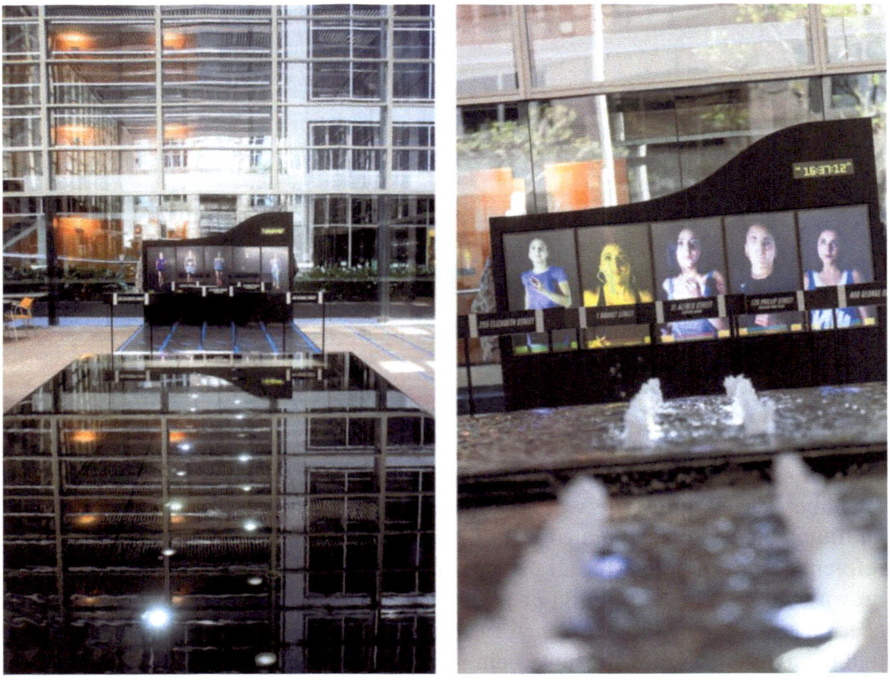

Building Run by Keith Deverell (Image credits: Keith Deverell (*left*) and Josh Hill Photography (*right*))

Building Run (2013) is a five-screen video installation work by Australian artist, Keith Deverell. It was first realised to visualise and metaphorise the energy consumption of five commercial buildings located in Sydney's Central Business District where it was also installed in September of 2013. It was commissioned by the City of Sydney and through Sensing Sydney, a joint initiative between Carbon Arts and the City of Sydney, that invited artists to bring sustainability data alive in ways that celebrate collective efforts to address environmental challenges. *Building Run* responded to this brief by providing a novel way for audiences to encounter the energy performance of their workplaces. It was produced in the context of the City of Sydney's carbon reduction goal to reduce emissions by 70 % by 2030, and it aims to illustrate the role that green buildings play, together with the communities that inhabit them, in achieving these goals.

Building Run was exhibited as part of the city's public art festival, Art & About, held between 20 September and 20 October 2013. The artwork was located in the foyer of Deutsche Bank Place, where project sponsors and building data participants, Deutche Bank and Investa Property, house their national offices. Unlike *Kilowatt Hours*, *Building Run* was a legitimised effort to draw attention to electricity consumption data.

The work animated energy consumption data from five buildings in the image of an elite athlete competing in a daily race for greater energy efficiency. Each building's performance was represented by a video avatar of this athlete, a role performed by an actress running through various stages of speed and fatigue. The installation tracked energy consumption every day over a period of 1 month, and each building's daily progress was measured against its own personal best. This in turn was based on an algorithm taking into account past performance and daily conditions such as the weather. This data was also displayed on the screens providing a public showcase of each building's performance.

There were three distinct audiences for this work – the general public, tenants of the buildings in the race and the building managers of those buildings. In-depth interviews with key project participants revealed some interesting observations about the impact of the artwork.

As a vehicle for incentivising other buildings to get on board with energy efficiency, data sharing and management, the City of Sydney, in particular, found that the artwork offered many opportunities to have meaningful conversations with partners in the sector, which allowed them to make progress in addressing energy efficiency challenges collaboratively. In addition, the very act of bringing the artwork to the building site involved coordination between different sides of Investa Office's business, as well as between tenants in the building. These often new connections reportedly served to cement relationships that are instrumental to driving a sustainability agenda, which by its nature requires collaboration.

The building managers, who were effectively pulling the levers behind the artwork and affecting the performance of the runners, reported that they felt that the work put them in the spotlight. However, competition between buildings was not overt, and generally the increased attention on the role of building managers was felt as a positive one. Bringing a mostly hidden, and potentially underappreciated, role to the fore gave greater emphasis to the work that goes into making tenants comfortable and buildings efficient in their use of energy.

Building Run created a focal point for the foyer of Deutsche Bank Place and for the lunchtime visitors frequenting the juice bar and café, differentiating it from other office buildings in the local area. As the Vice Chairman of Deutsche Bank stated at the launch event, the artwork served to remind the business about the high sustainability performance of the building and the reasons why they had located there in the first place. Investa Office reported that presentations about the artwork in the office activated a subset of employees who identified with art and/or sustainability. These staff brought enthusiasm to the project, expressing pride that these shared values were being put on display by the company. In this way, projects like *Building Run* can be contributors to employee satisfaction and employee retention.

A small, intercept survey conducted in the foyer revealed a mixed view of the artwork by building residents and the public. For some, attraction to the artwork was a matter of taste. For many, the presence of the artwork did not challenge them to change their patterns of behaviour in the foyer and move beyond, for example, buying a coffee, to spending time with the artwork in order to find out what it was about.

On the other hand, observations collected by the concierge revealed a strong curiosity about the work by tenants in its first couple of weeks of its showing. Over ten people a day asked the concierge about the work, what it meant, how it worked and who was winning. The overwhelming response was 'interesting'. There was a desire expressed by many to understand the mechanics behind the work and receive updates on performance in the race.

Meta Versus Matter

The discussion of these three case studies has been structured in two parts. First, it moves back from the realisation of the works themselves, through the processes behind their production and into the dark matter that they push against and reveal. This is the space of regulation, policy and bureaucracy. Second, it moves back in the other direction from artwork out into public space. How do these works reshape these public spaces and therefore notions of what publicness means? How effective are they as public envelopes for engagement? What assumptions are made in this artistic agenda for engagement?

Into the Dark Matter

Both the production and visibility of these three case studies play into the realm of policy, regulation and governance. *Amphibious Architecture* attempts to create an environmental legibility as a way of contesting river health and water quality. By revealing water quality information publically, this piece offers an intervention into public discourse around water quality and river health. It improves the audience's capacity to ask questions around why water quality might fall during a particular period and what conditions upstream have caused this. Although water quality data is available in NYC, the real-time and participatory nature of this work invites observers into an interpretation and synthesis of environmental conditions and leads to an engagement with a policy and regulatory landscape that would otherwise remain invisible.

Similarly, *Kilowatt Hours* and *Building Run* engage audiences with dark matter relating to energy consumption, efficiency and public access to this data. The anxiety around the installation of *Kilowatt Hours* reveals that although the CarriageWorks building (where the work was installed) is a public space, the energy footprint of the space was regarded as sensitive information at the time the work was installed. Yet the project also shows that there are possibilities for an artist to hack metering technologies and open up this data within the frame of an exhibition to an online audience. This widening technical capacity of non-experts also demonstrates how the falling costs of computation and its ready accessibility offer a rich potential for acts of quiet subversion such as this one.

Kilowatt Hours explores who is able to access information that describes collective behaviour and reveals a reluctance in some public institutions towards making data open and available. In this sense *Kilowatt Hours* can be thought of as a MacGuffin, an artwork that begins a process of opening up a data stream publicly. The ambiguous agenda enjoyed by art enables access to data that would otherwise be protected and masked by bureaucratic process. Similarly, *Building Run* continues pushing this agenda and serves as a useful counterpoint to *Kilowatt Hours*. Realised several years later in the same city, *Building Run* as a commissioned work by the City of Sydney shows the agenda of *Kilowatt Hours* to have been taken up widely by the city itself.

Also exploring *Building Run* as a MacGuffin, the major driving force in the commissioning and production of this work was the promise of seeing a real-time relationship between energy data and the behaviour of the runners on screen. Presenting the building's performance using a visual metaphor of competitive sport also necessitates a winner and attempts to motivate energy reduction using the driver of competition. To produce a real-time competition, it was necessary for the data used to be open and available in the public realm, a heretofore-unachieved outcome in Sydney and in the commercial sector where privacy of data is carefully guarded. As such, the work operates with a subtle agenda for shifting policy as it necessitates a more open approach to energy data. Data openness can offer potential for innovation and drive further energy efficiency gains.

Building Run's peculiar and novel design becomes a driving force for moving towards an open data society, and its realisation demands that the organisational, policy or regulatory environment be changed – in this case cajoling the corporate tenants of the buildings to open up their data. According to Hill (2012), choosing the MacGuffin and handling it carefully is the practice of design stewardship, and in this way *Building Run* drove progress towards open data. The work therefore was able to use the 'matter' of the artwork to address 'the meta' of unlocking energy efficiency in commercial buildings through innovation in open data communication.

Out into the World

The three case studies described here draw public attention to environmental and infrastructural systems. These are systems that define the material exchanges of our cities, assembling humans and non-humans into networks that move water, carbon or energy through places and forms. As such, these works bring the relatively abstract relationship between human action and environmental consequence into sharp focus and make public these processes that otherwise typically fall outside of our daily attention.

All three works use various aesthetic and performative strategies to do this. *Building Run* anthropomorphises energy consumption of architecture and its inhabitants in an attempt to generate a better understanding of peaks, troughs and patterns throughout a daily cycle. *Kilowatt Hours* aims to engage an audience via an emergent visual form that reveals relative use through time. It also attempts to reach an online audience through a social media stream. Both *Kilowatt Hours* and *Building Run* operate by giving a visibility to the energy infrastructure within the immediate built environment for the benefit of speculation. Presenting visual consequences of the building's electricity consumption in a public space also transforms the passive, unconscious relationship between infrastructure and its users into a more active and reflective engagement. By augmenting the visibility of energy infrastructure and its use and placing what would otherwise be private information into the public realm, these works reshape the viewer's relationship to the surrounding architecture, thereby politicising it as a 'public matter of concern'. *Kilowatt Hours*, and later *Building Run*, can both be seen as small references to the larger unfolding reality of society and our energy demands.

Amphibious Architecture also relies on a visual augmentation of the river and on public engagement through the use of social media. This work displays two distinct datasets, one measuring water quality and one measuring fish presence. The interplay between both positions the audience to make connections and hypotheses about how these datasets might be related. The audience participates in the interpretation of the installation as an interface to the river, a process that has the potential to expand their political agency.

Amphibious Architecture goes the farthest towards DiSalvo's (2012) notion of adversarial design via the production of what he calls an agonistic collective.

An agnostic collective establishes linkages among objects, people and actions to create open, interpretive and participatory spaces of contest, in which the elements gathered together are able to act out a plurality of conflicting practices, values and beliefs. *Amphibious Architecture* can therefore be approached as an agonistic collective as it produces novel performed calling cards for riverine species. Audiences are invited to enter into a technologically produced conversation with river creatures in order to produce a more empathetic relationship with them. The logic is that audiences will respond to their newly befriended urban cohabitants by advocating for better aquatic living conditions, thereby producing the contestation that is central to adversarial design. According to Benjamin and Yang (2011), participants did indeed engage with the work onsite, sending an average of three text messages per person.

In their 2011 paper, Fritsch and Brynskov discuss the challenge of capturing public attention in urban environments that are inherently transitory, discussing new media artworks that engage with climate change. The ephemerality of temporary works can be advantageous, delivering the micro-perceptual shocks inherent in the novelty of these installations. Yet Fritsch and Brynskov suggest that a more permanent integration of these works into the urban fabric could lend themselves potentially to a deeper form of public engagement. However, this also risks the opposite effect, where the work disappears into the fabric of the cityscape when its novelty has worn off. This tension between a work's duration and audience engagement raises an interesting question: If an artwork is intended as a conversation, how long can the conversation last? *Amphibious Architecture*, to be permanently installed in New York City in 2014, will provide a valuable case study for this question. Will the work require an ongoing programme of animation to keep it alive in the public's imagination thereby enabling it to deliver the desired shift in behaviour or consciousness? Or will its presence be subsumed as an ambient information display, called upon only when conditions make the data relevant?

How a work redistributes public attention is an important aspect of a broader agenda shared by all three of these works, which is to shift perception and motivate change. Yet, underlying this intention is the notion that there are straightforward connections between data-driven artworks, an individual's awareness and collective action on a concern, deeply held assumptions that exist at the heart of much public environmentally concerned public art. These assumptions raise questions of who the audience is. And, what agency and public engagement does this audience actually have in order to take some sort of remedial action? As Australian scholar Emily Potter (2009) observes, works such as these position the individual as 'the site of action', posited with the ability to shape government policy and inspire collective action (p. 1). Yet, this assumption can be problematic as it readily overlooks the very real political and social barriers that hinder progress towards better 'ecological governance' (Bratton and Jeremijenko 2008, p. 12). In light of this, and returning to Hill's strategic design framing, perhaps the most significant potential for change offered by this category of work is to understand artworks as powerful symbols and motivations for a small but influential audience consisting of the bureaucrats, the politicians and cities' governments themselves. The production of artworks such as these has the potential to soften up the boundaries of policies, regulations and data availability in a way that citizens often do not. As such, it is critical that artists and curators working in public space understand their practices as having this unique potential to be strategic design works that can reshape the dark matter of urban space.

Conclusion

Amphibious Architecture, *Kilowatt Hours* and *Building Run* are a small collection of case studies representative of a growing field of public art practice that see artists engaging with data streams in urban space. These works experiment with how data can be presented and performed in novel ways to inspire a more active engagement between their audiences and the systems from which the data is drawn. Using artistic strategies like creating visual novelty, anthropomorphising data and performing data through social media streams, these works attempt to draw audiences into closer and more intimate relationships with surrounding urban infrastructures and environmental health indicators. They draw the ongoing processes and conditions of these domains out from periphery and redistribute attention towards them.

Approaching these works through Hill's (2012) strategic design framework casts the negotiations required to install and realise them as equally significant to the final public outcome. The implementation of these works softens, reveals and potentially shifts existing regulations and bureaucratic norms, opening up new possibilities and questions as to what the smart city is and how it might be realised. These works give small but important voices to artists, curators and audiences who exist outside of the heavily corporate agendas of the large multinationals who are currently the dominant voices in the shaping of the 'smart city'. Given this, and in light of the limitations of the political agency of citizens, public art has the potential to provide an exploratory space and a cultural motivation for regulators and governing bodies to explore and give voice to other approaches to environmental data collection, presentation and synthesis.

References

Benjamin D, Yang S-i (2011) Amphibious envelopes. In: From Sentient City: ubiquitous computing, architecture, and the future of urban space. Architectural League of New York/MIT Press, New York City/Cambridge, MA

Brand S (1995) How buildings learn: what happens after they're built. Penguin, Harmondsworth

Bratton BH, Jeremijenko N (2008) Latent interfaces, suspicious images. Situated technologies pamphlets 3. The Architectural League of New York, New York

de la Peña B (2013) Embracing the autocatalytic city: how bottom-up growth, driven by citizens, trumps central command. http://www.citylab.com/politics/2013/03/embracing-autocatalytic-city/4909/

DiSalvo C (2012) Adversarial design. The MIT Press, Cambridge, MA
Greenfield A, Kim N (2013) Against the smart city, The city is here for you to use. Do Projects, New York City
Haque U (2012) What is a city that it would be 'Smart'? City Box 34:140–142
Hemment D, Townsend A (eds) (2013) Smart citizens, vol 4. FutureEverything Publications
Hill D (2012) Dark matter and Trojan horses: a strategic design vocabulary. http://www.cityofsound.com/blog/2012/08/dark-matter-trojan-horses-strategic-design-vocabulary.html
Hollands RG (2008) Will the real smart city please stand up? City 12(3):303–320
Mouffe C (2000) Deliberative democracy or agonistic pluralism? Department of Political Science: Institute for Advanced Studies (IHS), Vienna
Potter E (2009) Climate change and the problem of representation. Aust Humanit Rev 46:69–79
Townsend AM (2014) Smart cities: big data, civic hackers, and the quest for a new utopia, 1st edn. W. W. Norton & Company, New York

Chapter 3
Experiential Ecologies: A Transdisciplinary Framework for Embodiment and Simulacra

Grisha Coleman and Daragh Byrne

Abstract Within this chapter, we explore a conceptual framework of the simulacra for the ecological art project, echo::system. We will describe an arts-driven process which bridges practice and research to consider how diverse fragments of ecological information can be synthesized into emergent structures for exhibition, participation, and performance. We draw on our experiences of developing mediated simulacra of urban-desert landscapes through interdisciplinary collaboration and expertise, to qualify the challenges and opportunities of such a process. Further, we consider the impact of movement-based participation on the perception of the public. Finally, we attempt to disambiguate the roles of narrative, performer, audience, media, and computation in order to articulate how this framework supports the dynamic expression of real and imagined ecologies.

Introduction

Environmental art as a genre allows for a reorientation of perspective, to examine, reflect, and inquire into a relationship with the world of our surroundings. Its artists seek to "inspire, advocate and innovate, revealing and/or enhancing ecological relationships while modeling ecological values (Wallen 2012)." Robert Smithson's seminal piece "Spiral Jetty,"[1] Andrew Goldsworthy's site-specific land art (Goldsworthy 1990), and Walter De Maria's "The Lightning Field" (Baker 2008) all reflect a desire to transcend the boundaries of the museum in revealing the art already around us. By contrast, artists such as Ana Mendieta (Schneider 1997) created "earth-body" sculptures that explored metaphors of the

[1] http://www.robertsmithson.com/earthworks/spiral_jetty.htm

G. Coleman (✉)
School of Arts, Media + Engineering, Arizona State University, Tempe, AZ, 85281, USA
e-mail: grisha.coleman@asu.edu

D. Byrne
School of Architecture, Carnegie Mellon University, Pittsburgh, PA 15213, USA
e-mail: daraghb@andrew.cmu.edu

F.T. Marchese (eds.), *Media Art and the Urban Environment*, Future City 5,
DOI 10.1007/978-3-319-15153-3_3

human body, relationship to place, and identity through highly politicized use of materials (e.g., blood, earth) in durational performance. These eco-artists "*worked at re-defining and re-understanding the position that humans have in nature*" (Simon 2006). They do this by drawing their audiences to an unknown place, exposing the spectrum between art and nature, and by leveraging the interplays between four interdependent elements: location, time, material, and change (Krug 2002). In many respects, these works are examples of the foundational principles of the ecological art movement.

For many of these early eco-artists, this form of art typically "involves nature and 'natural' materials" and is often "the result of the artist simply rearranging materials and/or objects found in nature in a specific way, minimally interfering with the environment where the artwork is situated or placed" (Olivier 2007). However, the "environmental art" (encompassing land art, earth art, and eco-art) of the 1960s and 1970s in the USA has continued to intensify, increasingly motivated by the current environmental crises we face as a planet.

In our current digitally proliferated culture, eco-artists may ask different questions about how we understand the "natural world" and engage the public, either by bringing the natural world to the audience or by bringing audiences into *other* worlds. Dimensions of scale, data, and the instantaneous speed of networked information have extended the tools of ecologically inspired artwork. Environmental art emerging in a contemporary context can educate and inform on the natural world or the "hidden social and philosophical questions behind the complex environmental crisis" (Faber et al. 1996) by facilitating "*the need to reflect on the context and boundaries of the issues at stake* (Simon 2006)." As art, the work can go beyond education to inspire, invite, or endorse activism by "*framing environmental processes both from ecological and biological perspectives and from more social, economic and political angles* (Simon 2006)."

Deeply embedded in the practice of ecologically motivated artwork is the use of crosscutting collaboration between art and science (Wallen 2012) to inform, enrich, and embody scientific evidence in the processes and outcomes of the artwork. While, traditionally, ecological art has emphasized a material connection to the natural, the advent of new digital tools and new mechanisms for online dissemination and participation has diversified the means for production of and engagement with these works. The digital offers new ways of revealing the ecological and creating interventions. For example, Brooke Singer's *Superfund365*[2] extends the scope of the work beyond an individual site by leveraging collective action. She invited citizens across the USA to participate by cataloging superfund sites and sharing their documentation online. Online and digital tools offered Singer a means for the scale of the work to greatly increase by cultivating communities of engaged participants, offering new modes of contribution, and aggregating diverse perspectives across many locales and sites.

[2]http://superfund365.org/

Digital media has enabled new narratives on ecological systems and new forms of distributed engagement with ecological art. Simultaneously, it has also created new mechanisms for representation. Increasing access to large, long-term data sources on natural phenomenon and tools for data analysis, modeling, and visualization have empowered aesthetic representations that consider the natural world at a scale much larger than a site or a city; it creates the possibility for ecological art that affords new perspectives on the world as a whole. Donna Cox's work on simulating and visualizing F3 tornadoes has not only had impact on scientific understanding but has delivered a potent format for general audiences to come to know the complex forces operating within them (Cox 2006; Patterson and Cox 2005). Wattenberg and Viégas leverage visualization as a mode to aesthetically render the "*delicate tracery of wind flowing over the US*"[3] using the National Digital Forecast Database and reflecting on this work note that visualization offers artists a tool to present "a forceful point of view" (Viégas and Wattenberg 2007).

The co-opting of visual analytics and scientific modeling toward artistic interpretations is undoubtedly a powerful method for the production of environmentally motivated digital art, especially as it draws on the already strong connection between art and science within the realm (Samsel 2013; Viégas and Wattenberg 2007). These examples successfully incorporate both scientific data and knowledge with aesthetic, emotional, and metaphoric resonance. Many have remarked that a metaphorical component is central to ecological art forms, to facilitating needed dialogue and in communicating ecological issues (Cox 2006; McClintock et al. 1997; Simon 2006; Wallen 2012). Metaphors "*create a space for understanding to emerge*" and offer a "*way to structure our understandings*" (McClintock et al. 1997). However, in this context, the metaphors are translational in that they are used to convey scientific, political, or economic knowledge to audiences. In this way, they are often deeply grounded in literal representations of data and information. These emphasize provocation rather than speculation and favor illustration over abstraction.

In a departure from this trend and returning to more traditional modes of eco-art, *echo:: system*[4] (Coleman 2013) explores simulacra, instead of simulation, as a metaphorical framework where the use of fiction, abstraction, and interpretation becomes a tool to raise questions about our sense of place. It is a response to our current global crisis caused by contemporary human inability to reflect upon our own impact on the natural world, which is conceived as a series of five large-scale, multimedia environments constructed for both live performance and interactive installation. Within this work, notions of ecology and nature are explored *without* directly drawing on the material objects of nature itself. The connection with the land is digitally mediated rather than materially mediated as is typical within ecological art. Technology is used in place of materials of nature to explore these five ecologies, metaphorically. Importantly, this work is not centered

[3] http://hint.fm/wind/

[4] http://echo-system.net

uniquely in the past or present of the ecologies but also includes prospective, speculative futures presenting opportunities for abstract, expressive relationships with the natural world. Additionally, it underscores the embodied and kinesthetic rather than the observational, as the vehicle for connecting with and experiencing place. Within this work, simulacra are used to construct an embodied experience of ecologies but also as a means to transcend the specificities of geographic location to activate metaphors of place and space.

We ground our discussion of simulacra in the most recent iteration of the *echo::system* work, *actionstation.2 – the desert*, which has been collaboratively developed with an interdisciplinary team of artists, researchers, scientists, architects, animators, and performers. The production integrates sound, video, architecture, and ground-based installations with computation, contemporary dance, and narrative to construct an imagined environment – a simulacrum of the desert – with an aim to promote reflection on how and where we live.

Developing such simulacra requires both a negotiation of a variety of disciplinary perspectives and knowledge from a variety of sources (arts, ecology, ethnography, literature). We recognize urban and natural ecologies as complex spaces that encompass a broad range of interdependent systems including weather patterns, flora and fauna, environmental conditions (pollution), human inhabitation, built infrastructure, socioeconomic conditions and historical context, etc. Thus, our challenge is to develop a space for installation and performance that can map the terrain between the urban and the natural, combining these complex, interlaced systems aesthetically, to evoke sensory and emotive signifiers and embody the rich and vibrant texture of the environment, as well as our impact upon it.

echo::system

echo:: system is a fusion of art installation, choreographed multimedia performance, and public engagement that looks to mediate a connection between art and science. The project is a series of five large-scale environments created for both performance and installation. Each *echo::system* "habitat" – called an "actionstation" – corresponds to a natural biome: #1 abyss, #2 desert, #3 forest, #4 prairie, and #5 volcano. These *actionstations* are developed as hybrids of real and imagined information, drawn from the cultural, historical, and ecological information of the given habitat (see Figs. 3.1 and 3.2). In the tradition of speculative fiction, the project creates alternative environments to promote reflection upon how and where we live. The goal is to examine intersections of art, environmental sciences, and technology; information and place; performance; and public engagement through the practical realization of the work.

The trajectory of research looks to understand how time-based art can play a critical role in the integration of abstract information and complex current events with our everyday lives through tangible, aesthetic, real-time experiences. Live performance, particularly dance, emphasizes knowledge based on the present physical

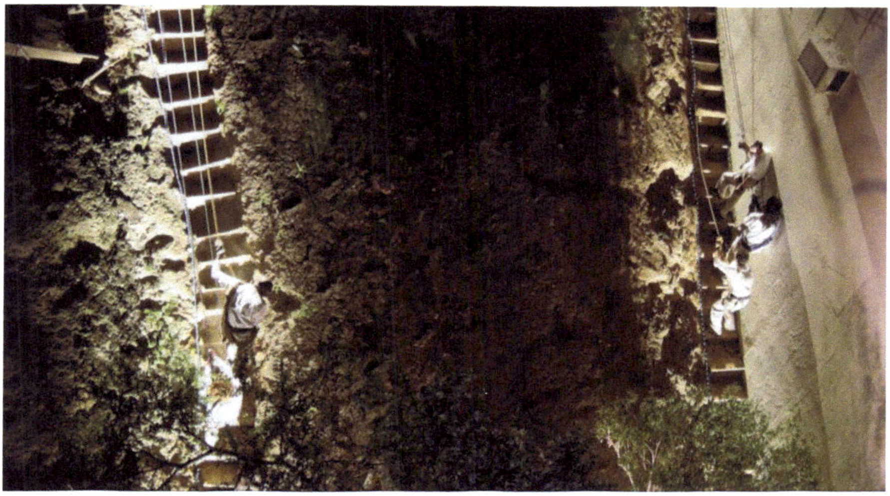

Fig. 3.1 ***actionstation.1 – the abyss***, *echo::system*. Performance at the California Institute of the Arts, Los Angeles, CA, 2005 (©Grisha Coleman)

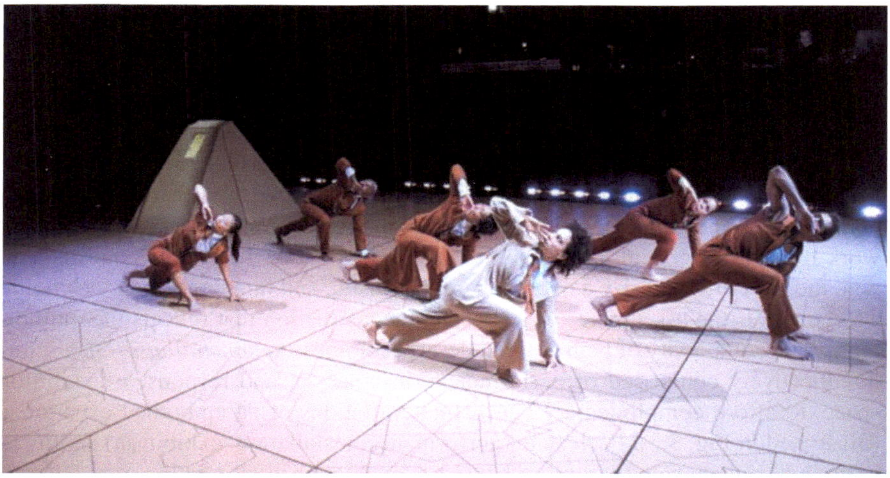

Fig. 3.2 ***actionstation.2 – the desert***, *echo::system*. Performance at Carnegie Mellon University/Studio for Creative Inquiry, Pittsburgh, PA, 2008 (©Grisha Coleman. Photo: Tim Friez)

moment, and expressing these moments in time is one of the strongest aspects of this form. The five constructed environments of *echo::system* mesh the visceral, phenomenal acts of contemporary dance with an interactive media-rich space, envisioning a meta-environment. Along with creating a site for performance, the space is built for participatory installation, populated with customized, augmented treadmill interfaces used by the public. These machines are reconceived to offer

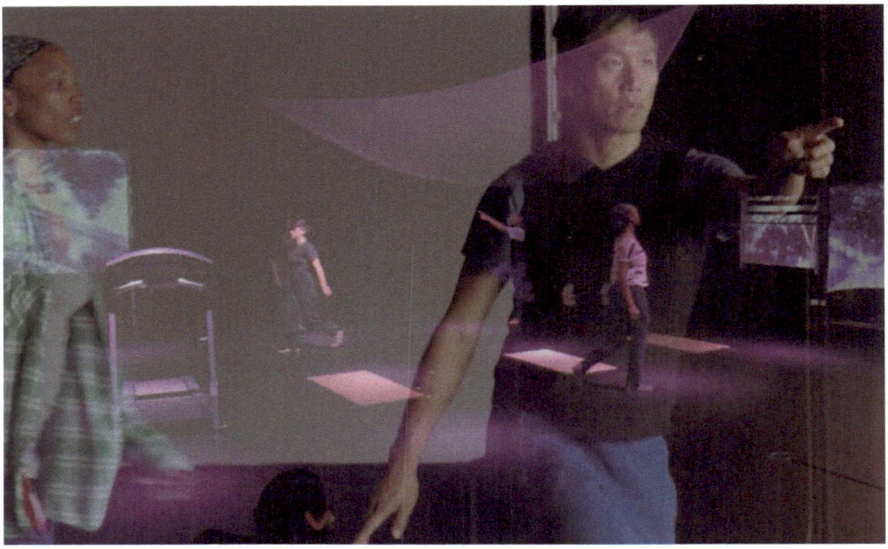

Fig. 3.3 *echo::system, actionstation.2 – the desert*. Performance at Ohio State University, Columbus, Ohio, 2014 (©Grisha Coleman. Photo by Tom Heban)

an explicitly physical engagement in which to navigate the landscapes of real and imagined place, to reconnect complex ecological data sets with lived experience and cultural narratives. The event moves from participatory installation to observable performance, leveraging the power of both kinesthetic experiences to dynamically communicate and connect sensory experience to information space. In doing so, this work looks to offer multiple perspectives on our relationship to the environment, with the goal to synthesize a more coherent, holistic, and interdependent viewpoint.

The aim is the assemblage of an aesthetic construction that reflects the multidimensional complexity of being in real desert ecologies. *Actionstation.2 – the desert* is collectively comprised of performance, installation, and responsive exhibition space, to afford complimentary but distinct variations of an embodied experience (illustrated in Fig. 3.3). Each state (performance, installation, exhibition) is offered as an independent *simulacrum* – a heuristic, qualitative interpretation of the desert. Individually, these present a faceted representation of experiential knowledge, reflecting human experience of the environment: through data, movement, first-person accounts, myth, abstract interpretation, and multimedia.

Simulacra of the Desert

In Lewis Carroll's *Sylvie and Bruno Concluded* (Carroll 1893), one of Carroll's characters notes some practical difficulties with a fictional map that had "*the scale of a mile to the mile.*" He comments, "*we now use the country itself, as its own map, and I assure you it does nearly as well.*" This employs a phenomenological humor to highlight the dangers inherent in the act of a simulation that tries to "out-real" the actual. Simulation is defined as the imitation of the operation of a real-world process or system over time (Nelson et al. 2001). Simulacra are copies that depict things that either had no reality to begin with or that no longer have an original (Goldman and Papson 2011). In his 1981 treatise *Simulacra and Simulation*, contemporary French philosopher Jean Baudrillard describes a contemporary existence in which it is no longer possible to assume that representation is derived from an "original" (Baudrillard 1994), due to the infinitely replicable and mutable nature of media, creating its own reference points, symbols, and signifiers in an ecology of knowledge. In the digital age, notions of real and copy, original and representation, are forever altered and perceived along a gradient of hybrid (virtual/material) experiences. We leverage the concept of *simulacra* to allow for a creativity that is inclusive of art production and practices as well as scientific knowledge and research methods. Simulacra allow for a sense of place that is not linked to a necessary geographic location but to a metaphysics of space and placemaking.

In *echo::system*, the container of the simulacra across all three "states" (installation, performance, exhibition) supports a range of expressive and researched data in a mixed-media, hybrid event. Through this container, the creative work is informed by actual ecologies, but is not a literal mapping of them, creating scenarios that are perceived as *plausible*, without having to be *possible*. In the tradition of speculative fiction, this framework acknowledges complexity, nonlinear time, and multidimensional space.

Unlike simulation which implies the approximation of real behavior, process, or systems through the modeling of scientific knowledge, simulacra offer a framework where aesthetic, experiential, and scientific knowledge can be blended. This provides an equally important mechanism for embodying and expressing complex interacting systems that acknowledges the influence of human belief systems on the perception and understanding of the same "outer"-world systems. The simulacra allow *a desert* to be made manifest through a range of sensorial constructs to more fully embody the experience of what it means to be in a place rather than offering a definition of the processes it contains. It recognizes the importance of metaphor while shifting from modeled, simulated, and literal knowledge (Cox 2006; McClintock et al. 1997) toward experiential understandings of ecological space, time, and process.

In approaching the work through simulacra rather than simulation, we seek the experience for a public to be revelatory rather than informative, by allowing one to have a personal, individualized experience that can allow for the variations of what it means to be in the desert. Predicated on informal, heuristic, and abstract knowledge, the artwork creates multisensory opportunities for one's experience to be read and reread, experienced and reexperienced, interpreted and reinterpreted by audiences. To achieve this, *echo::system* offers three distinct but coherently connected scenarios shifting the public among states of interactivity, participation, and engagement.

Experiencing the Desert with Performance

As a live performance, echo::system is expressed primarily through choreographed movement and composed sound, with a score that allows for improvised material to emerge. The performance places the highly visceral expression of contemporary dance in a media-rich space to develop a movement-based evocation of the desert. The choreography depicts an ambulatory narrative that explores the transitional space between urban and "country" environments by following a tribe as they embark upon a journey into a mythic desert. Themes and images that examine relationships between "desert" and "urban" inform the choreographic process and studio exploration. Examples include: the city in the desert, the parking lot, the grid, ambulatory, quicksand, locomotion through concrete vs. sand, verticality of the urban landscape vs. the horizontal vistas of the desert, thirst, mirage, oasis, etc. These themes manifest in the choreography by bridging more traditional generative dance strategies with other approaches, such as algorithms as a structure for patterned emergence.

By way of example, "36 Walk," a key sequence in the piece, is developed based on sets of algorithmic rules defining movement characteristics and performer interactions. The movement design is inspired by ironies embedded in mechanically assisted modern walking machines (i.e., treadmills, modern gym, etc.) where performers' movement is defined as a series of spatial strategies arranged on and off "the grid," metaphorically playing on traditional perspectives of space in cities. This is explored in more detail in Coleman and Byrne (2014). By creating the movement logic in this way, the structural underpinnings of the choreography are based on a translational language that uses a probabilistic score for highly composed opportunities for improvisational movement. In a strategy toward expression, intuition, and flow in the dance, performers work with numbers and spatial cues rather than psychological or emotional triggers, to guide the composition of their individual movements, and this is responsible for an emergent, dynamic improvisation. Human interactions pattern and accumulate to yield the narrative presented to the audience. The manner in which the performers react and respond to the algorithm and coordinate with one another results in intention, story, emotion, and psychology of the broader performance narratives.

Along with the development of movement, we also consider the relationship among movement, media, and computation in evoking the urban-desert metaphor. As part of exploring the performance environment as both physical and imagined space, we consider technology and choreography as interdependent, conceptually, symbolically, explicitly, and implicitly linking these hybrid worlds. The use of human-machine interaction is embedded in the theme of the environment as interface; our surroundings are the immersive medium in which we interact. Our fictional tribe members, seeking knowledge of the future of their species, *interface* with the land via the treadmill in their version of a modern walkabout. Yet, while searching for clues on how to live more gracefully and comprehensively in their desert environment, they are caught in a paradox of walking without

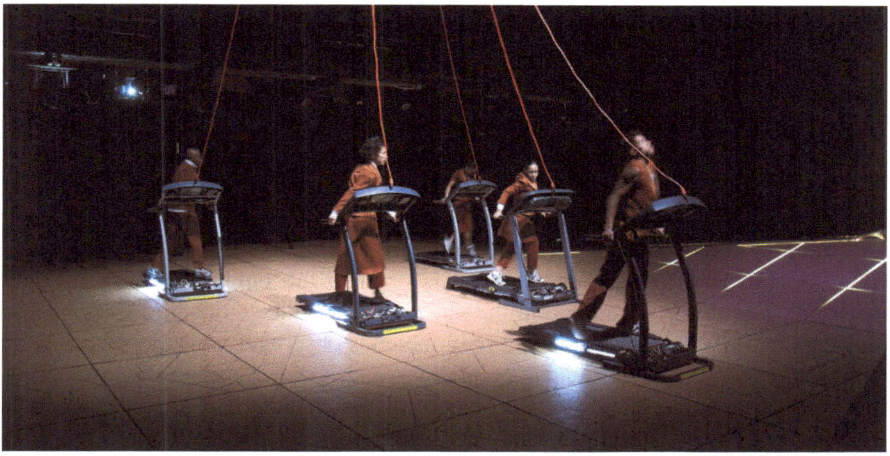

Fig. 3.4 Treadmills used by performers during *echo::system, actionstation.2 – the desert*. Performance at Carnegie Mellon University/Studio for Creative Inquiry, Pittsburgh, PA, 2008 (©Grisha Coleman. Photo: Tim Friez)

traveling and come to realize they have an inadequate understanding of their relationship to the land they traverse, their past culture, and its connection to their present state. By extending/reframing the interface in this most basic way, the choreography of performers on treadmills (see Fig. 3.4) conveys a parable of our contemporary use/misuses of technology. This "desert mythology" recasts the commonplace activity of working out in a gym as a surreal journey. The treated treadmills extend the metaphor of an interface with the land in that they activate memory and reveal histories of our relationship to the environment. The nonverbal, choreographic narrative describes an emotional landscape of misunderstandings that seeks expression but finds no resolution. As they run toward their future on treadmills, notions of contemporary progress are thrown into paradoxical relief. Symbolically, this activity also raises issues of our own energy expenditure that is no longer linked to hunting, farming, fleeing, or even moving toward one's destination. Added to this, the fact that the treadmill consumes energy to power itself only further parodies our larger energy economies that are slow to reckon with energy expenditures that have no (intended) feedback loop. The myth, as parable, depicts a fictional species attempting to navigate their surreal surroundings. Dancers perform in a space created with media, computational systems, and architectural elements. In the performance of *actionstation.2 – the desert*, the species imagines the treadmills to be functional replacements for walking on the earth. In the context of performance, treadmills become anthropomorphic and symbolic – seeming to drive the (movement) choices of the performers, who respond within an improvisational framework in a choreographic score.

Experiencing the Desert by Moving in Place

The treadmill not only offers a metaphorical prop (see Fig. 3.4) through which the performance plays out the tensions between moving in urban and desert environments. It is also repurposed as an interactive installation that provides opportunity for audiences to experience representations of actual, rather than mythopoetic, desert ecologies (see Fig. 3.5). Making meaningful connections between scientific information and expressive, live performance can prove obscure to the public. Consequently, the installation interface is open to the public in advance of the performance. This gives an opportunity to experience a level of subjective physical engagement that is lost if the event only includes observation of the performance. When coupled with the participatory, highly physical experience of the installation, the metaphorical yet highly kinetic nature of the performance thereby affords a rich opportunity to reflect on our relationship to the landscape through the body in movement.

Drawing on realms of embodied interaction (Dourish 2001) and somatic practice (Feldenkrais 1977), the treadmill interface emphasizes the affordances of human movement through a more explicitly kinesthetic engagement. This acknowledges and leverages interactions between system (computational) and system (physiological). Somatic issues of orientation, gravitation, timing, and the complex coordination of these elements in navigating our environment impact our perceptual experience and therefore the way we make meaning. Similarly, the treadmill in inviting audiences to physically walk through a virtual landscape is designed to similarly engage these somatic issues. Through this treadmill interface, we question

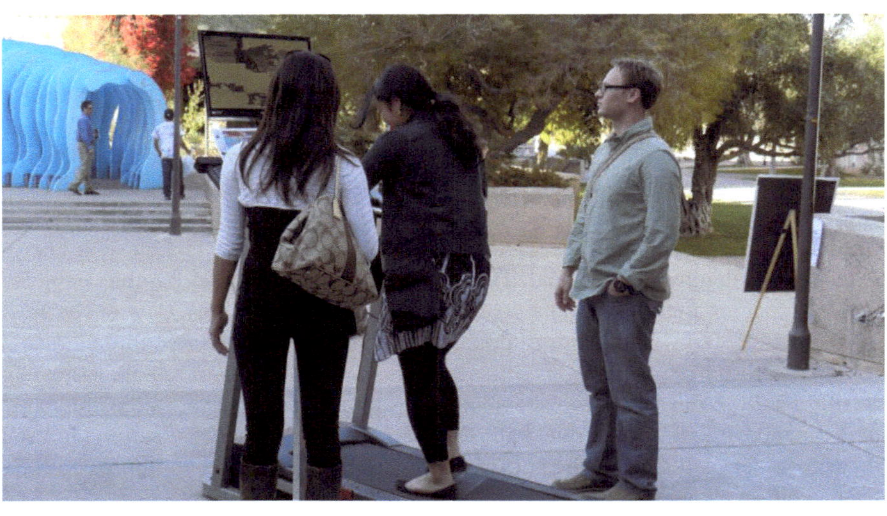

Fig. 3.5 Treadmills as participatory installations. From Emerge 2012 at Arizona State University (©Grisha Coleman, photo credit Matthew Briggs)

the way the complex, multidimensional experience of "taking a walk" could affect the way we understand an environment and our relationship to it.

Seeking to craft a sense of "being there," this embodied interaction is designed to craft a more authentic but situated simulation where the audience does not leave their environment to enter another but can experience *both* the real and the virtual together as a continuum. This is distinct from immersive virtual environments like the CAVE (Bowman and McMahan 2007; Kjeldskov 2001) in which the user is almost completely encompassed in the simulated environment. Within this room-sized infrastructure, virtual movement is often not dependent on real perambulatory motion. Conflicting with expectation, instead, the world typically moves around the user and is controlled through devices, joysticks, or gestures (Kjeldskov 2001). This overlooks the visceral experience of moving in space as foundational to how we understand, interpret, and interact in any world, real or virtual.

Instead, this installation integrates the familiar experience of walking on a treadmill with dynamic, interactive audio and visual information, displayed in response to the activity of walking. The walk is propelled by sensing and display technology to present a hybrid kinesthetic walk through desert landscapes in topologies local, distant, and mythological.

Two display surfaces surround each treadmill (Fig. 3.6): the first, a projection at eye level providing the main viewpoint onto the walk, and the second, a tablet affixed directly above the treadmill control panel is used to orient and coordinate the user through the walk they are taking. The walk is presented primarily as a frame-by-frame sequencing of an actual walk that progresses from a natural to an urban environment (or vice versa) and is matched to the speed of the treadmill. Binaural audio of iconic events in the place (e.g., train horns, traffic, fauna, etc.) are sampled and included so as to be independent of the speed of the walk. As one walks, the secondary display dynamically updates to indicate the user's "location" in the virtual space, as well as revealing land-use, air-quality, and socioeconomic factors through a series of on-screen data visualizations. This allows users to intuit "invisible" information relating to the real places they navigate and connects that sense of "being in the world" directly to the information itself by correlating the abstract information with the physical act of walking through a specific, highly identified, and geographically mapped space (Dourish 2001). Finally, a wireless sensor pack affixed magnetically to the underside of the treadmill transmits real-time information on the treadmill's use to drive feedback, synchronize playback, and automatically start and stop the walk. The interface is designed to foster a focused, multisensory individual experience without the need for immersive infrastructure (e.g., CAVE or augmented glasses). Additionally, this allows the installation to remain portable and versatile in diverse deployment scenarios and situations.

Presenting audiovisual accounts of real deserts in combination with interwoven ecological and ethnographic data, it invites the audience to consider their relationship to desert and urban environments. The present installation is focused around information gathered in the southwestern US desert region. In particular, research has drawn on ethnographic investigation, which folds local knowledge and lived experience into the walks. Participants are exposed to dynamic visual

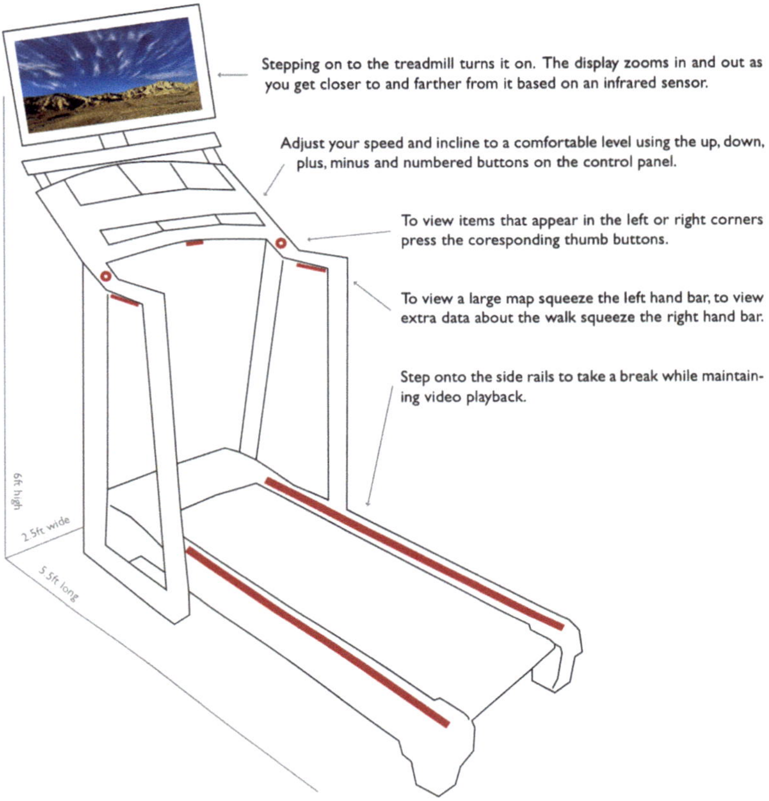

Fig. 3.6 Interface sketch of the treadmill installation

and sonic displays of a remote location while walking in place, on an augmented treadmill. The walks are curated to provoke reflection, from highly personal meditations of what it means to see and hear the land to issues of environmental justice and ecological understanding. The walks transition from urban to rural landscapes, showing the proximity of industrial areas, to residential neighborhoods, to regenerated wilderness in areas of South Phoenix.

The interface explores the potential for how highly physical interaction through a multimodal treadmill can influence an experience, as compared to reading, listening, or watching in a seated, stationary position. It allows users to make new correlations among the audiovisual digital display and the socio-ecological data provided on the secondary map display. It seeks to impart a sensory motor experience of being in the desert that augments a more abstract, reflective experience of watching the performance. In this way, audiences can integrate the real and the abstract to develop a fuller sense of the topology, society, and lived experience in the desert.

Experiencing the Desert by Being in Place

The performance and treadmill installation offer *two* modes of simulacra for experiencing a desert environment, one more receptive and one more active. A *third* simulacrum encompasses the first two and provides an exhibition space for the hybrid environment that is immersive and expressive. This space evokes both real and imagined ecologies of the desert as a layered, ambient, durational experience. This exhibition simulacrum is intended to give a spatial, multidimensional context for the performance and installation and speak to the underlying general themes of the echo::system project. Inspired by representational and participatory exhibition events such as Olafur Eliasson's *The Weather Project* (Eliasson 2003), Robert Henke's *Fragile Territories*,[5] and Louise Bourgeois' *Maman* (Morris 2007), *actionstation.2 – the desert* as exhibition combines architecture, media, and an interactive computational framework to evoke an emergent representation of a desert landscape. Distinct from the other modes of simulacra, it is widespread (encompassing, immersive), is of low intensity (easy to participate), and supports activation (responsive to activity, interaction, and participation).

To evoke different themes of the urban-desert relationship, a variety of designed architectural strategies are used in the event space (see Fig. 3.7). Around the boundaries of the space are louver walls. Typically used as a shade structure in

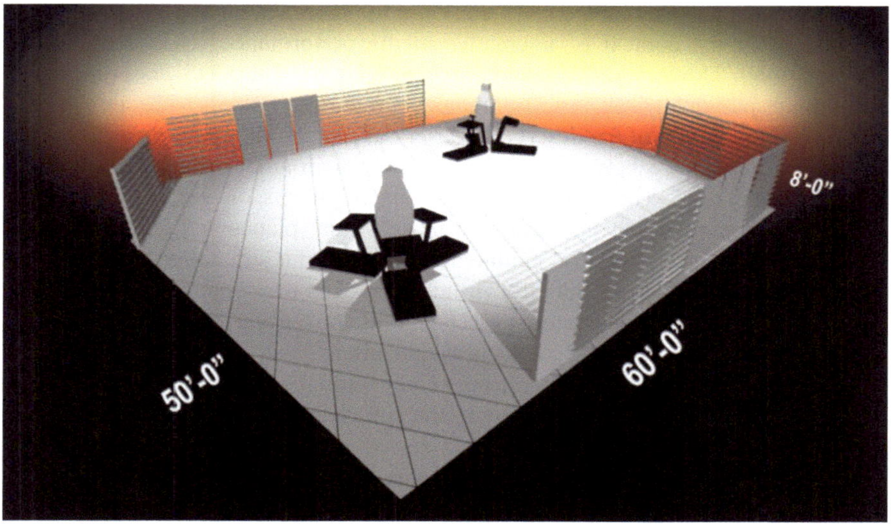

Fig. 3.7 Architectural designs and prototype for the exhibition space (see also Fig. 3.3 for current implementation)

[5]http://roberthenke.com/installations/fragile_territories.html

urban desert, the design team incorporated these structures as a means to describe a sense of space beyond the walls of the indoor exhibition space. RGB lighting was positioned behind each louver wall to emulate the natural light conditions of the desert. Designed lighting placement, color, and tonality cast shadows into the exhibition space and mimic the progression of light within real desert landscapes throughout the day. The objective is to create a sense of horizontality and horizon that evokes desert vistas. The louver structures serve a secondary purpose of providing an interactive canvas for media display. From the interior of the space, a series of projections fill the louvers' slats. The slats can then be adjusted to display the projection on its surface or allow the projection to pass through by rotating them. This gives opportunity for the audience or performers to interactively manipulate and recompose the presentation of media within the space. In addition and to reflect iconic objects of desert urbanity, other sculptural elements suggest large-scale telescopes, while projection surfaces reference solar panels.

While architectural elements such as solar panels and louvers create physical structures that evoke the desert and provide surface area for visual media presentation, the *Central Nervous System* [CNS] is a multimedia narrative engine which controls and coordinates the systems responsible for presenting media within the exhibition space: lighting, audio, video projection, etc. (see Fig. 3.8). This computational engine seeks to create an emergent media ecology. To achieve this, it draws upon a lineage of computational research into the emulation of natural systems through code (e.g., cellular automata (Codd 1968), Boid systems and flocking behavior (Reynolds 1987), etc.) and the more recent application of these techniques in the context of interactive media performance, as is described by Rowe (1999), Downey (2005), and Birringer (2008). The CNS disseminates an unfolding spatiotemporal narrative that explores desert themes through media-driven representation of the desert at a variety of scales and granularities to present the transitional space between urban or desert environments through a virtually modeled landscape. The engine places media fragments on virtual multilayered planes. Programmed "agents" navigate these virtual planes and activated media fragments they encounter. The engine next translates in situ action (e.g., walking on the treadmills) to direct both the virtual agents and the conditions of the model. Movement and media characteristics change as the virtual landscape transitions between simulated "regions" of natural and urban land. This further underpins concepts central to the piece by crafting another ambulatory narrative of movement in place and connecting conceptually to the simulacra of the treadmill and performance.

Media situated on this imaged virtual landscape is subject to transitions throughout the "day." Desert, via the media placed in the landscape, is represented at differing scales, for example, from the insect or animal scale to the human scale and beyond, and at different times, from dawn to high noon, through to dusk and night. The media is further organized into two distinct categories. Ambient or durational elements that create a sense of the larger conditions within the virtual landscape (e.g., weather, texture, lighting, etc.) are subject to slower transitions or shifts and have less opportunity to be influenced by activity within the space but instead

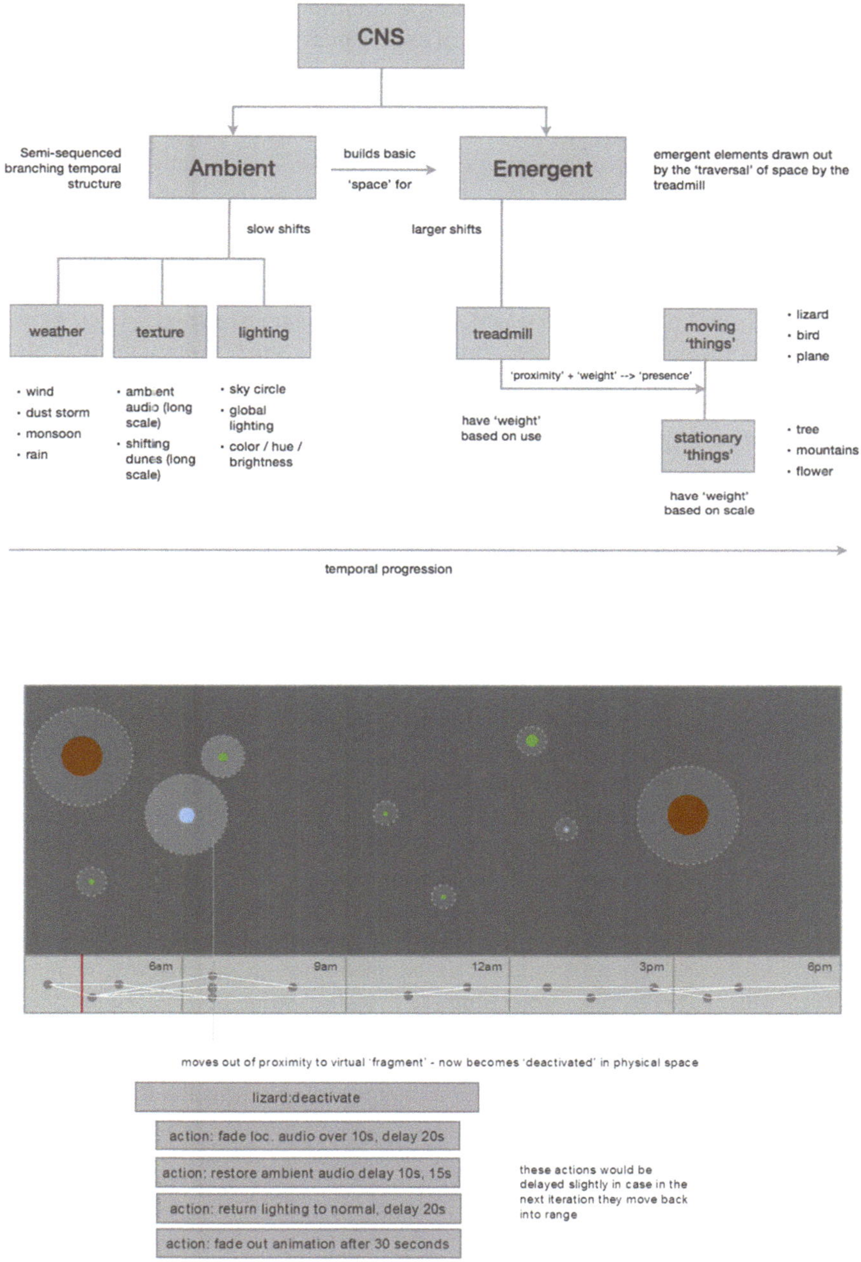

Fig. 3.8 Illustrative overviews of the Central Nervous System approach, which drives the desert simulacra exhibition space

transition gradually over time. By contrast, event-oriented elements are designed to encapsulate a moment or shorter-term experiences of the desert. These are more responsive to the state of the virtual landscape. Leveraging real-time input from the sensor-equipped treadmills, they respond to user activity within the space by triggering situated events in space. This couples the perspective of the treadmill walk with the larger environment. As a walker moves, the narrative engine adapts to feature imagery and themes related to the walk experience within the space. As more treadmills are engaged, the activation of the space is increased.

The virtual is made manifest in real space providing evocative proxies to the central concepts of the piece. The space and in particular the louver walls provide a canvas on which the desert can be evoked and expressed in situ through media. This media is coordinated through the CNS, which transforms that media and its attributed dimensions (temporal, spatial, nature, urban, surreal, real) into spatiotemporal spatio-temporal narratives that unfold in real-time score of a “day in the desert.” This simulacrum highlights the digitals potential to offer a durational, responsive, and experientially rich evocation that extends the performance and installation in two interesting ways: by extending the physical plane into a virtual landscape and by providing tribes of digital agents that can explore the desert at different granular movement scales and temporal trajectories and across poetic and literal interpretations.

The exhibition also acts to direct participation to the installations. As illustrated in Fig. 3.7, the architectural design places the treadmills as focal objects within the space. Arranged as a series of 3–7 treadmills surrounding a large central “hub,” the treadmill offers a mode of participatory engagement that can be experienced as a shared or individual experience. Activity on the treadmill simultaneously drives a first-person display of walking through the Sonoran Desert projected in front of the individual, as well as the activation of the space. The intensity of movement, as well as the number and location of treadmills being used concurrently, provides interactive input for the media-driven environment.

This media-driven environment is the context for both performance and installation. This simulacrum bounds the other two, situating them in a larger ecology, a larger landscape. These simulacra are coupled through interactivity, responsive to each other. The media, projections, and architectural objects are adapted to performance and choreography, providing the visual and acoustic decor to support the imagined mythopoetic landscape. Similarly, the space responds interactively to installation where audience participation interactively informs the dynamics of the media ecology which is revealed. While designed as independent simulacra, much as in the natural world, they also operate as connected subsets of a larger complex interacting system.

Collaboration and Interdisciplinary Expertise

The development of this project requires an iterative interdisciplinary process that connects a broad network of multidisciplinary collaborators from a range of fields to generate knowledge, representation, and form. Primary, ongoing participants include experts in movement and choreography, media arts, computation design, engineering, stage and lighting design, information visualization and 3D animation, ethnography, and architecture. However, short-term participation from scientific and artistic advisers plays a key role at junctures in the project development. The intention is to develop methods of transdisciplinary collaboration that connect diverse perspectives, knowledge, and expertise.

To develop an artwork that can create the conditions under which an authentic, integrated, and holistic experience of one's relationship to environment can emerge, team members are charged to understand and value both scientific and aesthetic perspectives. Scientific knowledge (weather models, ecological life cycles, pollution, heat dissipation) can drive data representations and visual schemas offering a more linguistic vision of a desert, while impressions, emotions, and that which is unarticulated quantitatively (i.e., aesthetic knowledge) comprise another mode of communication and engagement with a public. For the working team, communication across disciplines is coordinated around a series of production-oriented residencies. These are convened semiannually with approximately two residencies per annum. These residencies function not only as an opportunity for material development but also as forums for shared understanding and local contextual knowledge to be gathered and synthesized into production outcomes. More importantly, the residencies constitute a platform for understanding the process of production itself.

The development of this work has consequently become a tool for "research through production" (Koskinen et al. 2011; Mienczakowski and Morgan 2001) where the methodologies of development are explored in tandem with its production. In particular, it asks how scientific knowledge and content can be revealed and reconsidered within the context of a media arts performance. How is this knowledge interpreted, incorporated, and encoded within the experiential, kinesthetic medium of performance and installation? Through cooperative "production exploration" by the residency participants (e.g., the aesthetics of lighting design, the acoustic environment, choreographic creation, movement libraries, media projections, installation interface, etc.), this research explores the needs of the project and how participants satisfy them through knowledge acquisition, conceptual development, representational formats, and/or collaborative approaches. This perspective incorporates the goals of the project as a process-oriented inquiry where an unfolding understanding of the systems of making art and the systems of making science is continually interrogated through practice and iteration of the work as it develops. Key to this research is the understanding of how collaboration between the individual participants of diverse knowledge domains can work synchronously.

A key challenge has been coordinating the integration of diverse knowledge formats (scientific, experiential, aesthetic, ethnographic, etc.) into the work. Over the course of several residencies, the team has explored multiple strategies for approaching the fusion of knowledge into simulacra, and one particularly successful approach has been the use of *motifs* or thematic encapsulations that can incorporate diverse information in implicitly heuristic and humanized containers. We have found that the use of *motifs* provides a valuable translational framework that effectively coordinates multidisciplinary understanding and connects each of the three simulacra on aesthetic, representational, and informative levels in tandem.

Encoding Knowledge Across the Simulacra

Motifs frame and situate the choreographic score for performance, encode experiences for the installation's walks, and index media to be revealed in the mediated space. Importantly, they are also a mechanism by which hybrid practice and diverse knowledge can be coherently and powerfully synthesized to enable the creation of dynamic, emergent simulacra of environmental biomes within a mixed performance media-exhibition space.

This synthesis primarily occurs during the biannual production residencies that gather the design and movement team together. The residencies incorporate a diverse array of methods to facilitate the design and integration process. They combine the Feldenkrais Method's™ "Awareness through Movement" lessons (Feldenkrais 1977), coordinated walks through desert locations in the region, and meetings with scientific experts, along with the production of prototypical outcomes. In this way, the outcomes and knowledge produced during the residencies underpin movement, landscape, and the human relationship to space and place from a diversity of perspectives.

A clear example of the use of motif to encapsulate scientific knowledge in our qualitative framework comes from a residency in summer 2013. Early in this residency, the design team met with three scientific experts on desert ecologies. Each dealt with the relationship between natural and urban biomes at differing degrees of scale and complexity within the greater Phoenix area. The geologist dealt with the urban heat island effect (Anthony J. Brazel, Department of Geography, Arizona State University) and shared insight on the difference in heat dissipation from natural and urban spaces as well as the strategies and interventions for heat dissipation incorporated into current and future urban planning within the Phoenix area. An urban ecologist (Elizabeth M. Cook, Ecology) actively explored the relationships between native and nonnative flora, the introduction of new species, and the effect of urban and residential landscaping on plant species. The final expert (Jeffrey W. Ackley, Biology) presented work on the life cycle of lizards within the desert. This discussion uncovered the high sensitivity of flora and fauna to heat within the desert and some strategies for adaptation (e.g., migration of lizards avoiding activity at high temperature times of the day; having light colored skin to reflect light and heat), the scales of movement within the desert (e.g., large-scale movement of particulate matter through water and wind, to the small-scale movement of animals – where a lizard may move less than a kilometer in its lifespan), and shifts in activity and movement across the various times of the day. During this time, the scientists and design team discussed concepts around scales of movement, urbanization, temporal rhythms, and the complex relationship between heat, movement, and time. A seminal moment in this discussion emerged when the scientific advisers spontaneously expressed "the desert is shiny" in reference to the effect of heat on the lizards and their ability to radiate heat away from their bodies. This conceptual abstraction not only simply encapsulated and represented a body of scientific knowledge from the experts but also provided an abstraction that was well suited to aesthetic experimentation and artistic expression. This also highlights how "motifs" provide a mechanism to leverage scientific knowledge and sensibilities of environmental data but avoid being directly translational of these phenomena. Such a thematic approach allowed for scientific knowledge to be incorporated into an aesthetic, malleable, and abstract process while still recognizing the importance and specificity of these insights in crafting plausible, possible, and legitimate evocations of the desert. As such, "motifs" allow for inclusion of science research and knowledge without being representational and give the design team a vehicle to incorporate heuristic "impressions of science" in the experiential simulacra.

In developing motifs for evocative encapsulations of desert experience and phenomenon, they are applied at various levels of specificity, from the general to the particular. The most broad motifs include large themes such as "seeing the sun" (embodying heat, shock, melting, reflection, retention), "the desert is shiny" (iridescence, metal, mirage), "the tribe" (multiplicities, nomadic movement, myth, species, pattern), and "in the lee of the dune" (hidden activities, unnoticed change, shade). Other motifs focus on aesthetic qualities of the desert: materials and surfaces (concrete vs. sand, grass vs. glass), decay and vibrancy (bones, carcasses, blooming of plants), natural-urban objects (lawns, cacti, dwellings, etc.), and activities and actions (behavior of animals, traffic, human walking and other modes of locomotion, sandstorms).

Similar to Krug's four concepts of interdependency in ecological art (Krug 2002) and in addition to a thematic encapsulation, motifs are also organized and expressed along a number of distinct dimensions. This allows them to be mapped to a variety of times, places, and experiences:

Temporal: Motifs must consider change and transition of desert ecologies over time. Noting that the aesthetic texture of the desert as well as the presence of animals, objects, and levels of activity will vary greatly over the course of a day from dawn, noon, dusk, and through to night, as well as visibly shifting over weeks, months, and years, we consider how motifs are situated in "time."

Spatial: The motifs can also be observed at a variety of granularities from the large (environmental) to the human and molecular scales. Consequently, each motif is expressed at different scales to convey different intents to an audience, e.g., the desert is shiny might express the heat island effect at the city scale and also be demonstrated at the scale of the lizard.

Naturalness: A real desert may be experienced as the urban built environment and the structures that have been produced to support life there, but it may also be experienced as a natural "untouched" landscape in which that city is situated or as the wilderness beyond it. Similarly, there are parts that bleed through and hybridize that experience. Consequently, a motif may express as a hybrid urban-natural experience.

Real/Imagined: The project combines the lived experience of real deserts with a mythopoetic narrative of an imagined one. As such, this meta-dimension allows for motifs to transition between more literal and more abstract expressions of the desert.

By incorporating these dimensions, motifs build on the nonlinear nature of how we experience real spaces. For example, within the mediated simulacra space, each motif is mapped to sets of multimedia content (video, image, audio) and annotated to indicate the dimensions to which it belongs. In so doing, the content could be dynamically remixed and juxtaposed to develop a rich spatiotemporal narrative for each motif. For example, "the desert is shiny" might be expressed in radically different arrays of media for any combination of night versus day, real versus abstract, and natural versus urban. Through this recombinance, unfolding interactions between media emerge, allowing for spontaneity and indeterminate combinations in a rich expression of the complexity found in real desert natural environments.

Ongoing Work

Various iterations of the echo::system project have been presented; however, the work on this phase of the project will culminate in the production of an exhibition, installation, and performance in late 2014 and early 2015. Future residencies are already planned and as the project moves toward these milestone outcomes, we will continue to document and observe the collaborative process and outcomes. We anticipate that reflection on the totality of the project's development will reveal rich insight for media arts production, particularly where there are ecologically motivated underpinnings. Beyond exploring the methodologies and best practices for fostering interdisciplinary exchange and the integration of the qualitative and quantitative in media arts performance, we will also continue to document technical and design strategies for the multimedia spatial narrative of the simulacra. Once prepared, we foresee the context of the exhibition/installation/performance offering many opportunities for situated arts and human-centered contributions. Several explorations into the technical, experiential, and communicative affordances of the simulacra are already planned. These will explore the success of our emergent strategy for the expression of human and ecological experiences.

Conclusions

We explore the prospect of *simulacra* – rather than the more currently favored method of *simulation* – as a means to coherently encapsulate, embody, and express the variegated phenomenon and experiences of natural/urban environments. Using simulacra as translational framework, the work is informed by science and lived experience in parallel. *echo::system* explores the benefits of arts-led research through production that can incorporate science knowledge in an aesthetic experiential event. A plurality of perspectives is an essential part of both conception and production in *echo::system*, and a critical component of its development is the integration of this multiplicity of voices and expertise from artists, scientists, and researchers. This approach considers the complexity of desert ecologies, but challenges the production to integrate this diverse knowledge assembled across the domains of the aesthetic, human, scientific, and the ecological. Like the more representational simulation, simulacra acknowledge as well the complexity of these scenarios but afford a more experiential, abstract interpretation. Within the work, we demonstrate three distinct approaches to simulacra – in the mythopoetic performance, in the participatory installation, and in the spatiotemporal media narratives of the exhibition – when experienced together as a single event, they assemble a multifaceted and interwoven interpretation of the environment from the real to the imagined.

References

Baker K (2008) The lightning field (Dia Foundation), Dia Foundation series. Yale University Press, New Haven

Baudrillard J (1994) Simulacra and simulation. University of Michigan Press, Ann Arbor

Birringer J (2008) After choreography. Perform Res 13(1):118–122

Bowman DA, McMahan RP (2007) Virtual reality: how much immersion is enough? Computer 40:36–43

Carroll L (1893) Sylvie and Bruno concluded. Macmillan, London

Codd EF (1968) Cellular automata. Academic, Orlando

Coleman G (2013) Listening as the land talks back: ecology, embodiment and information in the science fictions of echo::system. Leonardo J 46(3):204–210

Coleman G, Byrne D (2014) 36 Walk: a case study of reciprocity in movement and computation. In: Proceedings of the 2014 international workshop on movement and computing (MOCO '14). ACM, New York, p 55

Cox D (2006) Metaphoric mappings: the art of visualization. In: Fishwick P (ed) Aesthetic computing. MIT Press, Cambridge, MA, pp 89–114

Dourish P (2001) Where the action is: the foundations of embodied interaction. MIT Press, Cambridge, MA

Downie M (2005) Choreographing the extended agent: performance graphics for dance theater. Ph.D. thesis, MIT

Eliasson O (2003) The weather project. Exhibition catalogue. Tate Modern, London

Faber M, Manstetten R, Proops J (1996) Ecological economics. Concepts and methods. Edward Elgar, Cheltenham

Feldenkrais M (1977) Awareness through movement: health exercise for personal growth. Harpers & Row Publishers, New York
Goldman R, Papson S (2011) Landscapes of capital. Polity Publishers, Cambridge
Goldsworthy A (1990) A collaboration with nature. Harry N. Abrams, Inc., Publishers, New York
Kjeldskov J (2001) Interaction: full and partial immersive virtual reality displays. In: Proceedings of IRIS24, University of Bergen, Bergen, pp 587–600
Koskinen I, Zimmerman J, Binder T, Redstrom J, Wensveen S (2011) Design research through practice: from the lab, field, and show-room, 1st edn. Morgan Kaufmann Publishers Inc., San Francisco
Krug DH (2002) Teaching art in the context of everyday life. In: Gaudelius Y, Speirs P (eds) Contemporary issues in art education. Prentice Hall, Upper Saddle River, pp 180–197
McClintock D, Kersten S, Ison R (1997) Emergent practice in agricultural research. Using metaphors and dialogue to create a space for understandings. In: Stowell FA et al (eds) Systems for sustainability. People, organisations and environments. Plenum, London
Mienczakowski J, Morgan S (2001) Ethnodrama: constructing participatory, experiential and compelling action research through performance. In: Reason P, Bradbury H (eds) Handbook of action research: participative inquiry and practice. Sage, London, pp 219–227
Morris F (ed) (2007) Louise Bourgeois. Exhibition catalogue. Tate Modern, London, p 170, reproduced p 171
Nelson BL, Carson JS, Banks J (2001) Discrete event system simulation. Prentice Hall, Upper Saddle River
Olivier B (2007) Ecological 'art' and the transformation of the world. S Afr J Art Hist 22:24–34
Patterson R, Cox D (2005) Visualization of an F3 tornado within a simulated super-cell thunderstorm. In: ACM SIGGRAPH 2005 electronic art and animation catalog (SIGGRAPH '05). ACM, New York, pp 248–249
Reynolds CW (1987) Flocks, herds and schools: a distributed behavioral model. ACM Siggraph Comput Graph 21(4):25–34
Rowe R (1999) The aesthetics of interactive music systems. Contemp Music Rev 18(Part 3):83–87
Samsel F (2013) Art – science – visualization collaborations: examining the spectrum. In: Proceedings of the IEEE VIS arts program (VISAP). IEEE, Piscataway
Schneider R (1997) The explicit body in performance. Routledge, London
Simon S (2006) Systemic educational approaches to environmental issues: the contribution of ecological art. Syst Pract Action Res 19(2):143–157
Viégas FB, Wattenberg M (2007) Artistic data visualization: beyond visual analytics. In: Proceedings of the 2nd international conference on online communities and social computing, 22–27 July 2007, Beijing, China
Wallen R (2012) Ecological art: a call for visionary intervention in a time of crisis. Leonardo 45(3):234–242

Chapter 4
Uncultivated: An Evolutionary Drama in the Urban Environment

Lynn Cazabon

Abstract This chapter will describe *Uncultivated*, a multifaceted public art project documenting wild plants in urban landscapes consisting of georeferenced photographs, commercial public displays (billboards, transit posters, etc.), a website, and community events. Commonly known as "weeds," the project encourages an empathetic view of these reviled plants and the natural environment we share in common while reflecting upon the changing biodiversity of urban landscapes due to the ongoing effects of global climate change. The project uses QR codes along with a dedicated website to provide contextual information on all the plants appearing in the photographs. The public displays are designed to deepen awareness of the immediate surroundings of the viewer by displaying a photograph taken in close proximity to the billboard or shelter. Started in Baltimore, MD, in late 2010, the project is designed to grow each time it is exhibited in a new city.

Uncultivated is a multifaceted public art project focusing on wild plants in urban landscapes. The project consists of georeferenced photographs, a dedicated website that pairs each photograph with information on all the plants featured in them, commercial public displays (billboards, transit posters, banners), exhibition prints, and community events. Begun in Baltimore, MD, in late 2010, the project is designed to grow each time it is exhibited in a new city.

The plants that are the focus of *Uncultivated* are the rats of the plant world: they are plants that most urban dwellers do their best to ignore, except when we want to get rid of them. Commonly known as "weeds," the project encourages a more nuanced and empathetic view of these reviled plant species and the natural environment we share in common while reflecting upon the changing biodiversity of urban landscapes due to the ongoing effects of global climate change.

The project asks why we privilege certain parts of nature over others, why, for example, we might throw our trash on the ground on a city street but not do so in a park.

L. Cazabon (✉)
Visual Arts Department, University of Maryland Baltimore County, 1000 Hilltop Circle, 21250 Baltimore, MD, USA
e-mail: cazabon@umbc.edu

F.T. Marchese (eds.), *Media Art and the Urban Environment*, Future City 5,
DOI 10.1007/978-3-319-15153-3_4

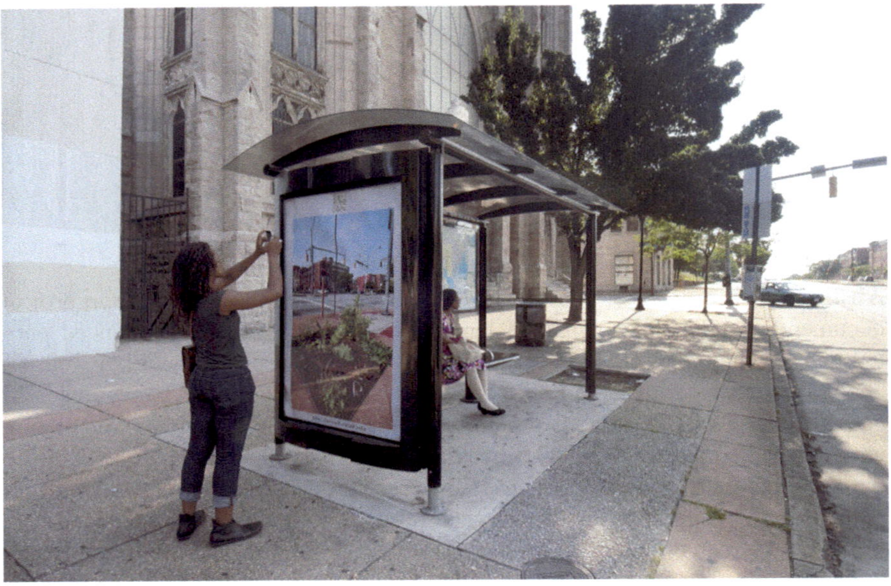

Fig. 4.1 Lynn Cazabon, Uncultivated_11.05.22 (39.311265, −76.616325), bus shelter installation, Baltimore, MD, July 2011

Uncultivated is also a story about biodiversity, its continuing loss, and coevolution: the side-by-side and codependent relationship between plants and humans. It is a slow-moving drama about survival, the plot of which stretches across time. *Uncultivated* pauses to consider how the way we choose to live in cities affects the plants that thrive there.

Photographs by nature are fragmentary: they present us with an image of reality suspended from time and place. I seek to contextualize my photographs, to place them as much as possible in the context from which they are torn. To that end, each image in *Uncultivated* has a corresponding web page, on which detailed information on each plant species appearing in the images is provided. Printed images are displayed with a QR code, which when scanned with a mobile device leads the user to the image's unique web page (Figs. 4.1 and 4.2).

Uncultivated represents a collaboration of art and science on many levels. In addition to the numerous sources I use for the information on the project website, the project relies on the contributions of Christa Partain, an independent horticulturalist, who has worked with me to identify the plants in the landscapes I choose to focus on.

The photographs in *Uncultivated* show wild plants in their natural setting within the urban environment. The project points the viewer toward a different way of seeing and understanding what is there, encouraging a transition from seeing just "weeds" to perceiving and naming what is there. For example, in the case of this

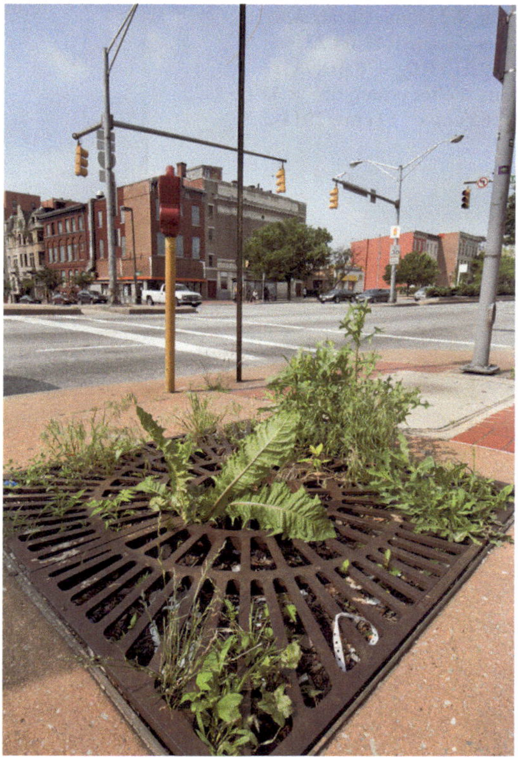

Fig. 4.2 Lynn Cazabon, Uncultivated_11.05.22 (39.311265, −76.616325), pigment ink-jet print, 45″ × 30″ (114 cm × 76 cm), 2011

image taken on North Avenue in Baltimore, MD, the plants in the foreground are annual bluegrass, spotted spurge, and common groundsel (Fig. 4.3).

Naming is a key strategy in *Uncultivated*: it is the first step toward making people take notice of plants in their surroundings. Plants found within each photograph are listed on the project website, which provides both scientific and common names for each plant.[1] Many plants have several common names, which are regional and historically specific, and these are highlighted on the website. A list of common names for *Taraxacum officinale*, for example, dandelion, lion's-tooth, blowball, cankerwort, pissabed, priest's-crown, swine snort, monk's head, fortune-teller, Irish daisy, peasant's clock, and fairy clock, comprises a historical inventory of human use and regard for this plant. Common names also indicate cultural differences in attitudes about plants. *Ailanthus altissima* is commonly called tree of heaven in its native Asian range, while in the United States it is often referred to as ghetto palm because of its unwelcome abundance in neglected urban settings.

[1] Scientific naming (also called binomial nomenclature) is a stable formal system for naming species of living things throughout the world, composed of the genus and species, using Latin grammatical forms as well as words from other languages.

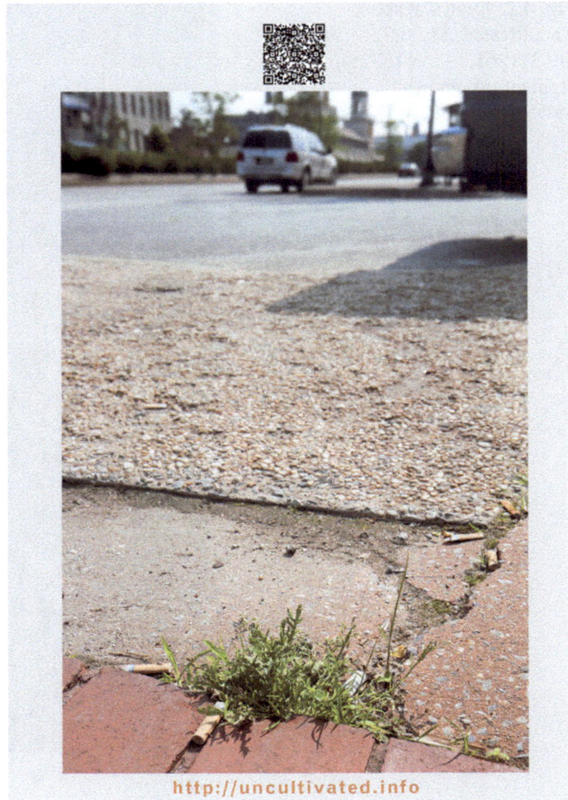

Fig. 4.3 Lynn Cazabon, Uncultivated_11.05.22 (39.310710, −76.618933), bus shelter poster, 65″ × 47″ (165 cm × 119 cm), 2011

Many of the plants included in *Uncultivated* are nonnative,[2] such as this group of plants (Fig. 4.4): white mulberry (Asia), spiny amaranth (South America), tree of heaven (Asia), and smooth crabgrass (Europe), alongside the plastic English ivy (European) in the shop window. However, there are also many native species that do well in urban habitats, such as this pokeweed growing near a dumpster in Washington, D.C. (Fig. 4.5).

Whether native or nonnative, these are the plant species that can survive in spite of us, plants that thrive alongside us, and because of the disruption caused by the way we treat the land in cities.

Uncultivated is a "live" project, ongoing, and continuously updated via the project website (http://uncultivated.info). The website was designed simply to work equally well on computer and mobile devices. Each image is georeferenced and is

[2]I use the term nonnative purposefully as opposed to the popular usage of *invasive* to designate any species thriving outside its native habitat. I find that the word *invasive* obscures the role that humans play in transporting plant species around the world, shifting blame on the plant itself instead.

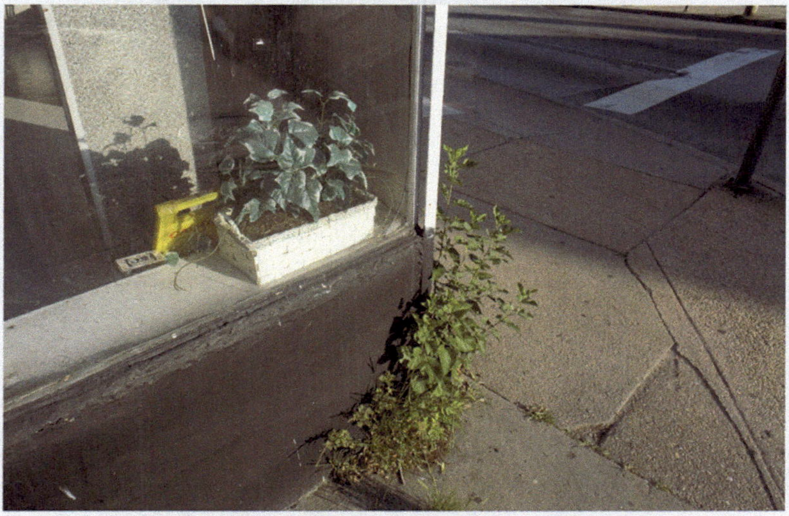

Fig. 4.4 Lynn Cazabon, Uncultivated_10.06.25 (39.337131, −76.633139), pigment ink-jet print, 22″ × 26″ (56 cm × 66 cm), 2010

listed with the decimal form of its latitude and longitude, more commonly known as GPS coordinates. Google Maps is used as an extension of the project website, showing where each image was taken on an interactive map (Fig. 4.6).

Each site page shows an image chosen to represent the location, the date it was taken, a list of the plants found there, and a link back to the project map (Fig. 4.7). Each plant name is in turn linked to its own page, which provides detailed information about the particular species, including its place of origin, urban habitat, ecological function, and cultural history.

Uncultivated is a hyper-local public art project. Commercial display spaces are used as a way to bring the project onto the streets of the urban landscape. Billboards, transit posters, and public banners show images taken within a short distance of the display location. These displays are designed to draw attention to the presence of wild plants within the immediate environment of the viewer by decontextualizing the familiar and representing it on a larger scale (Fig. 4.8).

Uncultivated is designed to continue to grow organically across time and space, with each new city constituting a chapter as the project is disseminated in new locations. In July 2011, six images were displayed as posters in bus shelters along

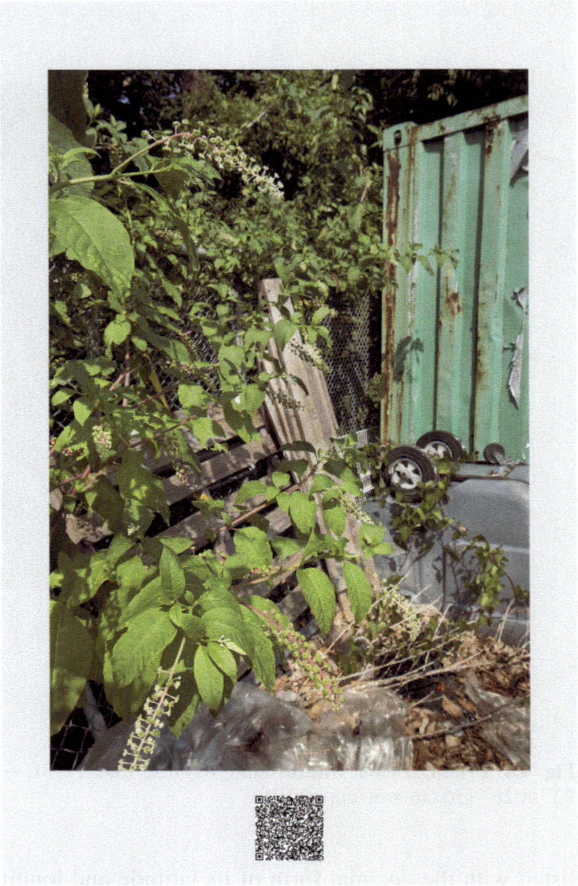

Fig. 4.5 Lynn Cazabon, Uncultivated_12.07.07 (38.909627, −77.077665), 96″ × 60″ (241 cm × 152 cm), pigment print on water-resistant fabric, 2013

North Avenue and near Penn Station in Baltimore, MD. For each shelter, the photographs displayed were taken within 25 ft of the shelter location (Figs. 4.9 and 4.10). The bus shelter posters were displayed with QR codes, offering the possibility to people waiting for the bus to learn about the plants underfoot.

In the fall of 2011, extended image sequences were displayed on digital billboards in New Orleans, LA, and Chicago, IL.[3] In each city, the project was customized for the location, with the images displayed taken within 500 ft of the billboard site. The location of the digital billboards along interstate highways enlarges the scale of the project, portraying intimate detail about plants that drivers normally speed past in their cars (Figs. 4.11 and 4.12).

---

[3]Sponsored by the Billboard Art Project, Richmond, VA.

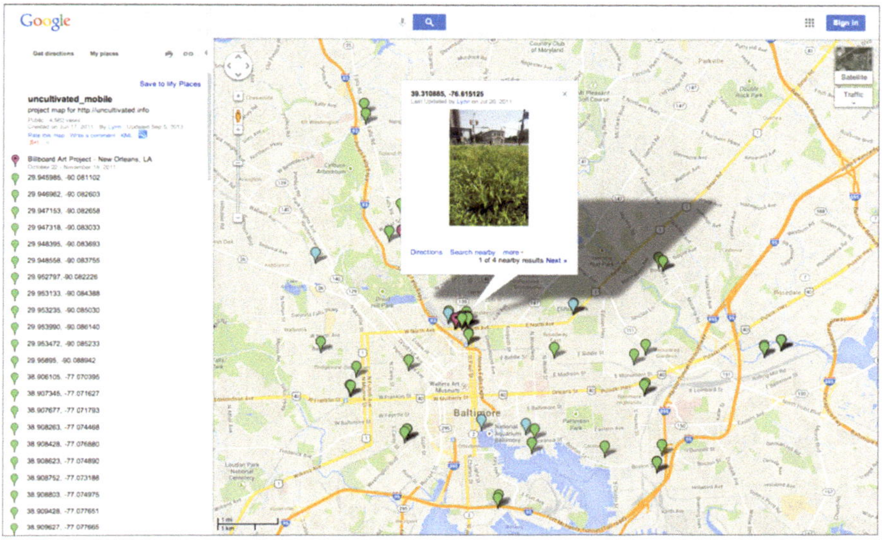

Fig. 4.6 Lynn Cazabon, Uncultivated.info, project website

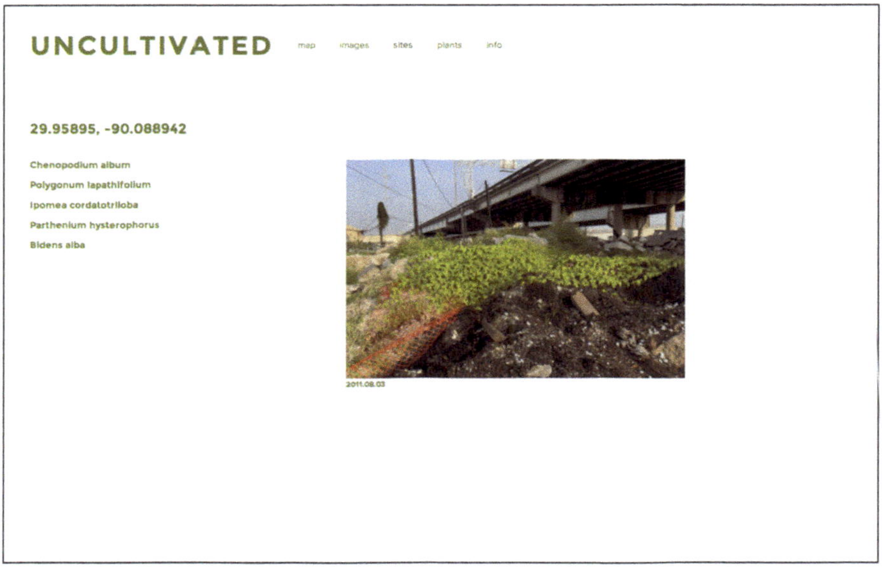

Fig. 4.7 Lynn Cazabon, Uncultivated.info, project website

In 2012, I created a set of 13 images on the campus of Georgetown University as part of an exhibition of the project in 2013 sponsored by the Spagnuolo Art Gallery. Six of these images were printed as large banners and installed in Regents Hall, the

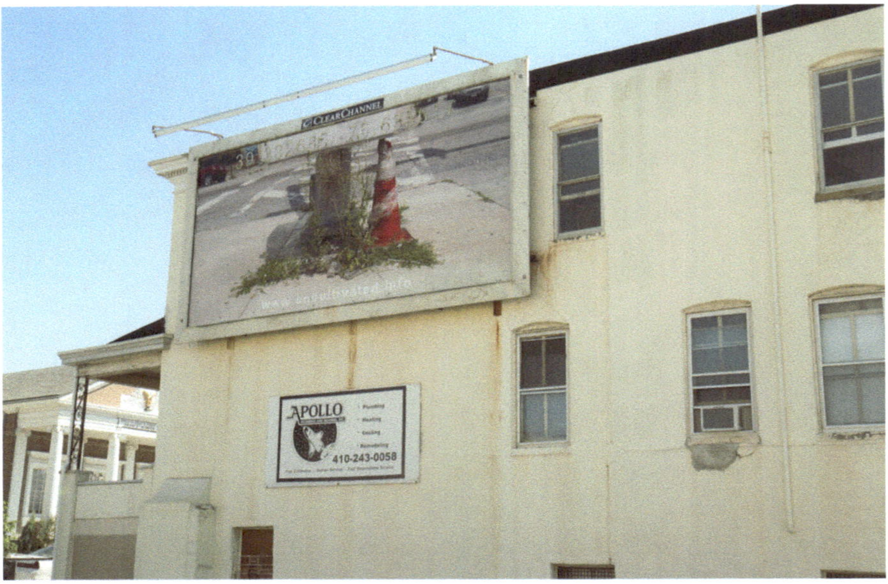

Fig. 4.8 Lynn Cazabon, Uncultivated_10.07.22 (39.332685, −76.635617), public billboard, Baltimore, MD, 2010

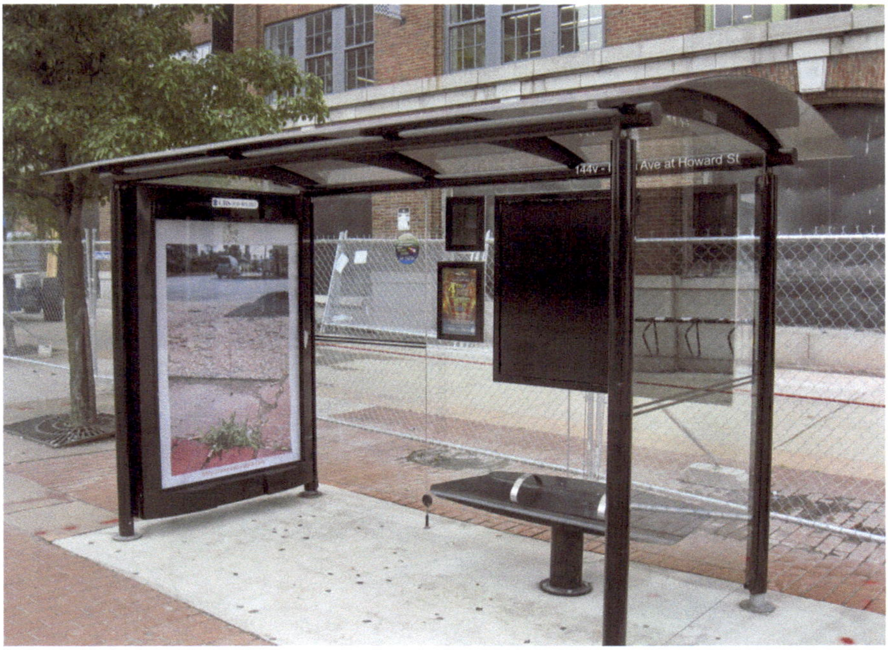

Fig. 4.9 Lynn Cazabon, Uncultivated_11.05.22 (39.310710, −76.618933), bus shelter installation, North Ave at Howard Street, Baltimore, MD, July 2011

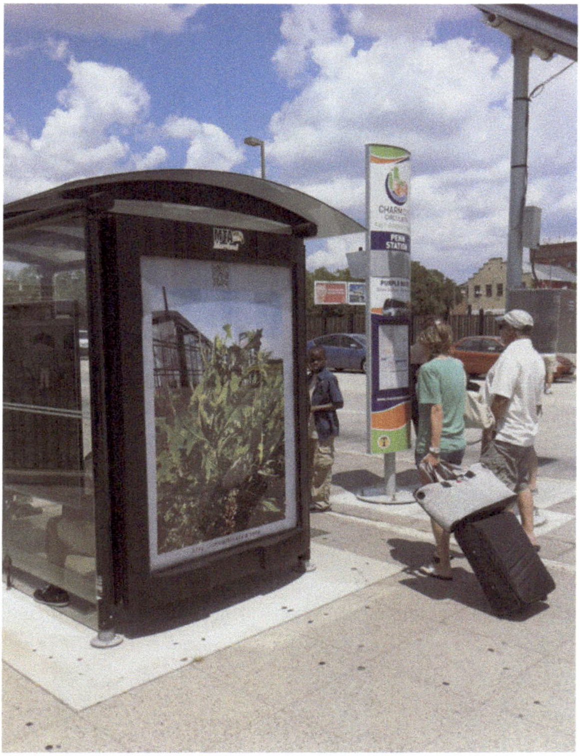

Fig. 4.10 Lynn Cazabon, Uncultivated_11.05.22 (39.306953, −76.614937), bus shelter installation, St. Paul Street at Penn Station, Baltimore, MD, July 2011

main science building at Georgetown, (Figs. 4.13 and 4.14). The highly manicured Georgetown campus is a pedestrian-scale, urban landscape situated within a mixed residential and commercial section of Washington, D.C., and contains numerous small habitats in which wild plants thrive. The banners displayed familiar parts of the campus from unfamiliar vantage points, as well as lesser-known locations where the campus borders the urban environment. During the run of the exhibition, a public plant walk was held which guided members of the Maryland Native Plant Society to sites I had photographed the previous summer and fall. At each site, we collectively tried to identify the plants we found there. Plant walks have also taken place in Baltimore to locations that people might not otherwise look for nature, such as under highways, near construction zones, and in refuse areas.

In 2014 for the multisite exhibition *inClimate: Climate Change Solutions, Awareness and Action*, I partnered with The Point, a community organization dedicated to the revitalization of the Hunts Point section of the south Bronx, NY, to hold a workshop for local kids which culminated in the creation of several "weed" gardens consisting of edible plant species that thrive in the particular conditions of the local environment (Fig. 4.15).

At its most basic level, *Uncultivated* is a series of landscape photographs; however, it updates the landscape genre in many ways, aesthetically, and in its

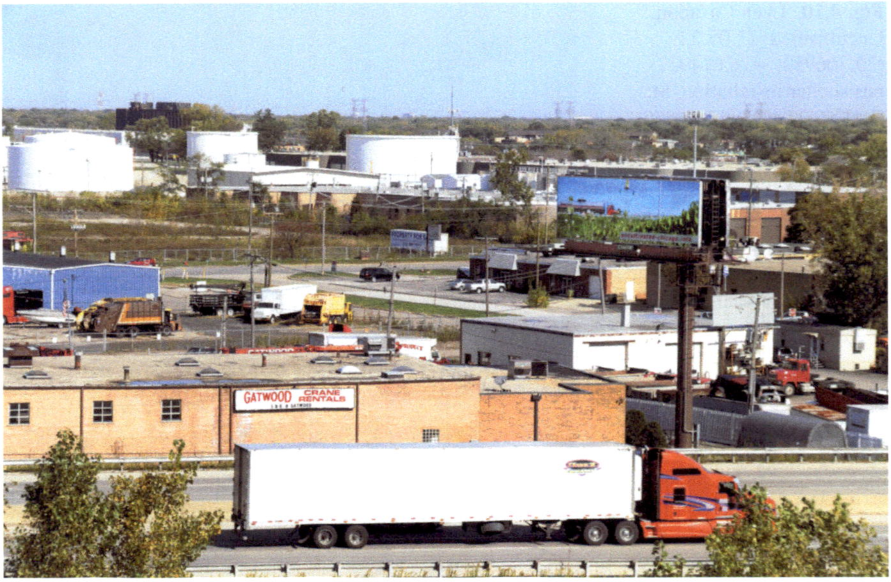

Fig. 4.11 Lynn Cazabon, Uncultivated, sequence of 20 images on digital billboard, Chicago, IL, 2011

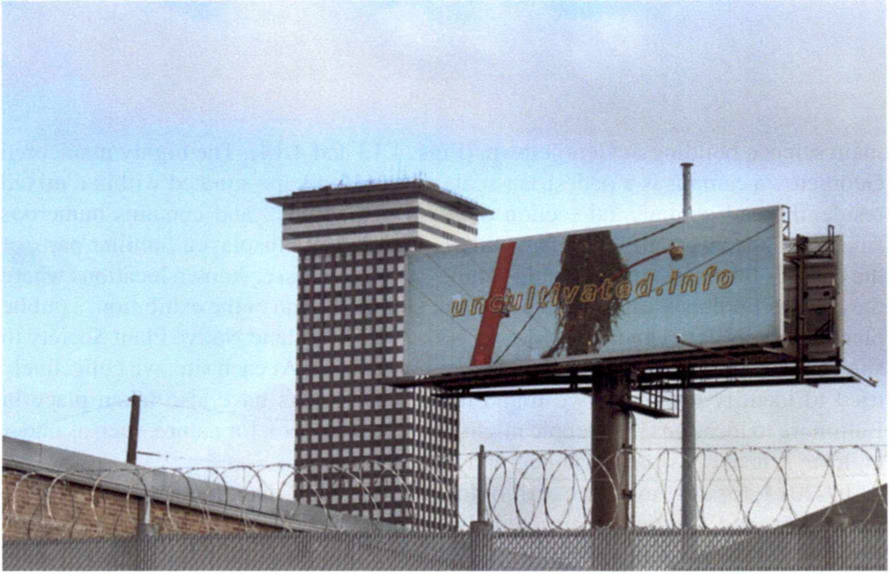

Fig. 4.12 Lynn Cazabon, Uncultivated, sequence of 12 images on digital billboard, New Orleans, LA, 2011

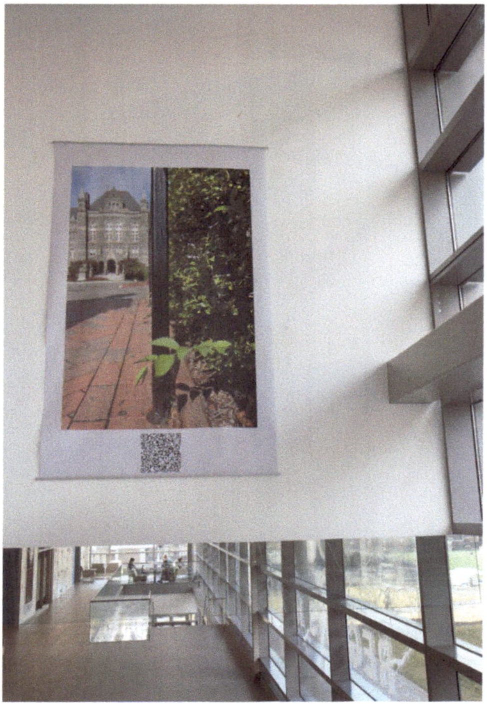

Fig. 4.13 Lynn Cazabon, Uncultivated_12.08.12 (38.907677, −77.071793), installed in Regents Hall, Georgetown University, Washington, D.C., 2013

use of new technologies and incorporation of community events. These are not the pristine or operatic landscapes of Ansel Adams. Instead, the images in *Uncultivated* are often taken from a low perspective, from the point-of-view of the plants, and juxtapose natural and human-made elements. Historical precedents for *Uncultivated* can be found in the photographs of Lewis Baltz, particularly his *San Quentin Point* project from 1985.[4] However, the use of color, contemporary geolocation tools, and the Internet in *Uncultivated* intentionally shrinks the aesthetic cushion that black-and-white photography allows between the viewer and the represented landscape.

Looking for the "wilderness" within the urban environment has led me to view our relationship to the nonhuman world in an inverted way – "wilderness" (in the traditional or historical sense) has become a largely human-made managed and created phenomenon, cultivated and preserved in parks: a bubble within the larger human-colonized world; whereas the plants I am photographing are unintentional, and therefore one could argue, truly wild. It seems that at the heart of the cultivated world, the city, there is a wilderness.I see wild plants in urban landscapes as living

[4] George Eastman House, http://www.geh.org/ar/strip87/htmlsrc2/baltz_sum00005.html#85:1265:0001

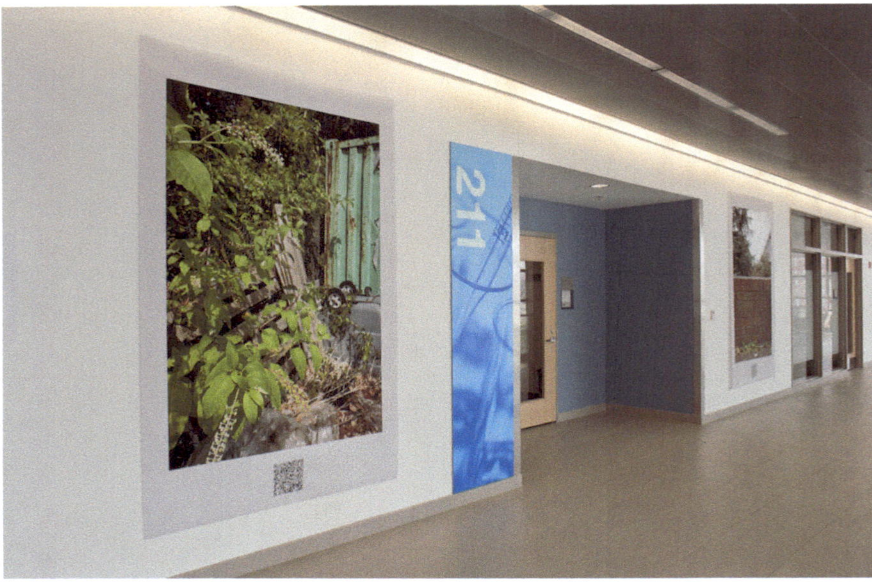

Fig. 4.14 Lynn Cazabon, Uncultivated_12.07.07 (38.909627, −77.077665), installed in Regents Hall, Georgetown University, Washington, D.C., 2013

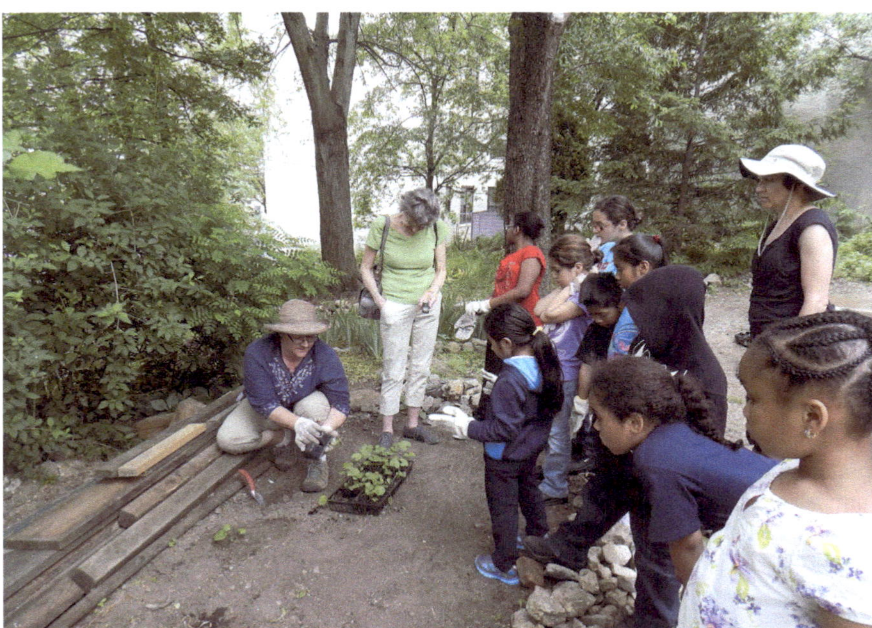

Fig. 4.15 Uncultivated: Edible Weed Garden Workshop, The Point and Bryant Community Garden, Bronx, NY, June 2014

expressions of the unintended consequences of conscious human activity on the land. These plants thrive on the disruption we create and, like a shadow, trail behind us wherever we tread.

Uncultivated seeks to break the habit of privileging certain parts of nature, inside national parks or wilderness areas, over the nature that we live with on a daily basis, because this habit leads to irresponsible behavior toward the nature that surrounds us every day. As the climate continues to shift, the landscape of our cities will change, but these shifts occur below the threshold of most people's perception. Photography is well suited to record change over time, and *Uncultivated* serves as a benchmark for the coming changes in the urban landscape.

Chapter 5
Alone Together in the Dark: Horror-Based Artworks and Fan Participation in Urban and Extra-Urban Space

Jillian McDonald

Abstract My artworks are inspired by stories from horror films. I avoid extreme violence and gore in favor of stripped down narratives and familiar archetypes like ghosts and zombies. This chapter outlines several projects in which horror fans and others perform live or on camera. Urban and rural locations, populated by local actors, are sites for conversation about monstrous, environmental, and social transformation.

Introduction

The horror film genre is appealing in part because of the robustness of its massive fan base, its litany of recycled mythologies, its distinct and faithful archetypes, and its rapt attention to setting or landscape. My performance and video artworks are motivated by these factors and their interconnectedness. From brief encounters with passersby in public places to more lengthy engagements where casts of amateur actors populate the visual stories, these works depend on participation.

In *The Adoring Audience*[1] Joli Jensen identifies two types of fans, "the obsessed individual and the hysterical crowd," a comparison which can also be used to differentiate vampires and zombies. These two fan categories and horror archetypes in particular interest me, literally and metaphorically. Collaboratively, I work with horror fans to develop performances and produce a transformation that is physical and sometimes perceptual. The participants have an experience in the making of the artwork, related to their niche interests, and the resulting artwork is intended for a gallery, performance, or screening audience.

[1]Jensen (1999, 2001), p. 9.

J. McDonald (✉)
Art Department, Pace University, 41 Park Row, #1205A, 10038 New York, NY, USA
e-mail: jmcdonald2@pace.edu

F.T. Marchese (eds.), *Media Art and the Urban Environment*, Future City 5,
DOI 10.1007/978-3-319-15153-3_5

Mythologies evolve, but generally it is the steadfastness of the lore that attracts fans. Exploiting human fear of the dark, the woods, getting lost, entrapment with no escape, being eaten alive, and becoming the monster, to name a few, many of these stories remind viewers that people are no match for nature or the supernatural.

Although science fiction is more typically associated with global disaster tales, horror features archetypes that undergo horrendous transmutations, endure cautionary tales, and sometimes suffer catastrophic extermination.

We know a horror setting when we see it – whether bleak or formidable it lies somewhere along the axis of industrial ruin porn and sublime gothic landscape, luring unsuspecting victims like poison poppies in *The Wizard of Oz*. Settings like apocalyptic cityscapes, midnight wildernesses, and desolate landscapes are all suggestive of horror. They suggest not only the pathos of humanity lost in the environment but also the uniquely human disregard for it.

Zombies in Condoland

In the mid- to late 2000s, an explosion of zombie movies caused an avalanche of zombie-themed products and a soaring undead popularity that included zombie-filled proms and weddings, zombie figurines, toys and books, zombie walks and flash mobs, and fan pilgrimages to sites such as the Monroeville Mall outside of Pittsburgh, the set of director George Romero's *Dawn of the Dead* (1978). Romero is the highly venerated godfather of zombie films. He is commonly credited with the contemporary definition of "zombie" as a being brought about through plague or supernatural means rather induced artificially as a means of ensuring slavery,[2] though he did not originate or even use the word "zombie" when he made *Night of the Living Dead* (1968) and subsequent films. The zombie is an unsettling (though timeworn by now) metaphor for unsustainable practices: for extreme consumption, for the working class, for the disenfranchised, and for the "99 %" championed in the Occupy movement of the early 2010s. Zombie mobs have been known to protest shopping, "brainless" TV watching, selling of violent video games to minors, unethical corporations, mining, and oil drilling. Their relationship with outbreaks, contagions, and viruses are also allegorical.

In the mid-2000s, New York City Council passed a City Planning Department rezoning proposal, relegating much of Williamsburg, Brooklyn's manufacturing structure to the archives, and paving the way to over 200 new blocks of residential condominium towers. The crowning jewels of this period perch at the waterfront, while whirlwind tearing down and building up practices continue to radiate outward in every direction. This is the story of many metropolises expanding upward in city centers and outward to swallow suburbs and bedroom communities at their edges.

[2]Boon (2011), pp. 5–6.

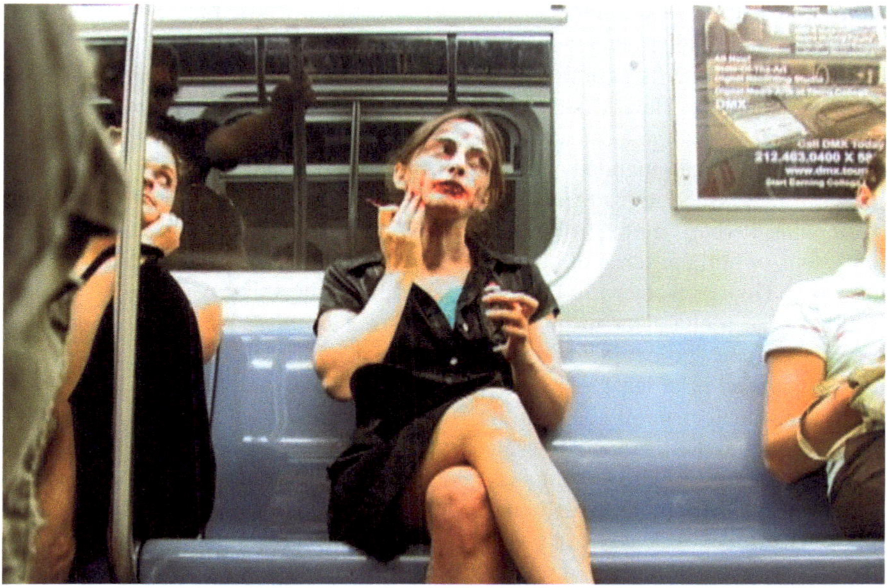

Horror Make-up, performance and video (detail), 2006

While studying amateur horror makeup videos on YouTube and watching the makeup artists' featurettes on horror film DVDs, I took notice of women applying makeup on public transit, specifically subway trains. In a filmed performance, I applied makeup to my own face on the L train in New York – a crowded commuter line that runs from Manhattan to Brooklyn and passes through Williamsburg – transforming myself into a monster rather than improving my appearance. The original event was a live performance; a second-generation audience can now view a documentation video, titled *Horror Make-up* (2006). In the video, the camera is hidden, faces the performer, and records only her and the subway riders in adjacent seats who steal glances and in some cases move away, keeping her in sight. "Ewwww" is heard from off screen as her fingers smear fake blood across her lips and chin. Like a contagious uncontainable disease, the zombie is abhorred.

Live and video audiences may read that we are all consumers – zombies commuting on the subway in an urban underground featureless world, which is no way to feel alive. Flocking like wildfire to an over-gentrified neighborhood that is growing skyward while consuming itself, transplanting the communities who once lived there and can no longer afford to, stretching thin the resources and capacities of the land and moving on.

Toronto's upwardly mobile skyline has also transformed radically in the past decade as suburbanites reenter the capital and climb skyward.[3] The city has a zombie legacy, inspiring the first zombie walk for a mere six participants in 2003[4] that was commandeered by Thea Munster, the consummate fan girl who today runs year-round events like zombie car washes and film festivals to raise the profile and funds for a massively well-attended annual zombie walk. It is hard to tell why the zombie walk format, which spread quickly to several continents and countless cities, is so popular. Perhaps more people identify with the zombie's condition and what it represents or find the sense of belonging to something universal and strange simply attractive. George Romero left gritty Pittsburgh for the shiny new city of Toronto in 2004, like a character in his own *Land of the Dead* (2005) who heads north where there are "no people to mess things up." The Canadian North is portrayed as a safe haven in at least one other film, *World War Z* (2013), where Brad Pitt's character's nuclear family ends up in Nova Scotia, a seemingly zombie-free zone. But the zombies are there, not missing out and colonizing exponentially more space.

Zombies in Condoland (2008) was an all night performance commissioned for Nuit Blanche Toronto, a 12-h nighttime citywide art event with attendance near one million.[5] The artwork was a participatory performance simulating a zombie film set situated in College Park, an urban plaza-cum-condominium tower construction site. It offered a chance for everyone to join in and be monsters for a moment, an hour, or all night. In a bygone era horror filmmaking meant low budgets, DIY productions, and casting from the ranks of friends and acquaintances; today more grandly conceived contemporary productions solicit extras from their tremendous fan bases or generate them with sophisticated software. Zombies are plentiful and also expendable in most zombie movies, and mobs of them are needed for effective storytelling.

During the performance, though all participants signed a release form, only documentary footage was recorded. There was no resulting movie. There were no stars and no dialogue but rather a series of representative scenes filled with legions of extras, enacted repeatedly in sequence while even more willing volunteer actors of all ages were churned out of makeup and costume tents ready for action. The undead rose from the ground, chased victims, swarmed a moving vehicle, surrounded survivors, and tried to enter a building – over and over again. A film crew followed them around while a "director" called out instructions from a megaphone and throngs of passersby were urged via signs, volunteers posted at the perimeter, and earlier, a blog, to slip into makeup and costume tents and be cast immediately. No one willing to be "turned" (pun intended) was turned away. Die-hard fans showed up already outfitted in elaborate zombie makeup and mangled attire, distinguishing themselves quickly from the regular Nuit Blanche attendees. This event was clearly in their comfort zone, and they helped others get into characterand hone their

[3]CBC News, June 28, 2012. www.cbc.ca/news2/interactives/before-after/toronto-condos/

[4]Lauro (2011), p. 208.

[5]Statistic from www.scotiabanknuitblanche.ca/about/event-history.html

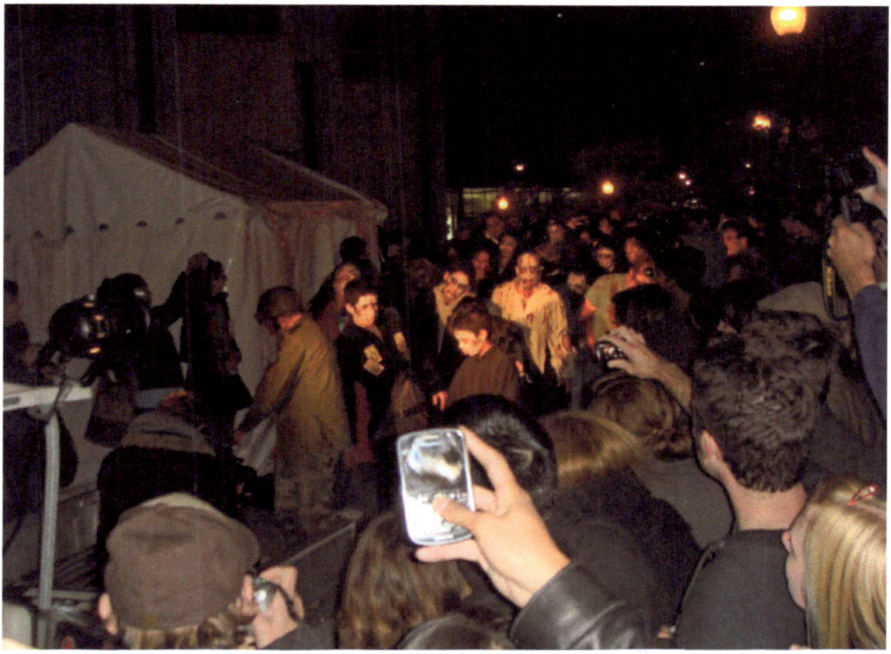

Zombies in Condoland, performance at Nuit Blanche Toronto in College Park, 2008

walks and looks. A sizable core group stayed all night long in a marathon of zombie activity. Heroic and creative makeup artists worked zealously all night. Thea Munster and Sarah Juliet Lauro, a California-based zombie scholar, joined the hordes. As the night turned to dawn, zombies in clusters hung about drinking coffee and discussing condo development, worker exploitation, and how the city is consuming its suburbs. Zombie activist mobs staged unscripted protests, calling an end to waterfront development and the community changes that come from gentrification and unsustainable urban progress. Splinter groups dragged themselves off into the night, reportedly hunting event attendees across the city above ground and under, in subways. And when the sun rose, the die-hard fans shuffled off, both invigorated and tired.

Alone Together in the Dark

Wilderness is and has long been a site of retreat from city life but equally a source of profound fear – innumerable fairy tales, fables, and horror films are sited outside of doors, and for good reason. Though rural residents may findsolace there, city dwellers are more often alarmed by the countryside, especially in the dark. Things snap and scurry and creep and scratch and howl when night falls; nature is cold and offers little protection from its "wildness." There is no end to the list of dangerous things lurking there. From *Little Red Riding Hood* and *Hansel and Gretel* to *Evil Dead* (1981) and *The Blair Witch Project* (1999), the forest itself, its potential for getting us lost, and the monstrous unknown that lies therein are all parts of the problem. In contemporary vampire and zombie stories, much of the action takes place in nature. An impossibly lush northwestern spectacle of forest is the playground for vampires and werewolves in *Twilight* (2008), the somber Southeast's featureless backwoods provide a setting for the loathsome zombies stumbling through *The Walking Dead* (2010–present), and the steamy swamps of Louisiana are home to *True Blood's* (2008–present) tantalizing supernatural figures. In these contemporary instances there is a subtext of excess, of nature's (including the supernatural) dominance and tendency to destruction, and of humanity's blithering ignorance and near-complete helplessness.

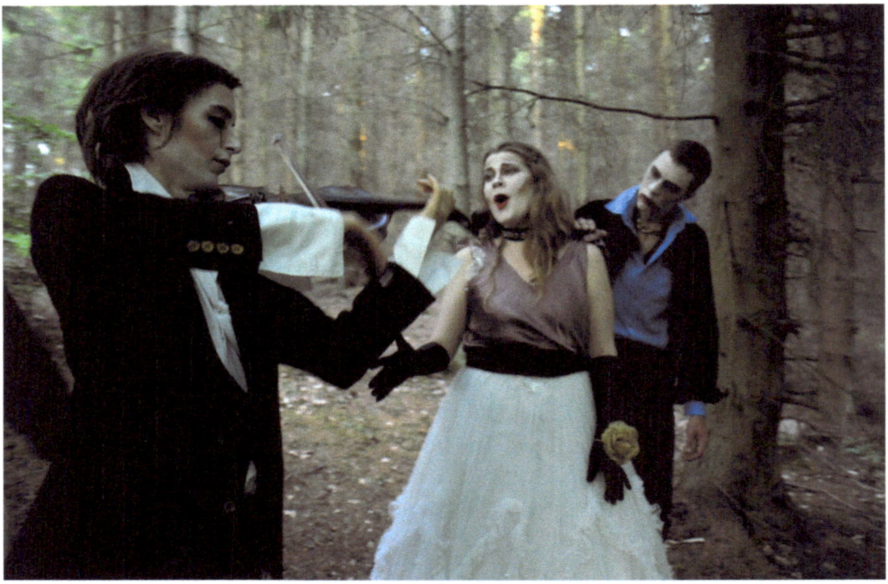

Undead in the Night, performance (detail), 2009 (Photo: Petter Pettersson)

Undead in the Night (2009) was set in a forest at the edge of Malmö in Southern Sweden: a live performance collaboration with Lilith Performance Studio, featuring 100 actors from Malmö, Copenhagen, Lund, and nearby cities who responded to urban poster campaigns that read, "Wanted: Vampires and Zombies." Five hundred showed up for auditions, and the selected actors and musicians committed to rigorous schedules.[6] Details of the event were kept secret. Tickets were instantly sold out. The press was rabid. Audience members were shuttled by minibus in groups of eight to the undisclosed location, to the tune of a commissioned soundtrack that set the scene for unspeakable events. Deposited in the woods without lights, mobile phones, or cameras, their sense of location and comfort was immediately destabilized by the dark and cold of the Swedish Spring nights. Small groups were led in silence along a 3-km path in the woods where 18 cinematic and chilling scenarios were set by artificial and real moonlight. Swamps, fern-covered clearings, ponds, thickets, hills, and valleys were populated by vampires, zombies, and victims who, every 15 min for three consecutive nights, reprised their roles for a new audience.

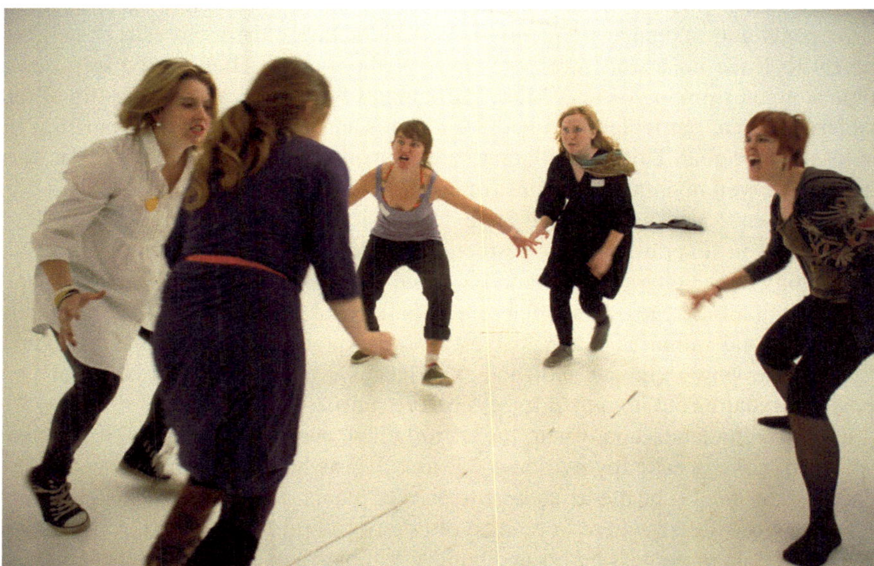

Undead in the Night, rehearsal, 2009 (Photo: Petter Pettersson)

The event of being thrilled and terrified simultaneously was lost on neither the audience nor the cast. Eavesdropping in a cameo role as a ghostly white spirit in the woods brushing past each small group, I witnessed breathless nervous titters on the path at sunset escalating to stifled cries from couples clingingtogether in fear.

[6]Each actor who walked into the Malmö rehearsals already self-identified as vampire or zombie; those who weren't sure were cast as victims. After 6 weeks of meticulous rehearsals, the actors, even those with no prior acting experience, wholly inhabited the spirits of those terrible monsters and bloodcurdling victims.

Screams cut through the crisp silent nights. Only days after the performances ended, a body was found nearby in the forest, bringing into awful reality the ability of the forest to serve as secretive and grim hiding place.

At the dress rehearsal I interviewed the actors about why they wanted to participate and give so much time to this production. Among the reasons cited was a desire to meet new people, having free time, feeling curious, participating in horror fandom, and seeking adventure. One man in his mid-30s, who traveled from a nearby town to attend rehearsals and performances, said he wanted to be a vampire (in real life) and the experience had brought him closer to that goal. Nearly all cited the collective ecology, expanding their social circles and getting to know the forest as the highlights. The urbanites in particular savored their daily visits to the forest and said that getting to know intimately the paths and the trees made them feel closer to nature. During and after the rehearsals and performances, participants and audience frequently called the setting "their" forest. They were careful to leave no trace, to respect and fear the woods. They stayed in groups; some stopped smoking on site. Many participants took protective ownership of the wilderness area, particularly around their own scenes. Some cited stewardship of parks as a new goal. This particular park, though outside an urban environment, was remarkably well maintained and inspirational.

In residence at The Arizona State University Art Museum in Fall 2009, I participated in a series of artist residencies called Social Studies, responding to the ecology of the city and community. *Alone Together in the Dark*, my photo and video installation created in a collaborative gallery/lab environment, features bands of zombies and vampires in a Wild West showdown at dawn in the desert. The title signifies the way a cinema audience experiences a film and the situation of two separate creatures out in the wilderness before sunrise.

The Arizona landscape was to me entirely new and astonishing. I felt stunned, unable to move or take my eyes away at first. It was hot and I moved slowly. The place itself rattled – brittle as its hostile prickly plants. I've never been so parched as I was in the desert, where I dreamed of thirsty vampires prowling through rocky caves in search of hapless hikers and hungry zombies scavenging for bits on bones. Both creatures fierce, gnarled, and dried like the land. At the time of my visit, the environment was in crisis and the global economic downfall in full swing, and yet Arizona's desert was a paradox. Golf courses abounded, and neighborhoods were decorated with bright green grass, nonnative plants, swimming pools, and indoor skating rinks; air conditioners hummed everywhere; water tanks perched on residential streets like cars full of gas. Hopeless bedroom communities were modern ghost towns, dotted with foreclosed homes where the desert was reclaiming the land, nudging the margins. Tumbleweeds swept through yards overgrown with thorny brown flora. The monumental Hoover Dam loomed nearby, holding back a seemingly impossible deluge.

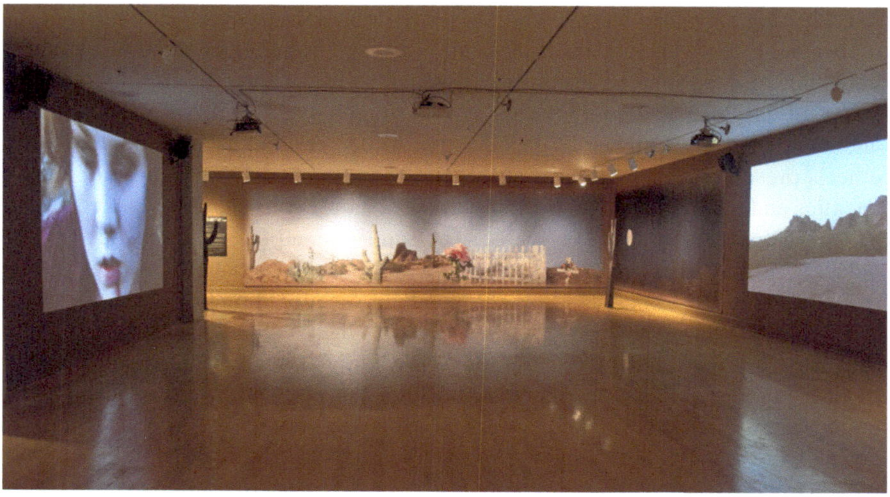

Alone Together in the Dark, installation at Arizona State University Art Museum, 2009

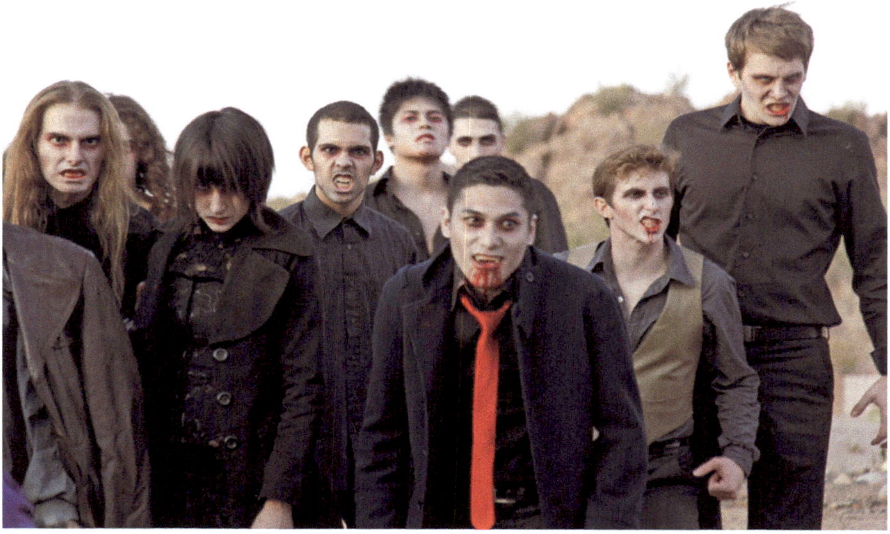

Alone Together in the Dark, production still, 2009 (Photo: Steve Gittens)

The Western, another film genre which proved impossible to ignore in this artwork, hinges on the West as more than landscape, as an essential character drenched in golden light. Good guys and bad guys wear white and black, respectively, tote weapons and are doomed to a final showdown. In *Alone Together in the Dark*, the shot sequences are modeled on Western filmmaking styles – starting with the inhospitable landscape and its details, unfolding and lingering over the conflict with long and close-up shots of the adversaries, and ending again with the landscape, to come full circle. The conflict in this case is simple, as it usually is – the vampires and zombies[7,8] each come out of their own wilderness and become aware of each other's presence via smell and sight. Since vampires and zombies are not natural enemies and have no need for each other, the standoff consists of prolonged gazing and snarling at one another en masse. The sun rises and the vampires flee (true to their mythology of aversion to sunlight) with unearthly screams to the hills. The zombies are the last ones standing, the good guys signified by their white clothing. They slowly disperse in every direction, and then the whole drama repeats.

The participants were all local residents including a car mechanic, math teachers, models, photographers, students, security guards, and artists. Anyone with a serious intention was accepted at auditions, and although some were uncomfortable in front of the camera, and implausible as characters at first, the actors persevered together to overcome these limitations. Many were horror film fans, and after 5 weeks of intensive rehearsals, a group performance with over 50 participants, honed and meticulously enacted, was captured on video. Installed in the gallery amid photographic murals depicting the desert, the showdown was spectacular. The soundtracks were composed to reflect each archetype and pay tribute to the Western via instrumentation and composition.

The location is municipal parkland, a place called Papago Park straddling Phoenix and Tempe with its distinctive red rock formations that are millions of years old. Hikers and cyclists, picnickers and tourists frequent the park. The metaphor of thirst and hunger was not lost on participants. Riverbeds were dry and cracked, rock formations crumbling. Ravenous and desperately thirsty creatures must live in the caves and hills. Collectively, we forgot to bring water on the day of our film shoot, and when our water supply arrived, we drank gallons, aware that the species that store water survive. We visited urban canals, desert graveyards, and ghost towns.

[7]The two separate mythologies of vampires and zombies are squarely seated in the pop cultural canon. The folklore occasionally shifts, but the deeper monstrosities remain unchanged: shuffling zombie swarms in *Night of the Living Dead* (1968) are dead ringers for the infected who race through *28 Days Later* (2002), and the crooked Count Orlok of F.W. Murnau's *Nosferatu* (1922) is fundamentally the same as *Twilight's* (2008) glamorous vampire clan, the Cullen family. If zombies are the metaphorical stand-in for the lowly consumer and disenfranchised everyman, vampires represent the powerful, aristocratic, or boorish CEOs. Zombies are perfect savages, and fans that identify with them recognize they represent the monstrous side of themselves and their worst fears. Zombies lack individualism and are rotting, soulless, unconscious, unfeeling, unattractive beings, though not inhuman. Vampires are the movie star victims of celebrity culture, trapped and bored in immortality, sometimes-androgynous soul suckers, bitingly cruel, and hopelessly alone. Post-humans might be thought of as capable of existing in an apocalyptic condition, evolving and adapting to radical environmental changes, and synthesizing human and nonhuman perceptions and responses. Vampires and zombies, despite their radical departures from humanity, both retain some degree of human sensitivities.

[8]Lauro (2008).

Economic and environmental collapse was evident in Arizona, increasing at its edges. A small group held informal story circles in the desert park, spinning ghost tales of lost hikers, watchful desert animals, poisonous and burned things, quick sand, hills that have eyes, oases, and mirages created by hazy heat. We spoke of northern desert tundra where the polar bears move to cities for food. We listened for the cries of coyotes.

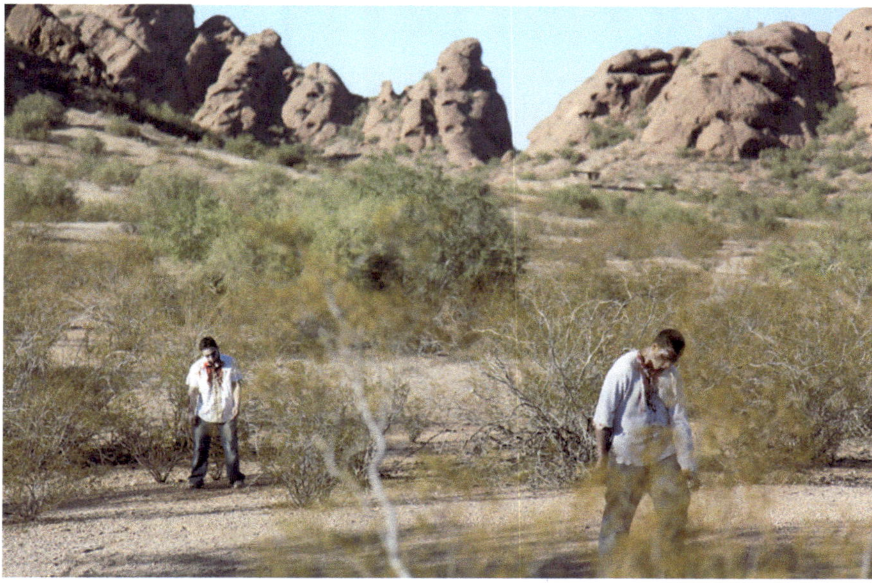

Alone Together in the Dark, production still, 2009 (Photo: Steve Gittens)

Fields of the Dead and Undead

Ghosts, spirits trapped in the netherworld, and hybrid creatures can epitomize or warn of unresolved histories.

Filmed at Fort Barry in the Marin Headlands in Golden Gate National Park, California, over 6 weeks, *Field of the Dead and Undead* includes 25 actors, each responding to simple directions, "walk into an open field on camera, "die," and stagger off as the undead." Each actor interpreted these words and was filmed separately. Later the actors were laboriously composited together into one scene using animation software, and the way the ghostly figures slippast each other in the fog, alone together in this field, is otherworldly. Some are sunlit, some are in shadow, some are transparent apparitions, and they inhabit collectively the murky gray field they walked in solitude, lost in an unrelenting loop. *The Shining* (1980) and *Poltergeist* (1982) are just two examples of horror films where the undead hover in tormented states between life and death, sometimes at locations of past cultural unrest such as disrespected burial grounds.

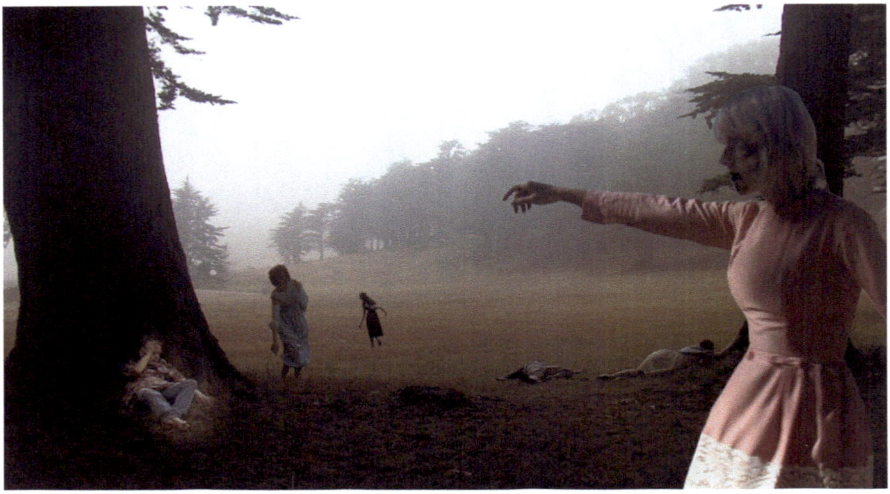

Field of the Dead and Undead, HD video (detail), 2011

The surrounding landscape of Fort Barry houses centuries of seacoast fortifications dating from the Civil War to the Cold War, built to protect the Headlands from enemies arriving by sea. Adjacent to this field is the only restored Nike missile site in the country, originally built as one of three hundred for Cold War defense. The field is the heart of a former military settlement, where visitors are expected to tread lightly on the land and to displace nothing. Permission is needed to film there in respect of the natural habitats and populations. Eucalyptus trees, seals, river otters, egrets, rattlesnakes, coyotes, and mountain lions are among those which call the setting home. National Park services educate visitors about responsible park use with signage and rules of recreation.

Most of the actors were visitors to the park and surprised by its microclimate (often visited by entirely different weather than San Francisco, a few miles away), many attracted by its context. In interviews before and after filming, they spoke of the landscape and history of the place. The ecology of the protected park was paramount, as well as its history of wartime occupation, the gold rush,and ancestral sites of the Coastal Miwok who were eradicated during Spanish colonization by disease and labor.[9] Actors were mindful of this complex history and haunting of the setting by past, present, and future inhabitants.

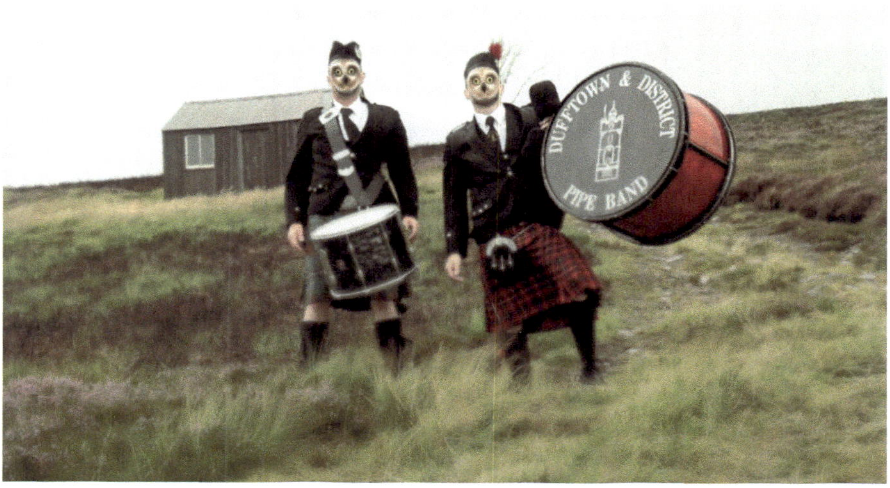

Valley of the Deer, HD video (detail), 2013

Valley of the Deer (2013), a film featuring 50 actors in modified highland dress as masked predators and prey amidst jaw-dropping Scottish landscapes, was produced during a 9-month extended artist residency at Glenfiddich Distillery in Dufftown. Shimmering phantoms of live animals like white reindeer and orange sheep appear among figures inspired by local legend and tradition like a cannibalistic family, a washerwoman who washes blood from the clothes of a future victim, and the Brollachan or shape-shifter who appears as whatever his victim fears most. Each shot or scene is a separate composition and a nearly still image where little movement, save hair in the wind or an animal moving its head, betrays the stillness. Slight movement becomes remarkable and disconcerting. The landscapes are as bewitching as the actors, made anonymous by their masks and swaying as if to a preternatural rhythm.

Masks in horror narratives are ubiquitous. *The Phantom of the Opera*, Leatherface from *The Texas Chainsaw Massacre* (1974), Jason Voorhees from *Friday the Thirteenth* (1980), and small gangs of anonymous killers found in *The Strangers* (2008) and *You're Next* (2013) all wear masks to hide monstrosity and/or the wearers' identities. Scotland's own *The Wicker Man* (1973) depicts a ritualized hunt of an outsider by townspeople in homemade animal masks. Inrecent years they can

[9]http://www.nps.gov/goga/index.htm

be found littered throughout mainstream ritual and performance, in weddings, music videos, and fashion shoots. They remain uncanny, anonymous, naïve, frightening, and hybrid.

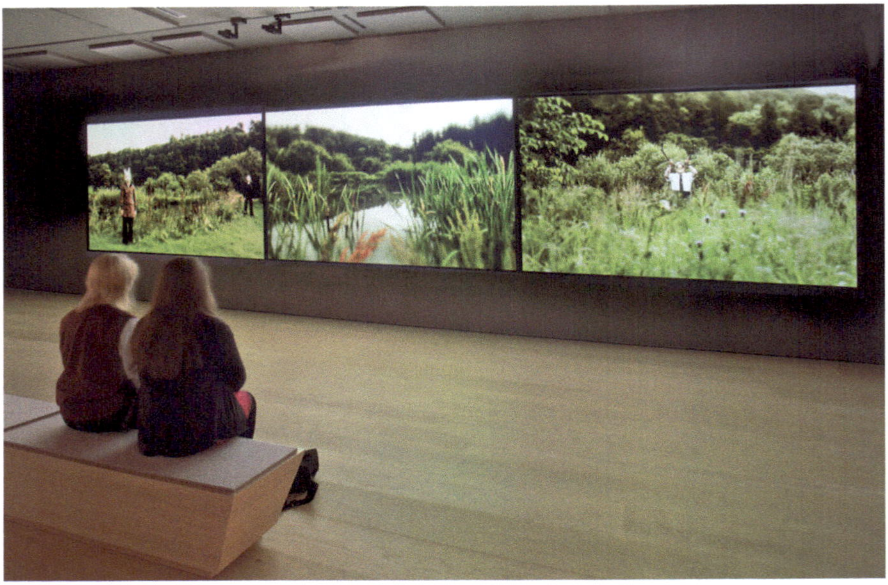

Valley of the Deer, installation at Esker Foundation in Calgary, 2013 (Photo: John Dean)

From the Valley of the Deer (2014) was commissioned by Turbulence.org[10] as an accompanying augmented reality artwork. A series of videos, images, and sounds originating from the film are launched on mobile devices via Layar, an augmented reality application. Characters are transported from the highlands and programmed to local GPS coordinates when the work is exhibited in urban galleries and are overlaid onto the device's live camera feed. They are discoverable as mapped "points of interest," in augmented reality terminology. This is not unlike touristic points of interest such as roadside vistas on the highway, indicated by signage and often, a camera icon. Virtually speaking, the actors travel across the world to tell their tale, leaving a trace or a copy of themselves in each new place. Set on radial paths from each exhibition venue, the characters may be encountered on a planned walking tour or stumbled upon by passersby, hauntingeach new location

[10]It was made possible with funding from the Jerome Foundation. www.turbulence.org/Works/fromthevalleyofthedeer

with the spirits of a distant valley. Clicking on the characters launches short video clips, glimpses into the world of *Valley of the Deer*. To date, this artwork exists in Québec City, Brooklyn, Calgary, Saskatoon, San Francisco, Montréal, Dufftown, and Newcastle.

The actors in *Valley of the Deer* are local craftspeople at the Glenfiddich Distillery, who work in grounds keeping, bottling, the gift shop, and the art gallery. They are distillery guides, production managers, coopers, and administrators. One was a recent retiree who worked at the distillery for 50 years and still lived within walking distance from the cooperage where he had finished his career. They were also members of the local Pipe and Drum band, health care workers, artists, cafe owners, a fireman, workers at other distilleries in the area known as the malt whisky capital of the world, and others. The soundtrack was produced from sound recorded on site: distillery noises like barrels being filled, corked, and emptied; animals such as shriek owls in the woods near my house; bagpipe laments and drum marches played by the local band; and a solo performance of Loch Tay Boat Song, a traditional song sung by a ceilidh singer and recorded from a well-worn armchair in her centuries-old farmhouse at the foot of Ben Rinnes.

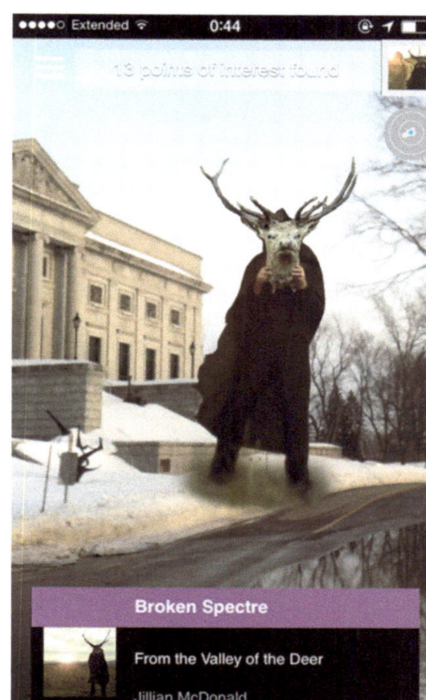

From the Valley of the Deer, Broken Spectre, at Musée National des Beaux-Arts du Québec (screenshot), 2014

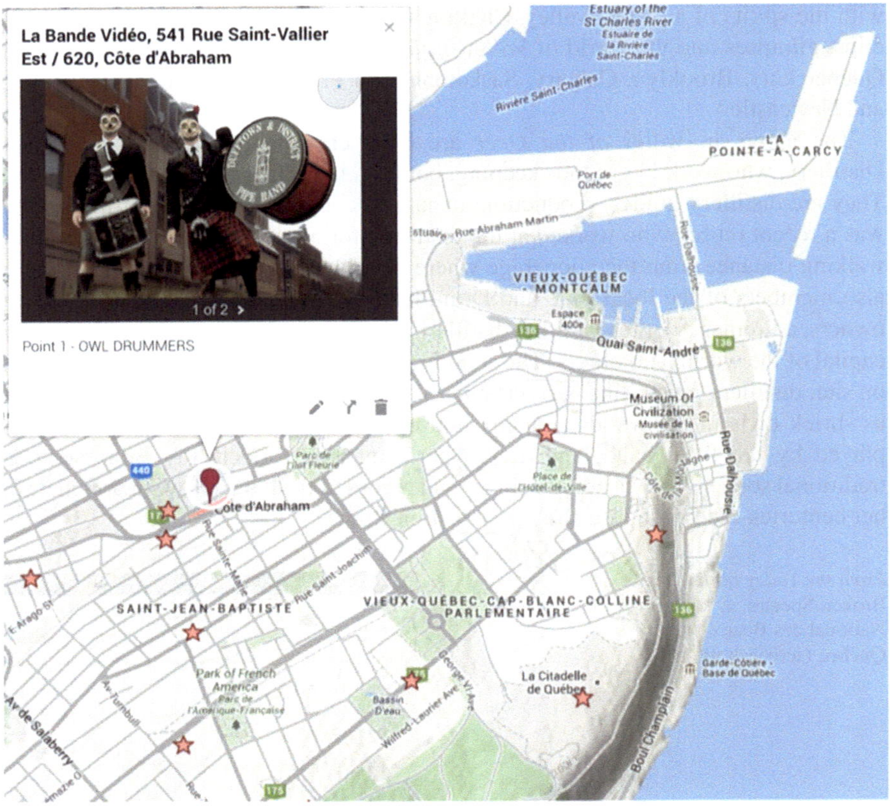

From the Valley of the Deer, Google map of Québec City locations (screenshot), 2014

Most, when questioned, did not consider their homeland remarkable. Some had never traveled away from Scotland or even the northeast.

Rural Scotland is a place where things that are local and time-honored are held dear. The pending referendum for Scottish independence from Britain was on the minds of many, yet some of the key factors in the argument for separation are not visible without a closer look, and without staying a while to get to know the place and the people. As "untouched" as the land may seem in the film, the narrative landscape betrays the reality of the local ecology. Many of the actors work in industries that contribute in some way to environmental decline. Disputed "Scottish oil" has been dredged from the North Sea since the 1970s, and Goliath wind energy mills produce impressive shares of alternative energies but cause interruption of wildlife habitats and obstruction of natural sightlines. Numerous distilleries and their euphemistic "angels' share" (evaporated alcohol) are likely responsible for a black mold that thrives in ethanol-rich environments and covers surfaces like buildings and trees that stand nearby alcohol warehouses. In some areas where I filmed, poverty levels are high and steady employment is rare. Some residents travel great distances to work offshore or where they can find employment.

The actors spoke of faeries, thought to be an ancient and peaceful race of smallish humans driven to the mountains by stronger peoples, and whose lingering presence in folklore and culture gradually changed them into magical spirits of the natural world. Some took me to sacred and mysterious places like standing stones and stone circles, faery rocks and cairns, churches with carved runes, and woods full of bones. One took me to a faery knoll in Aberlour that had been recently fenced in on private property. Residents don't look kindly on this move. These mounds of turf and peat and stones are not meant for private use, because many believe that the souls of ancestors dwell there. Defiantly my guide leaped the fence, and I followed; we walked on the knoll, though we took nothing from it. The Scots are proud of and determined to exercise the ancient "freedom to roam" that gives non-motorized universal access rights to cross and be on land in Scotland, responsibly, for recreation and education. Actors and others calmly took me walking in and around castle ruins, farmland, privately owned forests, and abandoned fields and houses.

Some saw their animalistic roles in the film as prophetic of a local and global need to find again a reverence for animals and land instead of destroying them, before they turn on humankind in an unscripted apocalyptic scene. They wondered whom was the symbolic prey falling without a whisper in our climactic on-screen moment when the masks are pulled aside. A fragile human deer is killed in a drone of bagpipes and rushing water and the video ends with a rainbow, nature's most beautiful sign, signaling victory after the fall.

Final Thoughts

The projects discussed above, and others that have come afterward, are examples of artworks with a core of active participants and horror fans. Each local community presents a unique opportunity for discussion about particular ecologies. "The abnormality thus gives imaginative form to anxieties about being human under evolving conditions, in a de-familiarized world... figuring the plight of modern subjects "at two with nature", with our circumstances, and with our selves."[11]

[11]Comaroff and Ker-Shing (2013), p. 30.

References

Boon K (2011) And the dead shall rise. In: Christie D, Lauro SJ (eds) Better off dead, the evolution of the zombie as post-human. Fordham Press, New York
Comaroff J, Ker-Shing O (2013) Horror in architecture. ORO editions, China
Jensen J (1999, 2001) Fandom as pathology: the consequences of characterization. In: Lewis L (ed) The adoring audience: fan culture and popular media. Routledge, London
Lauro SJ (2008) A zombie manifesto: the non-human condition in the era of advanced capitalism. Boundary 2 35(1):85–108
Lauro SJ (2011) Playing dead: zombies invade performance art... and your neighborhood. In: Christie D, Lauro SJ (eds) Better off dead, the evolution of the zombie as post-human. Fordham Press, New York

Chapter 6
Mobile Maps of Chameleonic Cities: Urban Cartographies and Methodological Procedures and Experiences

Pedro Marra and Carmen Aroztegui Massera

Abstract Every day, hundreds of thousands of people circulate throughout the center of Belo Horizonte, the capital city of one of Brazil's largest states. The physical space of this city changes constantly, and it is the site of conflicts with the government and among the population. Understanding the constant process of change is essential to performing appropriate ethnographic fieldwork. However, difficulties may occur when dealing with the temporal dimension of observation, when working with passersby, and with the current technological nature of recording artifacts. Traditional academic fieldwork has had trouble capturing this chameleonic city's changing characteristics and its continuous expansion of urban images and representations. This chapter discusses the experiences related by The Urban Cartographies Research Project, a research group who dealt firsthand with such methodological difficulties. First, this research explores how temporality and spatiality are constructed in contemporary cities and tries to evidence the temporal mobility engendered by global capitalist fluxes, especially on developing countries and within the context of mega events, such as the FIFA World Cup. Next, this chapter describes and reviews the specific work of the research group, highlighting its methodological practices. Then it discusses some examples of urban interventions that focus on urban art experiences. Such experiences reveal ways of exploring and making academic findings visible, as well as inviting the city's inhabitants to take part in participatory knowledge-based construction strategies. Finally, the research discusses the group's recent experimental methodological

P. Marra (✉)
Centro de Convergência de Novas Mídias – UFMG, Avenida Antônio Carlos, 6627, 31270–901, Fafich, Pampulha, Belo Horizonte, MG, Brazil

Universidade Federal Fluminense, Rua Miguel de Frias, 9, 24220–900 Icaraí, Niterói, RJ, Brazil
e-mail: pedromarra@gmail.com

C.A. Massera
Centro de Convergência de Novas Mídias – UFMG, Universidade Federal de Minas Gerais, Belo Horizonte, Brazil
e-mail: aroztegui@gmail.com

F.T. Marchese (eds.), *Media Art and the Urban Environment*, Future City 5,
DOI 10.1007/978-3-319-15153-3_6

procedures and outlines future actions that might allow for a better and more participatory understanding of an ever-changing and mobile city, thus allying the social sciences and various artistic practices.

Introduction

Every day, hundreds of thousands of people circulate through the center of Belo Horizonte, the capital city of the Brazilian state of Minas Gerais. The location is in a constant state of change. Daily, new buildings emerge, city traffic is reorganized, and seasonal vacations and celebration days create boosts in the circulation patterns of the people. The city center encompasses different regions, mutating according to the presence of people, trees, benches, and other urban equipment and drawing and redrawing its own invisible and symbolic borders. Multiplicity is the key to understanding modern and contemporary cities, which often are characterized as chameleonic because the most prominent feature of such cities is their constant state of change, like the skin of an animal.

This chapter discusses the fieldwork experiences of a research group who faced methodological difficulties when dealing with these changes in urban inquiry. The group, called the Urban Cartographies Research Project, has been researching citizens' everyday practices in Belo Horizonte's urban space setting since 2004 in order to determine their roles in the city space configuration. Our research produced mobile maps using sound, video, photography, and texts to register peoples' uses of the space.

Traditional academic fieldwork and artistic projects alike have trouble capturing a chameleonic city's characteristics, as well as the continuous growth of urban images and representations. The main difficulties occur when dealing with the temporal dimension of observation, the issues that emerge when working directly with passersby, and the current technological nature of recording artifacts. How can we contemplate these various changes in the appropriation of space over time? What can we do to keep from reducing our conversations with a city's citizens to "interviews with qualified informers?" How can we include both the researcher's own urban experiences and alternative recording methods which incorporate digital media such as soundscapes, image sequences, urban installations, and social networks? In an effort to answer these questions, this article discusses some of the strategies inspired by contemporary urban art interventions that were created and incorporated by the research group.

This chapter first explores how temporality and spatiality are built into contemporary cities, evidencing the temporal mobility engendered by global capitalist fluxes, especially in developing countries, as well as within the context of mega events such as the FIFA World Cup. Next, we describe and review the work of the research group of which we are a part, The Urban Cartographies Research Project (UCRP), pointing out the methodological procedures we call upon so that we can address the issues discussed in the first section. Then, we discuss certain experiences of works of art in an effort to elicit the esthetic procedures that help us to explore and make our findings visible and invite the city's inhabitants to take part in participatory, knowledge-based construction strategies. Finally, we describe and analyze the methodological procedures with which we have recently been experimenting to try to outline future methodological procedures we may need to understand our ever-changing and mobile city, in a participatory way, thus allying the social sciences and various artistic practices.

Understanding the Temporality and Spatiality of Belo Horizonte's City Center

Belo Horizonte is the capital of Minas Gerais, Brazil's second-largest state in terms of population and third-largest state with regard to economic relevance. The city was planned and built to serve as home to the state's government at the end of the nineteenth century, and today, it has a population of approximately 2.3 million people. It is mostly a commercial and services city, although neighboring cities (which include a metropolitan area with a population of approximately 5.2 million people) serve as home to a number of industrial plants.[1] The previous capital of the state, Ouro Preto, was a colonial city with narrow and curvy streets, tucked in among mountains mined for gold in the eighteenth century. However, Belo Horizonte was planned as a European-style modern capital with wide streets full of trees and avenues crossing each other along two orthogonal planes, all encircled by an avenue designed to allow for the rapid transportation of both goods and people. As Ermínia Maricato (2000) has stated, however, this modern urban tradition offers up the city only to those few who enjoy the rights of citizenship. One hundred and sixteen years after its inauguration, Belo Horizonte has outgrown its original borders, confirming Jane Jacob's view that "it's foolishness to plan a city's appearance without knowing what kind of innate and functional order it has" (Jacobs 2000).

The Seventh of September Square[2] is one of the main areas of interest in the city and is an excellent example of some of these contradictions. It is estimated that every day around 10,000 people travel the city's streets and sidewalks during peak hours (2:00 pm to 3:15 pm), because it is one of the main traffic articulators for the city. At Christmas time, it is estimated that 800,000 people per day circulate through this space (see Fig. 6.1). The space also serves as host to certain bank headquarters and government buildings responsible for issuing official documents such as identification cards and passports. At the same time, various commercial and recreational activities (e.g., street trading and services as well as artistic and circus presentations) can be found in the Seventh of September Square. The Square is also

[1]Most of the industrial factories in Belo Horizonte are dedicated to automobile manufacturing, food production, and mining; the city is situated on the biggest iron reservoir in the country.

[2]The square's name celebrates Brazilian Independence Day.

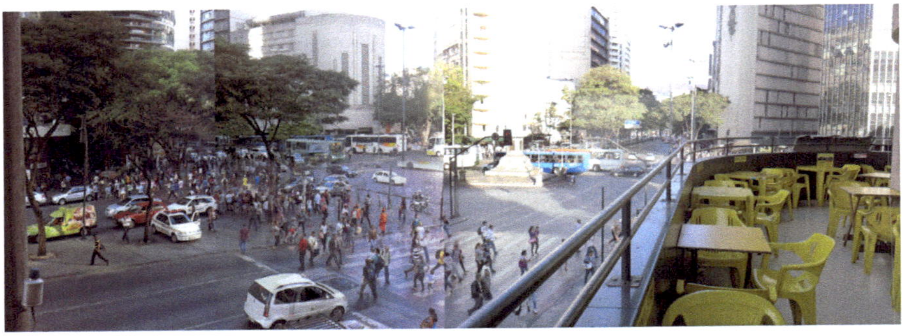

Fig. 6.1 Seventh of September Square, Belo Horizonte, Brazil

known as a place for criminals to acquire illegal guns and false documents. All of these factors together demonstrate what Milton Santos has called the hallmarks of a complex place "which generally coincide with metropolises, [where] there is a profusion of vectors: from the ones that directly represent hegemonic logics to the ones opposed to those. They are vectors of all orders, searching various objectives, sometimes external to themselves, but intertwined by a common space" (Santos 1996).

Thus, the contemporary city is the location of a vast array of disputes focused around its space, meanings, and resources and what configures it as a territory. Rogério Haesbaert conceptualizes the notion of territory as "always and concomitantly an appropriation (at a symbolic meaning) and [a] domain (at a more concrete political-economic approach) over a socially shared space" (Haesbaert and Limonad 2007). These contests over urban space are evidenced by the space's various uses of place that introduce into it the different temporalities and spatialities. The street, from the point of view when inside a car, is different from the point of view of a pedestrian; the citizen in a hurry, on his way to his job, experiences the sidewalk differently from the one who is there painting or selling his work.

> A city is also, simultaneously, the mutant presence of a series of events in which we take part as actors or audience, and that have made us live that certain urban fragment on a certain way that, when we cross it again, it reactivates that memory fragment. (Canevacci 1993)

The new technologies of information and communication, in a globalized context, speed up this changing process by making it easier to connect every city in every country to worldwide capitalist financial fluctuations. As a result:

> [T]he local space started to compress in it the whole world, to offer to its inhabitants, mainly in the big city, the multiplicity of times/speeds that represent practically a synthesis of the whole diversity of rhythms on the transformations in a planetary level. (Haesbaert 2006)

And, as Doreen Massey puts it, "for there to be multiplicity, there has to be a space" (Massey 2004), which in the case of a contemporary city such as Belo Horizonte means its features are constantly changing.

However, work is necessary in order to produce this kind of city. As Milton Santos has said, “the production of fluidity is a joint venture by the government and the private sector” (Santos 1996). While the private sector provides private networks within which circulates both information and structuring commands to the whole production, the government is in charge of providing, either directly or by grant, the technical territory for these flows. The contemporary city is part of this technical territory because it represents a field in which infrastructures are built. Therefore, it needs to be prepared by governments to receive what is necessary to accomplish those networks in order to make it a part of this whole system. In recent years, economic prosperity and the projection of mega events such as the World Cup and the Olympic Games have provoked adaptations of the city center, resignifying Belo Horizonte’s urban space under the sign of urban requalification, real estate speculation, and strategic planning. These interventions mostly include urban equipment repair, street widening, traffic reorganization, and the rebuilding of public spaces.

These dynamics are dramatically changing the city’s meanings – a process which was already multiple, within the frailty of visual, sound, and textual signs such as propaganda, traffic noise, and graffiti, among others – trying to coin two different images to two different kinds of public, since it “is an important factor in consumption-oriented development generally” (Burbank et al. 2002). On the one hand, it tries to forge an insider’s point of view, attached to invented traditions (Hobsbawn and Ranger 1997) through which the city’s inhabitants are able to express their individual identities. On the other hand, it aims to sell to foreigners a desired image that can be used to structure future investments and business opportunities in certain chosen economic areas. Mega events such as the FIFA World Cup are seen as moments:

> intended to spur local economic development by attracting tourists and media recognition for the host city. Pursuing a mega-event is a risky policy, however, because it requires a substantial outlay of resources [...] and the potential benefit to the city may be largely intangible. (Burbank et al. 2002)

The front page of Belo Horizonte’s strategic planning booklet, released in 2009 and entitled “The City We Want,” is an example of this image-coining process. It shows an imagined city that connects different city sites, buildings, and natural resources that couldn’t be shot in a single picture. In the first plane, you see Santa Tereza bridge, a construction preserved from the earliest days of the city, which is still protected today as part of the city’s heritage. Behind the bridge, the image shows the city’s skyline, shrunken by the monumentality of the bridge. In the background, there are the Curral Mountains, the main local element of the natural landscape. Lines are drawn along the image’s contour, as if the city is being redesigned. The small skyline camouflages the older buildings, and the mountains hide part of the city and part of the slums, which are rendered nearly unrecognizable. Figure 6.2 shows to foreigners what the city administrators “think” they want to see; the image hides what has been decided should not be shown: the not modern, not very tall buildings, the slums in the mountains, and the back slope, devastated by

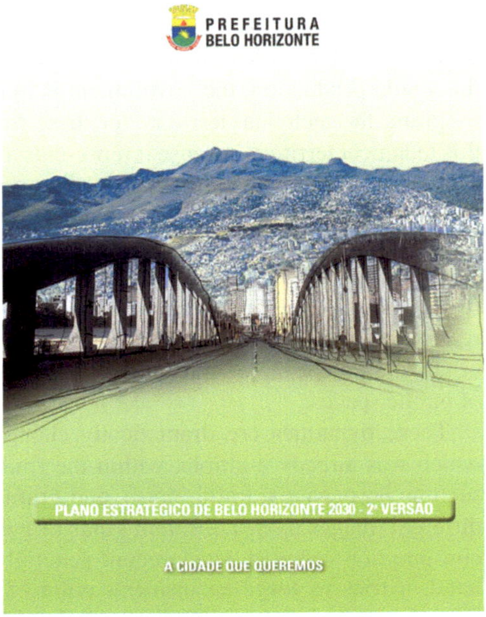

Fig. 6.2 Cover page of the "Strategic Plan of Belo Horizonte 2030 – second version." Published by the Belo Horizonte City Hall, 2008 (Use by permission. Link: https://bhmetaseresultados.pbh.gov.br/sites/all/themes/metas/pdf/planejamento_2030.pdf. Visited at 04/10/2014)

mining activity. At the same time, Belo Horizonte's urban planning for the FIFA World Cup in 2014 includes only reforms addressing the city's stadiums and airport, as well as the building of car and bus highways to connect those stadiums and airport to the city center, with no work being done to the regions of interest to its citizens. At the same time, the lack of accommodations in the city has been addressed with the construction of big hotels owned by international chains. New construction has been seen in various regions of the city, including the central area around the bohemian zone, in an attempt to disrupt and replace the low-cost prostitution currently found there with more lucrative forms of business tourism. These urban interventions mimic models conceived of and implemented during Barcelona's preparation for the Olympic Games in 1992, which was widely considered to be successful "in physically transforming the metropolitan area, including run-down inner city areas, as well as the waterfront and harbor and bolstering its economic competitiveness, particularly as a tourist and convention center" (Sharp et al. 2005). The result:

> may have provided extraordinary leverage for financing public projects [. . .] to transform the lagging city into a metropolis prepared to face the challenge of a unified Europe [. . .] at the expense of a hinterland particularly ill equipped in terms of urban policies and proper amenities and on which fall, increasingly, the problems that the central city displaces rather than resolves. (Sola-Morales 1993)

In order to implement such marginalizing policies, and just before those urban interventions occurred, a series of derogatory news stories about downtown Belo

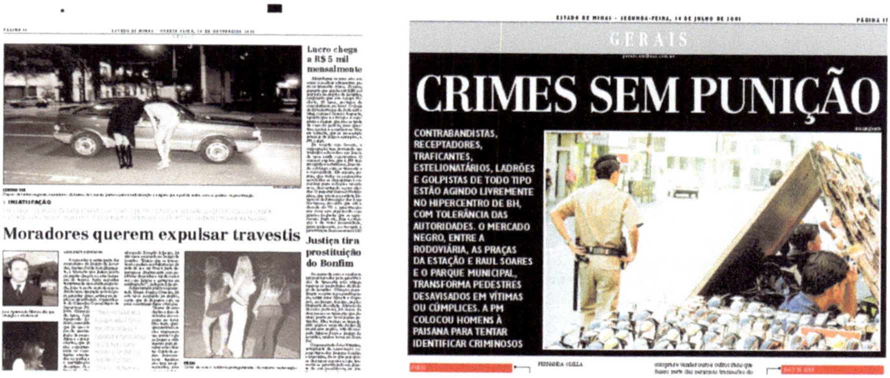
Lucro chega a R$ 5 mil mensalmente

Moradores querem expulsar travestis

Justiça tira prostituição do Bonfim

GERAIS

CRIMES SEM PUNIÇÃO

CONTRABANDISTAS, RECEPTADORES, TRAFICANTES, ESTELIONATÁRIOS, LADRÕES E GOLPISTAS DE TODO TIPO ESTÃO AGINDO LIVREMENTE NO HIPERCENTRO DE BH, COM TOLERÂNCIA DAS AUTORIDADES. O MERCADO NEGRO, ENTRE A RODOVIÁRIA, AS PRAÇAS DA ESTAÇÃO E RAUL SOARES E O PARQUE MUNICIPAL, TRANSFORMA PEDESTRES DESAVISADOS EM VÍTIMAS OU CÚMPLICES. A PM COLOCOU HOMENS À PAISANA PARA TENTAR IDENTIFICAR CRIMINOSOS

Fig. 6.3 Samples of newspaper portraying negative news about downtown Belo Horizonte (Used by permission)

Horizonte were published in the newspapers and broadcast by media, reinforcing an image of criminality, drug abuse, and prostitution already present in the city's collective imagination. This media campaign contributed to the arguments supporting urban reform (see Fig. 6.3). Sadly, in the field of academia at the beginning of the year 2000, a diagnostic study bolstering such arguments also took place. The study used a set of social science methodological procedures such as observation, surveys, and semi-structured interviews with people who lived and worked in the neighborhood to collect people's expectations and desires regarding what the region should become. Of course, these methodologies faced issues with regard to capturing movement. Therefore, most of the people who were either already downtown or who were in transit were excluded from the study. The outcome of this diagnostic research was a set of maps (Fig. 6.4) that oriented urban planning and requalification sites, but designed them to be distant from their "intervention objects" (Hissa 1998); the results were "plans far more rigid and inflexible and less responsive to changing circumstances" (McCoy 2003).

The Urban Cartographies Research Project (UCRP) questioned the value of these maps. On the one hand, the maps were a very specialized kind of text, readable only by experts and not accessible to the average city inhabitant. On the other hand, they froze the city's movement and vitality, tying some groups of people (or urban practices) usually in motion to a specific place. They also were initiated from a point of preconceived identities, so these maps did not achieve "a politic of exposing the maps of power through which the identities are constructed" (Massey 2004). An example of such rigidity is how these maps dealt with prostitution. According to the study, prostitution occurred only in one place, in some quarters of Guaicurus Street, and only at night. However, you can find different kinds of prostitution throughout the day and in several sectors of the city, including in other downtown areas. In the evening, prostitution can be found all along Afonso Pena Avenue, one of the city's most important streets. The UCRP also noted that these regions are

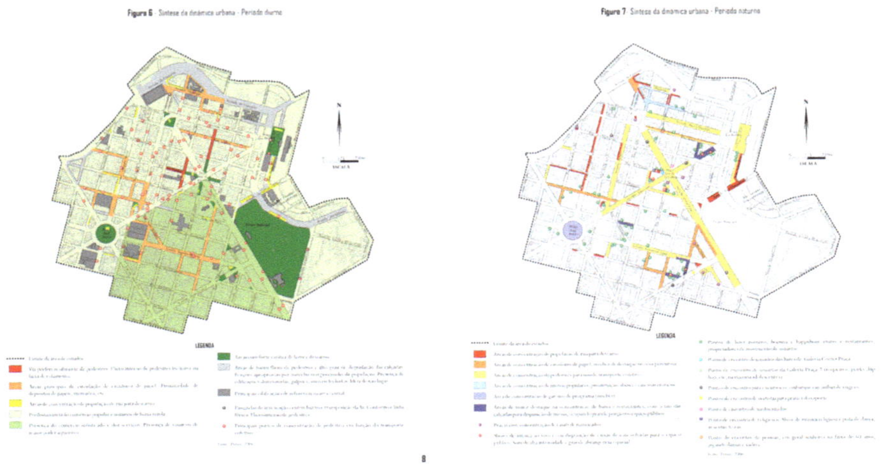

Fig. 6.4 Urban dynamic maps produced for the "Rehabilitation Plan of Belo Horizonte Downtown." Published by the Belo Horizonte City Hall, 2007, pages 11 and 12 (Used by permission. Link: http://portalpbh.pbh.gov.br/pbh/ecp/files.do?evento=download&urlArqPlc=plano_reabilitacao_hipcentro_p1.pdf. Visited at 04/10/2104)

not homogeneous in terms of the types of activities they house. There are many considerable differences among the practices surrounding this prostitution, from variations in the price rate to the types of prostitutes available (female, male, and transsexual, for instance). We also noted that, according to these maps, all homeless people do in the street is rest. Such a representation denies other activities such as informal car washing and the selling of parking tickets. The main objective of this kind of diagnostic study is usually to reinforce a derogatory image of a city in order to justify a perceived need to "clean up" the city center. The proposed reforms came as an effort to mint an alternative image of the city, a more marketable one, which would attract international investors and business and, thus, make the city more "sustainable."

What this diagnostic study missed due to its top-down planning is the everyday practices that arise from the inhabitants' uses of their city's space. One example of a practice that was totally ignored by the study is the street cry. Upon first look, the street cries sound like people screaming randomly within a large flux of pedestrians. However, a more careful study (Franco and Marra 2011) found that these street cries (which are performed by individuals holding informal jobs) are intimately connected both to the fluctuations of people and the noise produced by downtown urban dynamics, thus creating a site- and time-specific event. Therefore, what can be heard in the street cry is the coining, through urban experience, of a kind of intuitive understanding of the entire capitalist strategy for organizing the city. The criers usually interact with the forces manifested by traffic jams and pedestrian fluctuations. They find a fracture in the system in order to affirm their existence in public places and defy their situation of being underemployed. By ignoring city cries, this kind of strategic planning often fails to achieve true political recognition. As argued by Massey, "a more complete recognition of difference should recognize its contemporaneity, recognize that the 'others' real existence could not only be following us, but have their own histories to tell" (Massey 2004).

The Urban Cartographies Research Project

The Urban Cartographies Research Project aims to understand the city in all of its mutant manifestations and to perceive its inhabitants' acts which often are invisible to urban planning politics. The UCRP's objective is to help build up public spaces through the development of collaborative research projects and alternative diagnostic methodologies. It was created in 2004 in response to certain experiences with the Federal University of Minas Gerais' Cultural Centre, which is located in a popular commercial region in downtown Belo Horizonte, next to the train station and the bohemian zone with its small, dirty bars, and low-cost prostitution. This Cultural Centre used to present visual, scenic, and musical arts shows to a middle class and comparatively rich public, but this group grew increasingly resistant due to the Centre's "dangerous" location and lack of parking lots. In 2002, though, the Centre's occupation proposal changed in an effort to establish a dialog with the population who regularly lived, worked, and visited the area; the shift was toward a programmed diversification in hopes of encouraging the public to appropriate the Cultural Centre. Thus, the arts began to share space with a popular Internet access space, a library containing newspapers and magazines, young people from the city's suburbs, workers such as prostitutes, and marketers from the surrounding areas holding meetings with professional groups.

At this time, Belo Horizonte's downtown requalification, as described above, was just beginning. Thus, the Federal University Cultural Centre began to register this transformation through two media projects, R.U.A. and Concreto Sonoro. R.U.A. (Audiovisual Urban Register) produces short videos, mixing together a documentary and a poetic approach, about a variety of themes related to the space occupation in Belo Horizonte's center. Among these works, the video Relé[3] synthesized some of the issues that would later be developed by the Urban Cartographies Research Project, such as the several kinds of urban space occupation enacted by its inhabitants, the methods of urban mobility, and routes throughout the city designed and used by the population, connecting several far-flung regions. At the same time, Concreto Sonoro[4] (Sonic Concrete) has produced radio programs

[3]This video is available at: http://www.youtube.com/watch?v=O6-STW1_aqE. Last access: 4th of February, 2014.

[4]Some of these radio programs can be heard at: http://dajaneladomeuquarto.tumblr.com/post/3234283596/programacao-radio-le-1 and http://dajaneladomeuquarto.tumblr.com/post/4005368391/programacao-radio-le-2. Last access: 4th of February, 2014.

broadcast by a radio station in Belo Horizonte. Beyond a preoccupation with the city's sonic space formation, its noises, and voices, the programs began working with soundscape techniques as a means of understanding the place's temporality, spatiality, and rhythms.

These two experiences worked together within the laboratory of the Urban Cartographies Research Project; in 2004, the UCRP systematically began grouping together an interdisciplinary team with researchers from the fields of History, Communications and Media, Computer Science, Arts, Museology, Social Sciences, Psychology, and Architecture. Beyond the preoccupations previously mentioned, the research also intended to explore and elaborate upon alternate forms of registering, capturing, and visualizing the urban fluxes that materialize on various uses of urban space through the production of space- and time-mobile maps. After a series of discussions about the researchers' and the public's imaginary construction of Belo Horizonte, an intensive fieldwork period was organized in 2005. The International Situationist procedure of "drifting" was reinterpreted to produce a cartographic drifting. During these drifts, researchers walk along predetermined routes through Belo Horizonte's downtown, in order to register anything that might attract their attention. The group observations rely heavily on photographic and video cameras to register sound recordings and record field text notes. The researchers also employ searches of public archives to gather together both old and contemporary photographs, as well as short stories and print news published in the past, in order to deal with the collective imagination of the place.

In March of 2006, the group organized an exhibition at the Federal University Cultural Centre that brought together people from the surrounding area and the research results, in order to establish a dialog regarding what was produced. The rooms were organized in an effort to create an interactive environment in which visitors could manipulate the printed pictures, read parts of the short stories and other texts, and see and listen to the audio and video material captured during the research. A new set of cartographic drifting was made in 2007; this time, the walking occurred along routes that connected two bus stops in different regions of downtown Belo Horizonte. Based on an observation made during our first set of driftings, we concluded that the downtown area was a hub for bus transfers. The city's public transportation system uses buses that cross the central region and are arranged in a way that several bus routes together connect the two distant neighborhoods. Downtown has become a very busy port at which people transfer between buses to get to another region of the city. The bus stops were reinterpreted as arrival and departure gates, and the routes we researched simulated the population's everyday walking routes.

From this point, new cartographic driftings were organized around themes such as the city's bohemian life or its material heritage, circulating among monuments and around certain buildings. This research generated several academic articles that were subsequently published in journals and books and presented at symposiums and congresses focusing on the methodology that was developed (Silva 2009; Silva et al. 2008), the material heritage of the city (Silva 2007), collaborative mapping (Silva and Franco 2009), popular service cries (Franco and Marra 2011), bars and bohemian life (Garcia 2011), live music performances in the city (Garcia and Marra 2012), and soccer (Marra 2011), in addition to doctoral theses on the sociability that forms around newspaper and magazine stands (Fonseca 2008) and the types of visual communication that exist in the city (Gonzaga 2009). Besides these academic endeavors, the group actively participated in activist protests focusing on public space questions, urban planning, regulation of public forums, and academic extension programs leading from the city's public educational system. During 2007, Urban Cartographies research was also introduced on the city's Educational Public Network, in nine schools located in different regions throughout the city; students and teachers collaboratively mapped, diagnosed, and otherwise became more aware of their surroundings. This extension project was also used as means of introducing new technologies of communication and information into the classroom in a way that could be integrated into the school's pedagogical project.

The task of building up a database that would be able to gather and intertwine all of the pictures, videos, sounds, and texts produced by the research was a difficult one, and it turned out to be more arduous than we had originally thought. Our intent was to create a tool that would make possible the hosting of and access to the entire body of material produced and collected by the research team. This digital device would also make possible not only the uploading of new material but also the creation of a virtual ambient where memories and experiences could be shared by connecting data and targeting a broader audience in order to bring about discussions, collaborations, and interactions among the various academic fields. Some categories discussing the problems of mobility within the urban setting were constructed by the group, such as "fixed," for what could not be moved; "moving," for what could be moved by their own agency; and "movable," for what could be moved by means of the action of another being. We believed these categories to be relational enough to deal with the complexity of the problems we proposed ourselves. The task remains incomplete, though, for a number of reasons. First, we faced the subjective problem of this categorization, since it relies on the interpretation of the researcher to determine some situations (e.g., traffic cones would be classified as fixed because they are tools for organizing traffic, fixing routes, and controlling the fluctuations of vehicles, though they can be moved, making it also possible to classify them as movable). On the other hand, it was difficult to automate some descriptors and keyword criteria (e.g., how should one make a computer distinguish people from cars in a single photograph? How can one make a computer determine whether or not an object moves, either by its own power or by the action of another thing, since a photograph freezes every movement?) These were some of the problems that made the task difficult and required the manual work that was employed to construct the database.

Today, the group is currently developing a research project focusing on the four squares located in the center of Belo Horizonte. The investigation is attempting to compare the population's uses of the four places, analyzing (1) how the physical changes being made to these urban spaces relate to these processes, focusing on how people use public urban equipment and cultural heritage buildings at these locations; (2) the sociability aspects and space sharing related to these four locations, focusing on the disputes and division experienced by various social groups, in a self-organized way, present in the place and through mobile networks; and (3) verifying how these groups dialog with or otherwise face the global capitalist fluxes that take part in the physical interventions being performed on the city's spaces.

The group also continues to create procedures geared toward making visible this mobile and mutant character of the city in both space and time. Some of these techniques involve talking to people in the squares being researched, following people in their everyday routes through the city, making more generalized observations, writing in field notebooks, and photographing and recording sounds in these squares.

Learning from Urban Art Experiences: From the People's Everyday Practices to Event Disruptions

Faced with the need for methods that might adequately capture the chameleonic features of the city and the mega events approaching, the UCRP asked how art might provide new strategies of understanding urban practices, conflicts, and negotiations. The UCRP, conscious of the drawbacks of academic fieldwork, opened themselves up to the study of contemporary urban art in the hopes of understanding alternative ways of intervening in the city.

Whether or not urban art is capable of promoting a people's emancipation is quite a controversial issue. Several case studies (Sharp et al. 2005; Tornaghi 2008) reported on how art interventions could emphasize the gentrification processes, accentuate social exclusion, and become an authoritarian imposition upon locals. Other studies have claimed that such interventions "can help develop senses of identity, develop sense of place, contribute to civic identity, address community needs, tackle social exclusion, possess educational value, and promote social change" (Hall and Robertson 2001). Urban art interventions can provide "pointers to what, in public art terms, would define an inclusive city, as one giving expression to the multiple and shifting identities of different groups, as indicative of presence rather than absence, and of avoiding the cultural domination of particular elites or interests" (Sharp et al. 2005). In any case, the main concerns emerging from such studies not only deal with the concepts of inclusion, collaboration, and the promotion of social change but also, as Tornaghi has said, "the measurement of social benefits of the arts is one of the main challenging and arguable questions which animate an endless debate between critics and advocates of public art" (Tornaghi 2008). The way a local population relates to such intervention "shifts the emphasis from outcomes toward the processes through which public art is produced and how these can foster a sense on inclusion" (Sharp et al. 2005).

The term "city art" is ambiguous and comprises a large group of practices. However, the urban art interventions this research is interested in are those which articulate everyday people's practices, modify people's urban space appropriation, and point to participatory processes in urban space formation. In fact, the most provocative issues resulting from the conjunction of art and urban space emerge from three key features of art itself. On the one side, art deals with the concrete, material, and temporal dimensions of place. It proposes a site-specific intervention where the subject is not the mean citizen of social science's studies but an individual within his or her own intersubjectivity. Conversely, art proposes artifacts, either ephemeral or permanent interfaces that provoke modifications in the functional grid of the city; people's everyday practices are disturbed. As a third perspective, art may also evoke people's participation in the construction and experiencing of possible future worlds, as well as crystallize and highlight unequal social relations in order to make people aware of them.[5] Public art can be conceptualized as "different art manifestations – sculptures, video, music, and performances – that are located outside the conventional arts sites, such as museums and galleries, in public spaces" (Tornaghi 2008). Chiara Tornaghi (2008) points out that this public nature of public art is usually conceived in one of five ways: (1) as publicly visibility, in which the work is placed in key or strategic urban spaces; (2) as art installations in colonized public spaces, pointing to a gentrification process; (3) as critical pieces of art supporting discussions within the public sphere; (4) as process-oriented works, in order to explore the city inhabitant's participatory agency; and (5) as objects connected to places and populations in areas targeted by regeneration programs, in order to show local community roles in place making within these intervention actions.

The street cries, mentioned earlier, negotiate their sounds within the overall body of city noise. Such types of dialog permeate all urban events, from the everyday passing by of pedestrians to eventual political demonstrations. Sometimes integration becomes confrontational; sometimes it occupies the interstices of the city's rhythms. Thought of as a continuum, the word "event" contains different grades of determinacy. In the case of the street cries, the event is closer to the "the fortuitous, the accidental, transience, and unpredictability" (Doane 2002). This aspect of the event as contingent is connected to life, the concrete, and valueless practices. Perhaps this aspect is the one that relates best to Lefebvre's "lived space" (Lefebvre 1991) since it relates to our everyday lives, activities that actualize space and do not signify. Everyday practices such as the street cry are events not predicted as possibilities but rather events that simply happen. The social sciences approach to city planning ignores such nuances of time and space and the specificity of the events that occur as everyday practices for the citizens of Belo Horizonte. Different kinds of temporality occur within the urban space. For example, the sidewalk, which was intended to allow for a free circulation of people, actually features certain individuals who stand and offer products and thus invite passing pedestrians also to stop. This stopping slows down traffic; it works at the margins

[5]Such features – the site-specific, the disruption of the social grid, and the value of participation – all have a performative lure that approximates such practices with the drift procedure so extensively used by the research group.

of the general capitalist logic and confabulates against the mainstream imagery of consumerism. This situation presents a different scheme for understanding urban temporality as an intervention into a manifest creativity and of people finding gaps and taking advantage in order to introduce into the city different and singular rhythms connected to the inhabitants' respective subjectivities. This kind of event may confront the logic of capitalism, as Milton Santos states:

> The strength belongs to the slow ones and to those who do not hold the speed [. . .]. Who, in the city, has mobility – and can travel through it, or scan it – ends up seeing too little of the city and of the world. Their communion with images, frequently pre-fabricated, is their perdition. Their comfort, which they do not want to lose, comes, exactly, from their familiarity with those images. The "slow" men, for whom such images are a mirage, cannot, for a long time, be in phase with this perverse imagery, so they discover confabulations. (Santos 1996)

But those creative powers usually attributed to the poor are also the subject of appropriation by the capital, as Hardt and Negri remember, since "the poor [are] defined not by lack but possibility (. . .) [they] are completely within the global rhythms of biopolitical production" (Hardt and Negri 2009). They can renew the capitalist strategy of colonizing the city, as some city cries also show. Some criers are autonomous, but others – usually the ones announcing services and products related to telecommunications and dentists' offices – work within an underemployment scheme, structured by private business and agencies who make the work viable, bringing the participants drinks and lunch during their break time. The techniques developed by street criers have also been adopted by large stores located in other regions of the city center (e.g., in clothing shops at Paraná Avenue) where announcers emulate radio ads in order to invite people waiting at bus stops to enter the stores and buy. The challenge remains how to "find ways to translate the productivity and possibility of the poor into power" (Hardt and Negri 2009).

Urban art experiences deal with place, but, contrary to the quotidian actions of local people, the events constitute situations that interrupt everyday practices. Artistic approaches find and expose fractures in the way capital defines the city space. They seldom take the form of big panels, strange objects, or performances; sometimes they are mediated by technological means that try to disrupt everyday life in order to show the city's problems and point out possible worlds yet to come. However, art interventions also try to "eschew monumentalism as it was expressed in the nineteenth century with its thinly disguised appeal to elite interests" (Sharp et al. 2005) in what may reinforce power relations. The disruption of everyday life ranges from events that incorporate citizens' tactics, to those who question them or create new space fruitions. Urban art proposes a movement toward structure and the making of meaning.

Considering works that affect the soundscape of the city, the work *Evoke* by Usman Haque (2007)[6] exemplifies one of those strategies. In this event, colors

[6]Information on this work may be found at: http://www.haque.co.uk/evoke.php. Last visited on the 14th of February 2014.

and morphing shapes are projected onto a gothic British church in response to the sounds produced by the interactions of passersby. People scream, clap their hands, or sing into a microphone connected to a computer that captures the sounds, interprets them, and creates psychedelic images that are then projected onto the church's facade. This work of art creates an event that lures people into coming to a familiar place and seeing it differently. It challenges the definitive monumentalism of a socially significant building. Although *Evoke* has been produced in other places and therefore is not purely site-specific, it causes a memorable urban experience for passersby. The Gothic church, a significant and symbolic landmark in the city, becomes a screen of interactive painting, a strange spectacle of forms and colors. The proposed interaction, however, narrows down the possibilities for appropriation of the event into everyday practices. Haque's work evokes one very specific way of relating to the projected surface: the production of loud sounds such as screaming, singing, or clapping hands. As a result, the eye candy produced by this interaction has the power to make people re-signify this place, but in a narrow way, since the possibilities for interaction are few and somewhat predetermined, lacking the power of facing the various local forces implicated in that public space. Moreover, the event invades the place, ignoring and expelling quotidian non-planned uses of the site. Of course, these are risks artworks usually face. The effect an art intervention has on public space should be evaluated on a case-by-case basis in order to understand what impact it might have on dealing with its discussion and conformation. Certainly, such work refers to the most used sense of the word "event," connoting "a high degree of constructedness, as in the notions of a media event or social event" (Doane 2002). It leads to identifying a structure that makes an event meaningful, transcendent, a representation of totality. Such events could relate to everyday life activities, but they are made significant through a narrative. In *Evoke*, such narrative emerges from the memorable experiences imprinted on the passersby. Most importantly for our discussion here, such events interrupt the quotidian and do not articulate any aspect of the citizen's daily appropriation of urban space and could even create another kind of monumentalism, one that proposes as art the task and obligation of solving conflicts that belong to the capitalist production of place.

Another kind of artistic intervention that we find relates closer to our proposal is *Post Urbano* by the Argentine group WokiToki.[7] As stated by the creators, this intervention proposes a web platform that would allow for "urban signposting," creating an environment of collaboration between the dwellers and users of the city. Reedited in the city of Rosario, Argentina, a website visitor can create his or her own sign, marking the place on a satellite photo of the city and thus telling a story about an experience with the place. Later, the web posts are transcribed into posters and placed in the actual urban space, photographed, and then added to the initial web post. The experience of the territory, densely populated by stories and other significant moments from the lives of the individuals who visited there,

[7] Information on this work may be found at: http://post.wokitoki.org/. Last visited on the 14th of February 2014.

gets entangled with the web of "information flow." Their proposal "investigates the circulation, the value of information reconsidering notions of identity, territory, border and city."[8]

Synthesizing several experiences of geo-referencing using the API of Google Maps and other mobile technologies, the site allows for tagging various affections for stories about and impressions of the city. Different from sites that focus on building exclusively digital platforms for collaboration,[9] the project proposes a direct intervention into the city, creating a connection between the digital nowhere land of the Internet and the genuine experiencing of the city. The printed posters of 100×70 cm allow for a simple assemblage and therefore are easy to adapt to different spaces. Instead of mimicking the advertising strategy of poster repetition, the posters/signs place a particular narrative into a concrete territory. The project proposes that the user go to the specific places shown on the map and find the posters telling these everyday intimate stories (e.g., one such sign reads "this is the house in which I was born" and bears the author's signature). This kind of work inscribes into the city the subjective experiences of the common citizens who are able, then, to reclaim the immaterial possession of the place, thus creating affective urban territories. More than a question of offering visibility to less powerful people's points of view, *Post Urbano* invites people to experience, visit, and live in the whole city by encouraging them to search the places in the city where the posts are registered or insert into the city their own experiences with it. The result is the establishment of a discussion regarding the conflicts inherent in the meaning of place, because the posts can be erased or contested by other posters. WokiToki's proposal brings about the potentialities embedded in the evidencing of the dynamics of constitution of places and territories; it also serves as a way of taking part in the building of space.

Latest Work

Inspired by and evolving from these art experiences, the UCRP has developed experimental procedures to capture the dynamics of Belo Horizonte's city center. The focus of these procedures is to evidence the city's aspects of mobility.

The first procedure, named "audiovisual postcards," was produced through panoramic photographs of certain spaces and audio recordings made during this photographic registration. Focusing the procedure on transitional spaces – semipub-

---

[8]Information on this work may be found at: http://post.wokitoki.org/. Last visited on the 14th of February 2014. Cited translated by the author: "La propuesta indaga sobre la circulación, el valor de la información reconsiderando las nociones de identidad, territorio, frontera y ciudad."

[9]Several projects focus on the potential of web map applications for developing people's collaborations in the territory. Developed by the UNISINOS http://portoalegre.cc/, the main objective is to create "a space for citizen collaboration, where you can meet, discuss, inspire, and transform the city itself."

lic balconies hosting bars and other kinds of commerce and services – these postcards show both an internal and a street ambience. These transitional spaces, in their architectural design, articulate the tensions among private and public spaces. The audiovisual postcards turn these tensions visible by showing both internal and external aspects of place. Cars can be heard and seen at the same time, as can people crossing the street, people talking, music being played, etc. Serial postcards, taken at different hours of the day, show the various different uses of the space by different groups of people. For example, in Fig. 6.5 the postcard shows a balcony in the late afternoon when the space is occupied by people who are fresh from work and looking to spend some time drinking beer as they wait for the traffic to abate. The same place, later in the evening, becomes a funk party where people come to dance and flirt.

Another procedure, called "10 × 10x1," consists of a 10 min recording of a public space's soundscape, coupled with one photograph taken every minute. These photographs and sounds are then edited into a 10 min short film, with every photograph shown for one complete minute. Such a procedure shows the changes in a place over a short span of time. Figure 6.6 shows a "10 × 1" taken in the Seventh of September Square, at one of the sidewalks often used for political and activist protests. On this particular Friday, at about 6 pm, there was a student protest against the arrests of certain young people during the Independence Day Parade the week before. The audio track communicates the sounds of the protest, while the images show that the demonstration didn't disturb pedestrian movement in the square. One of the main arguments against political and popular manifestations in the city center is that they cause traffic jams from which emerges "all this desire of suppressing every obstacle to free circulation on goods, information, and money, in order to guarantee free competition and assure market leadership, becoming a global market" (Santos 1996). Of course, a big demonstration from out of nowhere bursts into the streets can make the traffic worse or even paralyze the city. However, they are not the cause of traffic jams, since mobility in the city is already difficult due to the large number of cars in the streets and the poor public transportation system. This record, then, could be used to show detractors that this argument against demonstrations is not necessarily valid.

This particular 10 min recording also shows what kind of people this type of event attracts or repulses. Surrounding the square are the headquarters of several financial institutions, such as banks and credit funds. The pictures were taken at a time when the individuals who worked at these institutions were scheduled to leave work. However, you can't see anyone wearing a suit in any of the pictures. They may have taken other routes so as not to cross the youth demonstration. On the other hand, later in the audio track, one can hear two people talking about one of the young men who was arrested and had not yet been released. Again, the repetition of these registers during other hours of the day serves to capture different situations. Groups, urban equipment, and people's practices all tend to attract certain people while repelling others, according to the particular social group's preferences, experiences, and affections.

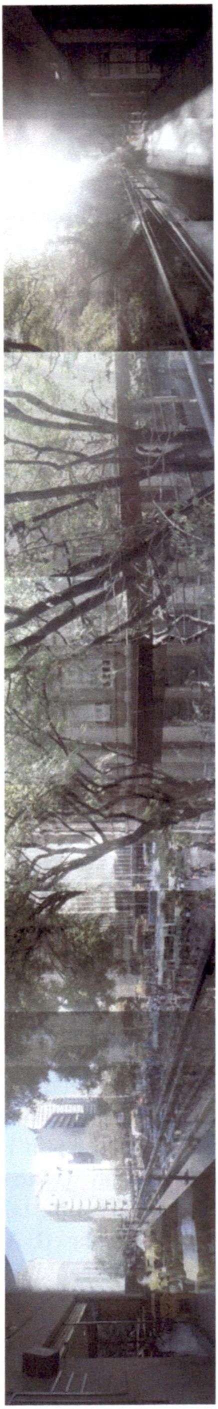

Fig. 6.5 A balcony facing the Seventh of September Square (To see the video, go to: http://vimeo.com/76094751. Produced by Maíra Oliveira and Pedro Marra)

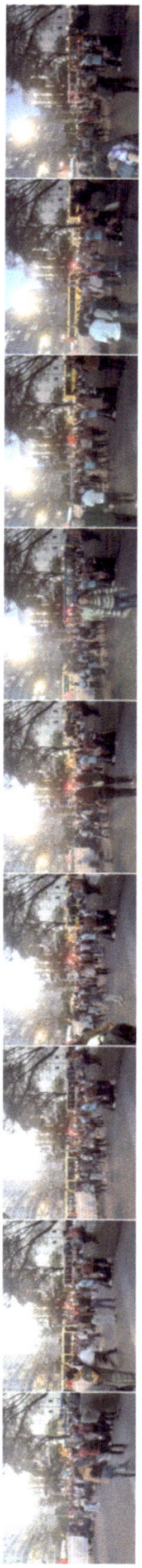

Fig. 6.6 Image showing a frame of the video produced through the "10 × 1 method" (To see the video, go to: http://vimeo.com/76092900. Produced by Maíra Oliveira and Pedro Marra)

These kinds of methodological procedures can help witnesses to grasp such mobility in both space and time. The citizens of our cities possess forces that, although not so strong as capitalist fluctuations and mega event interests, are potent powers in urban space dynamics.

One of the drawbacks the UCRP observed with their initial cartographic drifts was the immense quantity and diversity of data collected. The visualization of the data, the recording of the researchers' interpretations, and the broadcasting of the results all were performed insufficiently. Therefore, the challenge we face today is to create different ways of organizing the material, tagging the data, and exposing the results to a wider public. We plan to explore such drawbacks promoting events in which the material produced by these methodologies is projected and shown to people in the squares being researched. Such events aim to verify the passersby's self-consciousness of the city's changes and to discuss the group's results with the city's inhabitants, finding out what these processes mean to them and also connecting them to a shared urban experience. We want to explore the possibility of projecting these images and playing the sounds in the city, using its buildings and streets to make pedestrians stop, so we can initiate conversations with them about these materials and about the place and their urban experience. By using a disruptive approach, the group intends to grasp the city's attention in a way that allows them to feel free to expose their thoughts and feelings to the researchers, without caring about what we want to hear; the goal is to have the city's inhabitants focus on what they have to say, in the informal conversations so common in everyday life. By doing so, we expect to reach these people not as mere qualified informers in an interview but as participants in the processes of urban building and rebuilding.

Planned for the future is another approach to data gathering, inspired by the experience made by Tia DeNora in her study of the impact of music on consumption behaviors in stores in the United Kingdom (DeNora 2004). She proposed "consumer shadowing expeditions" in which she "shadowed a volunteer shopper such that both shopper and shadower wore clip-on microphones: the shopper was asked simply to 'think out loud' and the shadower commented on the volunteer's activities" (DeNora 2004). We thought of following pedestrians, trying to observe their actions and the ways they walked through the streets: where they walked more quickly or more slowly, where they stopped, etc. We also hope to invite them to comment on their own actions, the way DeNora did, giving them recorders and asking them to think out loud about what they are doing, seeing, hearing, about the places they pass, etc. Besides allowing access to people's thoughts and experiences about the city, the experiment represents another way of inviting Belo Horizonte's citizens to take part in both the city's symbolic construction and the UCRP's findings. Also, these recorded thoughts can later be edited in order to evidence the diverse affections and meanings provoked by the city. The results of such an experiment could generate a locus of interaction between the social sciences and the arts, since data collected from an ethnographic methodological procedure could be used as raw material for sound artistic work.

Acknowledgments The authors would like to thank the Universidade Federal de Minas Gerais, the Fundação de Amparo à Pesquisa em Minas Gerais, the CNPq, CAPES, and our colleagues from the Nucleurb who belong to the Centro de Convergência de Novas Mídias (New Media Convergence Center).

References

Burbank MJ, Androvich G, Heying CH (2002) Mega-events, urban development and public policy. Rev Policy Res 19(Fall 3):179–202

Canevacci M (1993) A cidade polifônica: ensaio sobre antropologia da comunicação urbana. Studio Nobel, São Paulo

DeNora T (2004) Music in everyday life. Cambridge University Press, Cambridge

Doane MA (2002) Dead time, or the concept of the event. In: Doane MA (ed) The emergence of cinematic time: modernity, contingency, the archive. Harvard University Press, Cambridge, pp 140–170

Fonseca C (2008) Interações Comunicativas no espaço urbano: conversas e derivas em torno das bancas de jornal. Ph.D. thesis, Universidade Federal de Minas Gerais

Franco J, Marra P (2011) Som e complexidade urbana: apontamentos a partir de uma visao sistêmica das sonoridades do comércio popular no Hipercentro de Belo Horizonte. Ciberlegenda 24(2):146–159

Garcia L (2011) Derivas cartográficas: tecnologia aplicada à memória numa incursão pelos bares do Hipercentro de Belo Horizonte. In: Anais da Conferência sobre Tecnologia, Cultura e Memória, vol 1. LIBER/UFPE, Recife, 13–15 Sept 2011

Garcia L, Marra P (2012) Ouvir música na cidade: experiência auditiva na paisagem sonora urbana do hipercentro de Belo Horizonte. Revista Contemporânea UERJ 10:43–57

Gonzaga M (2009) Diálogos Públicos no Centro de Belo Horizonte: Mapas de Sentidos em Comunicação Urbana. Dissertation, Universidade Federal de Minas Gerais

Haesbaert R (2006) Territórios alternativos. Contexto, São Paulo

Haesbaert R, Limonad E (2007) O território em tempos de globalização. In etc, espaço, tempo e crítica. Revista Eletrônica de Ciências Sociais Aplicadas e outras coisas 2(1):39–52

Hall T, Robertson I (2001) Public art and urban regeneration: advocacy, claims and critical debates. Landsc Res 26(1):5–26

Haque U (2007) Evoke

Hardt M, Negri A (2009) Commonwealth. The Belknap Press of Harvard University Press, Cambridge, MA

Hissa C (1998) Geografia e planejamento: entre o puro e o aplicado. Geonomos 2(6):33–43

Hobsbawn E, Ranger T (eds) (1997) A invenção das Tradições. Paz e Terra, Rio de Janeiro

Jacobs J (2000) Morte e Vida de Grandes Cidades. Martins Fontes, São Paulo

Lefebvre H (1991) The production of space. Blackwell, Cambridge

Maricato E (2000) A cidade do pensamento único. Vozes, Petrópolis

Marra P (2011) O Som como elemento da Experiência Urbana do Futebol. Sonora 6:1–9

Massey D (2004) Filosofia e Política da Espacialidade: Algumas Considerações. GEOgraphia 6(12):7–23

McCoy M (2003) Tendências atuais de fragmentação nas cidades latino-americanas e desafios para a política urbana e o planejamento urbano. Iberoamericana, Nueva época 3(11):111–128

Santos M (1996) A natureza do espaço: Técnica e tempo. Razão e emoção. Editora Hucitec, São Paulo

Sharp J, Pollock V, Paddison R (2005) Just art for a just city: public art and social inclusion in urban regeneration. Urban Stud 42(5/6):1001–1023

Silva RHA (2007) Patrimônio, espaço e mídia: discutindo as intervenções na Praça da Liberdade. In: XII Encontro Nacional da Associação Nacional de Planejamento Urbano e Regional – ANPUR, 2007, Belém do Pará. Anais do XII Encontro Nacional da ANPUR, Anpur, São Paulo

Silva R (2009) Cartografias Urbanas: construindo uma metodologia de apreensão dos usos e apropriações dos espaços da cidade. Cadernos PPG-AU/FAUFBA 5:83–100

Silva R, Franco J (2009) Collaborative mapping projects: mapeando os espaços de (com)vivência nas cidades. Arte e Tecnologia: para compreender o momento atual e pensar o contexto futuro da arte In: Anais do 7o Encontro internacional de arte e tecnologia, Universidade Nacional de Brasília. Brasilia, 1–4 Oct 2008

Silva R et al (2008) Dispositivos de memória e narrativas do espaço urbano: cartografias flutuantes no tempo e espaço. E-Compós 11(1):1–17

Sola-Morales I (1993) Letter from Barcelona. ANY: Architecture New York, Writing in Architecture, May/June:60–61

Tornaghi C (2008) Questioning the social aims of public art in urban regeneration initiatives. The case of Newcastle upon Tyne and Gateshead (UK). http://www.ncl.ac.uk/guru/assets/documents/EWP42.pdf. Accessed 2 Feb 2014

Chapter 7
Electric Signs

Alice Arnold

Abstract *Electric Signs* is a documentary film about signs, screens, and the urban environment. The film takes us on a journey through a variety of urban landscapes, examining public spaces and making connections between light, perception, and the culture of attractions in today's consumer society. This chapter summarizes the research that underlies the film and extends the discussion of several key ideas. The film and this chapter are divided into six sections: New Sign Systems (the introduction, discusses the role that outdoor advertising plays in shaping public space and public expression in urban environments), Manufacturing Consent (the section looks primarily at Hong Kong and describes how networked LED displays are transforming public spaces in cities by creating a sophisticated level of mediated experiences; it makes connections between real estate developments, marketing strategies and technologies, consumerism, and public space), Sign Wars (this section looks at the political fight in Los Angeles between the outdoor advertising industry, politicians, and community activists who oppose more outdoor commercial signage; it also explores ideas about visual culture and the urban environment), Pale Daylight (a history section that connects the industrialization of light, outdoor advertising, and sign spectaculars), Media City (this section looks at the merging of the built environment and the media sphere; global cities; the etymology of the word screen; the connections between screens, surveillance, and data collection; and sustainability issues), and Urban Lightscapes (the conclusion, the importance of public space and people's ability to shape urban environments).

Introduction

New screen-based sign systems are putting TV-style advertising into the public domain in cities around the globe. These electronic signs, along with the innovative tall buildings they are integrated into, are reshaping urban environments and redefining areas of public space by intensifying the commercialization of the public sphere. In addition to the explosion of screens in public spaces, screens

A. Arnold (✉)
A2Studio.org, 28-30 Clinton Street #1A, 10002 New York, NY, USA
e-mail: alice@a2studio.org

F.T. Marchese (eds.), *Media Art and the Urban Environment*, Future City 5,
DOI 10.1007/978-3-319-15153-3_7

are ubiquitous in workspaces and in people's daily life activities. These seamless, illuminated electronic surfaces are becoming the devices through which we frame our experiences. Screens, whether in the form of video billboards, computers, mobile phones, ATMs, digital cameras, iPods, PlayStations, or other electronic devices, are our link to communications, our conduit to information, our means to online shopping and other commercial transactions, and the gateway to many of our entertainment activities. *Electric Signs* is a documentary film that explores this new screen culture as it unfolds in the global city.

In 2005, Times Square was just beginning to transform into the screen-rich environment that it is today. I was there often that year, filming for a documentary that I was making on street art, called *To Be Seen*.[1] Times Square is actually one of the least likely places in New York City to encounter street art, but I was there to film the advertising signs, which are part of the fabric of Times Square and have been since the 1920s, when Times Square evolved into New York's entertainment and media zone.

As discussed in *To Be Seen*, street art is unique in our society today because it is a non-commodity form of art – there are no price tags on the wheat-pasted posters, stencils, or other types of images that street artists (illegally) post onto city walls. The advertising signs in Times Square, on the other hand, are powerful representations of the big brands and the major corporations that control the majority of goods and services that the average US citizen consumes (this situation applies to the inhabitants of most postindustrial cities). The signs are produced by teams of imaging and marketing specialists; they are highly scripted, high-budget artifacts that command prominent locations in our urban environments. Street art, especially in the early 2000s, has a more ephemeral quality and also a more limited range – if you didn't walk by the spot where it was put up and notice it, among all the other visual stimuli of the street, you wouldn't see it. Nowadays, of course, more people are aware of street art, and it circulates widely on the Internet, allowing it to transcend the space and time constraints of a geographically specific audience.[2]

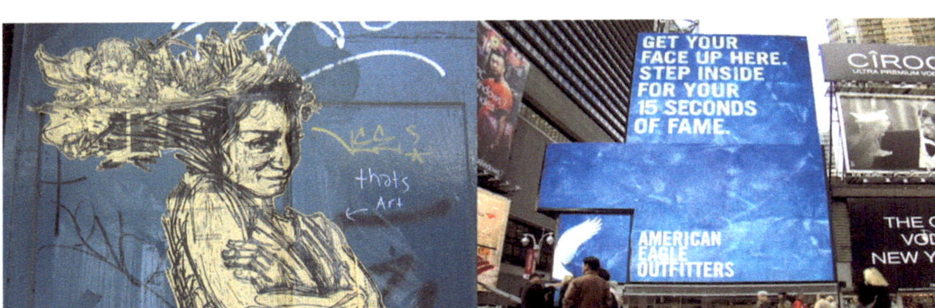

Street Art vs. Screen

[1]Distributed by Icarus Films: icarusfilms.com/new2006/to.html

[2]And several street artists, such as Swoon, now have a prominent position in the mainstream art world and are "brands" themselves.

At the time I was filming *To Be Seen* in Times Square, large supergraphic advertising signs and conventional billboards vastly outnumbered the digital signs, but the screens were so much more captivating. The sheer exuberance of these bright, large-scale images dancing among the skyscrapers conveyed a futuristic glow and inspired an effervescent mood. My visual pleasure in these signs, though, was tempered by layers of critical knowledge about public space and media power. I found this tension between visual pleasure and critical analysis an interesting space to inhabit and therefore a powerful place to make work from, so that lead me to my next film, *Electric Signs*.[3]

New Sign Systems

> Walking through the city I was reminded of the words of Walter Benjamin, who described the explosion of advertising as something that "hits... us between the eyes with things, as a car growing to gigantic proportions, careening at us out of a film screen." Now at the beginning of the 21st century, I wonder what he would think about the electric landscapes that I inhabit today? (Narrator, *Electric Signs*)

Electric Signs was conceived as a visual study to investigate this economy of signs and the new urban space of the media city. How are these screens, which add a sophisticated level of mediated experiences to the urban environment, changing the way people interact, utilize and experience the city? When public spaces become more highly concentrated with commercial messages, does public expression, an important element of democratic and open societies, contract and become less of a political force in the city? How do the screens (both the outdoor and the personal handheld ones), combined with new image and communication technologies, affect people's communications and spatial flows, impact visual culture, and shape people's perceptual experiences? And, as cities as well as devices become more networked, what are the connections between screens and surveillance? These are the research questions that propelled this project, which started with a Fulbright Fellowship in Hong Kong and then continued for many more years in other cities.

The film is structured as a documentary essay in the spirit of city symphony films[4] and features Hong Kong, Los Angeles, and New York, as well as Tokyo, Las Vegas, Shanghai, Vienna, Macau, Berlin, Seoul, Prague, and Kaohsiung. Also featured are a voice-over narrator, interviews, vérité footage, and vox populi style interludes. The interviews are with prominent lighting designers, advertising and marketing professionals, urban sociologists and visual culture experts, community activists, and a public space artist whose work offers alternative ideas about the use of media in the public sphere. The vox pop segments are stylized interstitials

[3]Distributed by Icarus Films: icarusfilms.com/new2013/elecs.html

[4]City symphony films originated in the 1920s, when film was still a relatively new medium. These films took the city as their subject and showed urban life in a poetic, nonnarrative style.

throughout the film featuring people in the city who walk, sit, work, shop, and daydream in these urban lightscapes.

The spine of the film is a voice-over narrative by a "city observer" who takes us on a journey through a variety of urban landscapes and weaves together the film's themes, cities, and various personalities. The narrator, who appears throughout the film as a shadow (the opposite of light), is modeled on the critic Walter Benjamin, who wrote about city life and media in the 1920s and 1930s. This was a time period, analogous to ours, when new forms of media and technology dramatically impacted society.

The vérité layer of the film utilizes super 8 mm time-lapse cinematography to translate the spatial qualities of the built environment and the unique lighting style of electric signs and media façades. The use of time lapse, combined with editing techniques that also affect time and pace, gives the film a subtle ripple of time manipulation, which helps convey the emotional effects of the intensification of urban life amidst the vast spaces beneath the skyscrapers.

Filming for *Electric Signs* began in Hong Kong, where I was a Fulbright Fellow in 2007. The project continued when I returned from Asia to the States in 2008. In Los Angeles, a sign war was erupting over digital billboards and supergraphic signs. In New York, the majority of signs in Times Square were now digital, and they were becoming larger and more spectacular. In Prague, the capital of the former Communist Czechoslovakia, digital signs were beginning to make an impact in the now capitalist capital of the Czech Republic. And in many other cities in Europe, Asia, and the Middle East, media facades were changing static cityscape views with architectural façades embedded with programmable LED lighting technology. In 2010, however, the real estate market crashed under the weight of speculation and the global economy spun into a recession. The Occupy movement gained momentum in New York City in 2011,[5] protesting the economic and social inequality that the recession magnified. The protestors' encampment in Zuccotti Park, a privately owned public space[6] in the financial district, helped call attention to public space issues and brought a new awareness to the public about the political importance and social value of public space.

[5]Building on the Spanish Indignants movement and the Arab Spring, Adbusters magazine proposed an occupation of Wall Street to call attention to increasing political, social, and economic inequality. See Adbusters call to action here: https://www.adbusters.org/blogs/adbusters-blog/occupywallstreet.html

[6]Privately owned public spaces are city plazas and spaces controlled by real estate developers but required by law to be open for public use. For more information, see http://www.nyc.gov/html/dcp/html/pops/pops.shtml

Manufacturing Consent

> My journey took me first to the brightly lit cities of Tokyo and Hong Kong, and to Mainland China, where the majority of the world's LEDs and flat panel displays are manufactured. This technology is allowing the visual image to be everywhere. (Narrator, *Electric Signs*)

The Pearl River Delta, which sits just to the north of Hong Kong on mainland China, is where most of the world's LEDs and flat-panel displays are manufactured.[7] LED stands for light-emitting diode. LED lighting technology is innovative because it is energy efficient and programmable, which means it can be used as a video platform. Many of the factories are controlled by Hong Kong companies, so in Hong Kong, LED technology is relatively cheap and readily available.

Ronald Lo, who is featured in the film, is the CEO of Join Merit Media, a marketing firm that operates a network of digital advertising signs in Hong Kong. Lo has charted the growth of LED signs in Hong Kong since the early 1990s, when his father, a businessman involved in LED manufacturing, field tested the first LED sign in Hong Kong. This experiment was short lived, but it paved the way for future signs. "Digital signage did not really take off until 3 years ago, until we put (one) on the site at Chung King," relates Lo. This screen is a 90 square meter sign on the façade of Chung King Mansions, a popular indoor shopping and low-budget hotel complex on a busy shopping street in downtown Kowloon. It is also famous as the location of Wong Kar-Wai's film, "Chung King Express." "That screen," continues Lo, "influenced a lot of real estate developers." That's because this building was sold at a high price shortly after the screen was installed. "So a lot of other developers actually picked that up, and ... right after that installation there were at least 15 or 20 screens built afterwards."

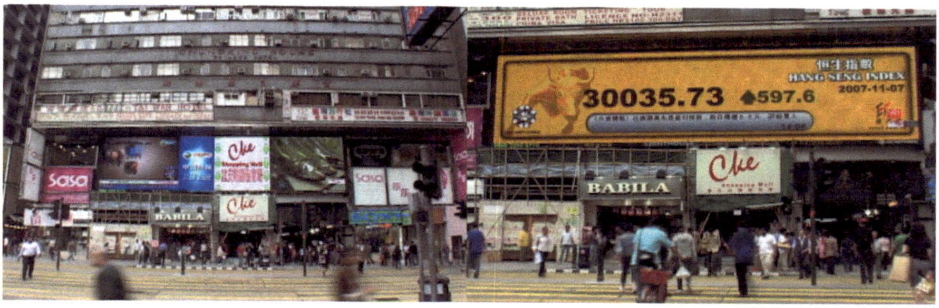

Chung King Mansions Screen

[7]The Pearl River Delta Economic Zone is often referred to as the world's factory because so many of the goods that people buy all over the world are manufactured in this region in southern China.

Ted Lo, Ronald's brother, who is a LED artist and lighting designer with offices in Hong Kong, China, and New York, adds that in addition to adding value to real estate, the screens also provide a better platform for outdoor advertising. He points out that digital billboards are more effective at capturing eyeballs than static billboard ads because the screens utilize motion and have richer colors. "So a lot of media and advertising people really try to push this projection technology to the mass(es)," observes Lo.

The development and implementation of this more elaborate, costly, and technologically sophisticated promotional apparatus is fueled by the need to capture consumers' attention in a marketplace that is saturated with entertainment, advertising, and marketing messages and populated by an increasingly mobile audience. For digital media producer Cedric Chan, who produces campaigns for commercial screens, they are "new business opportunities and new ways to interact or affect the public." He is thinking about strategies to link up mobile devices, the Web, and the public screens, "to get people more involved in what the client wants to sell." These screens are not broadcast television, he observes, so they're not truly mass market, but they are networked. "So that allows you to use web technology, to give you a level of control of your narrowcast. To say at this time of day let's play this specific clip because we expect people getting off of work to be a certain demographic. So who's being targeted? Everybody, everyone's fair game. It's public spaces. Broad and narrow at the same time. Now, we're getting to this level of control that we never had before. And that is a very different dynamic."

Hong Kong is a densely built commercial city with very little public space in the commercial areas of the city. It is a complex, speculative space, constantly changing, rebuilding, and expanding, and dominated by images. According to the Skyscraper Museum, Hong Kong's population of seven million residents live at an average density of 70,000 per square mile, which makes Hong Kong the most densely occupied city in the world.[8] Hong Kong's dense urbanization and concentration of tall buildings are due in part to its geographical and physical boundaries and in part to its status, dating back to the nineteenth century, as a free-trade city in the British Empire.[9] Currently, acceleration in urban density and growth is being stimulated in large part by the commercial and cultural competition Hong Kong faces from Shanghai and Shenzhen, two fast-growing free-trade zones in southern China.

Mirana May Szeto, Assistant Professor of Comparative Literature at Hong Kong University and a political activist, discusses Hong Kong's cultural policy and politics in the film. "The city is very subtly controlled," she notes. Key industries such as real estate and the financial sector have a lot of economic and cultural power because "... they are the backbone of the media. And [the] media, in order to survive, really cannot antagonize ... a consortium of this kind of financing."

[8]Skyscraper Museum: http://www.skyscraper.org/EXHIBITIONS/VERTICAL_CITIES/walkthrough_vertical_density.php

[9]For a comprehensive timeline of Hong Kong's history, see http://www.bbc.co.uk/news/world-asia-pacific-16526765

Professor Szeto cites the protest movement over the destruction of Queen's Pier (which she was involved in) as an example of how developers and politicians exert control over the city environment in Hong Kong.[10] Queen's Pier was a public space pier on Hong Kong Island that the government wanted to raze in 2007, so as to use this prime waterfront location to develop a shopping mall and highway. The protest opened up a critical dialogue in Hong Kong about the importance of public space and the lack of political power that ordinary Hong Kongers have over development decisions. Ultimately, however, the courts sided with the government, the protestors were forced out, and the pier was torn down. Professor Szeto draws the conclusion that the "... theory of manufacturing consent works extremely well in explaining the situation in Hong Kong... Media is really very highly and subtly controlled on all levels of its production."

Sign Wars

> The film Blade Runner depicts a futuristic, dystopian Los Angeles. Director Ridley Scott's vision for the film was influenced by his experience living and working in Hong Kong. And now his image of the city, created more than 30 years ago, is catching up to present day reality. (Narrator, *Electric Signs*)

These changes in the urban environment are taking root in cities around the world. But unlike Hong Kong, where outdoor commercial screens are part of the fabric of the city, in Los Angeles there was public outrage in 2008 over the installation of digital billboards and the changes they created in the visual landscape of the city.

The digital billboards were marketed as improvements over the existing ones, but they were installed without community district discussion, without zoning regulations, and under questionable legal conditions.[11] "The issues with signs is about money," states Dennis Hathaway, the director of the nonprofit organization Ban Billboard Blight. "These billboards are like slot machines for the industry. A single digital billboard can make up to 800,000 dollars a year in revenue. A big supergraphic sign in a high traffic location can make up to $50–100,000 a month." Hathaway believes that the visual environment of the city belongs to the public and that private interests who use this space should not be making enormous profits from a public resource. But in Los Angeles, collusion between the city and the billboard companies has financially benefited the outdoor advertising industry more than the public.

[10]This New York Times article explains the situation in more detail: http://www.nytimes.com/2007/07/30/world/asia/30iht-hong.1.6894441.html?_r=0

[11]LA Weekly writer Christine Pelisek investigated the issue and concluded that "the mayor and city council have let the billboard industry flout the law." See "Billboards Gone Wild: Is City Hall Corrupt or Inept?" Christine Pelisek, LA Weekly, 23 April 2008. Online link: http://www.laweekly.com/2008-04-24/news/billboards-gone-wild/

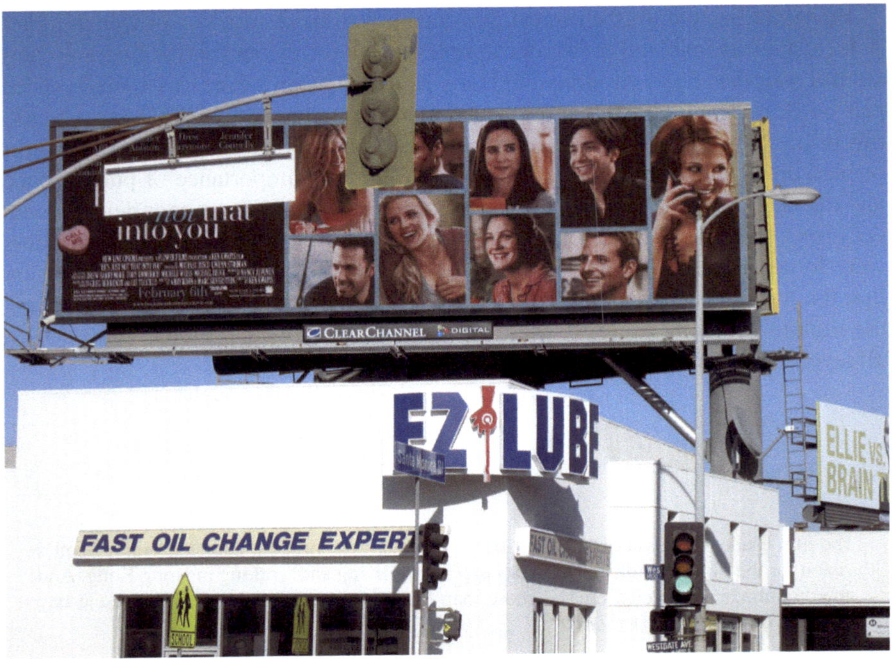

Digital Billboard, LA

Los Angeles is often described as a postmodern city because it does not have a unified center. Instead, it sprawls horizontally, networked together by its freeway system. Driving is part of the fabric of life in LA, and people spend a lot of time looking out at the city through their windshields, which is a very cinematic way to experience the city. And because people spend so much time on the road, they are a captive audience for the billboards.

But community groups have been working since the 1980s to reduce the number of billboards in LA, many of which were put up illegally. In 2002, the City Council passed a ban on all new billboards, but the billboard companies fought back with lawsuits, citing First Amendment rights for commercial speech. The city ended up settling one of these lawsuits by allowing three of the billboard companies to convert over 800 billboards to digital billboards. Pressure from community groups and media scrutiny forced the City Council to pass a billboard moratorium (2009), so that they could rewrite the city sign code. But the billboard companies responded with more lawsuits. Because of this tangle of lawsuits, plus the need for more revenue, the city is now considering allowing advertising in city parks and creating special sign districts.

Professor Dana Cuff is the founding Director of cityLAB at the University of California, Los Angeles. One cityLAB project involves rethinking planning regulations for the twenty-first century. In the film, Cuff points out the connections between city planning regulations and real estate development. Los Angeles, she notes, has a very weak planning department, which has allowed for a lot of vibrant development, but it also means that developers can bend or modify planning regulations. "When a big development comes along," Cuff observes, "a lot of the planning gets set aside so the thing can happen. And you see that in all of the big developments downtown." Additionally, the recent increase in commercial signage in real estate development projects in LA, as well as in other cities, is fuelled by selling the signage rights as part of the development deal that funds the project.

"The problem of course is that many of these screens or billboards are controlled by corporations, for whom it is of course a form of big business," adds Professor Erkki Huhtamo of the Design Media Arts Department at UCLA. "So the issue of access to those messages that surround us in public spaces is anything but simple. Public space is highly controlled space, so you just simply cannot put anything you want on that space, unless you chose to do it illegally."

Dennis Hathaway concludes the LA section in the film by discussing the psychological effect of the screens. "Being surrounded by all those screens becomes a problem because its sensory overload, and it just kind of robs you of any ability to take in and sense the larger surroundings. I know that some people don't really see a distinction between commercial advertising and say a piece of digital art. You know, it's all about the graphics, how it looks. To me, a major part of it is the message. The message from the commercial ad is be a consumer, buy. Whereas the screen with a piece of art could say to you, look at the way these colors interact. Having a visual experience without having the message to buy, you know, can be a good thing."

Pale Daylight

> Walter Benjamin thought deeply about mass media and the power of images. In the essay "This Space for Rent" he observed the power that advertising signs have to capture people's attention and imaginations. "It is not what the moving red neon sign says," he wrote, "but the fiery pool reflecting it in the asphalt." (Narrator, *Electric Signs*)

At the end of the nineteenth century, electric signs and street lighting began to light up city streets and spaces. New York and Hong Kong were two early adopters of electric lighting, which Hong Kong locals poetically described as "pale daylight." Electric lights lengthened the day for people. The workday was extended because quitting time was no longer dependent on when the sun went down, but leisure time was also expanded, as illuminated streets animated city life.

"During the nineteenth century, with the evolution of capitalism, the strategies of advertising were linked with what we call culture of attractions," explains Professor Erkki Huhtamo. By the late nineteenth century, cities were filled with signs and billposters battled one another for wall space, often posting on top of their competitors, which sometimes escalated into street brawls. In New York, the Municipal Art Society, a group dedicated to the betterment and beautification of New York City, sought to limit outdoor advertising. This led to a more regulated advertising environment, with people renting out the spaces on the walls, which eventually led to the billboard. "The whole space of advertising as it developed," continues Huhtamo, "was a place of struggle, was a place of tension, was a place of competition." The ubiquity of signs, however, has also contributed to urban literacy and, as Professor Zukin notes, "people like signs precisely because they are visual, they are immediate, they are in your face. They are a form of public art that we all grow up with."

Advertising's quest to capture the eye led to bigger, brighter, and more captivating signs, known in the trade as sign spectaculars. At the beginning of the twentieth century, electric signs were a new visual medium, lighting up the night sky with larger-than-life images. These immense glowing signs extended the commodification of urban space and helped to shape the modern city. By the 1920s, these giant electric signs had put Times Square on the map as the media capital of the world. Times Square became an urban spectacle, drawing huge crowds of people who came to look at the signs and be entertained.

From the 1920s to the 1950s, Times Square was an exuberant and popular entertainment center, but by the 1970s, the lights had dimmed. TV, shopping malls, and suburban homes had shifted affluence and influence, away from the city to the suburbs. New York City was on the verge of bankruptcy, and in Times Square, there were empty billboard spaces, vacant buildings, and illicit activities.

In the 1980s, redevelopment plans to make Times Square into a cleaner, safer, and more corporate business environment were proposed. The main feature was a set of skyscrapers on 42nd Street by Philip Johnson and John Burgee. These buildings, however, were widely disliked, and the Municipal Art Society, which in the early twentieth century had fought to reduce advertising signs, devised a plan to create a sign district in Times Square, to recapture the feeling of excitement that Times Square was known for. The Municipal Art Society argued that what made Times Square successful in the past was its unique populist character and that the city can reclaim and redevelop the space by keeping the visual character of the old Times Square. "So Times Square was made a unique zone in New York City," explains Professor Zukin, "where electric signs are required by law. Every building in Times Square must be covered on the outside by giant electric signs."

The sign district brought spectaculars back to Times Square, along with new investments and crowds of people. These signs have become a sophisticated form of visual capital – they are worth millions of dollars to advertisers, real estate developers, and city governments. And the sign district concept has been copied by cities around the world who want their own version of a Times Square district with sign spectaculars.

Media City

> I was walking by and it caught my eye because there were colours that were moving, and then I took a second look because I noticed it wasn't an annoying ad. (Chisa, Commenting on the Pixelator, New York, *Electric Signs*)

LED technology is creating new forms of programmable media that both inform and entertain us, and this is changing the role that an image plays in our lives.

The relationship between media and the public city space recently has become more complicated, as walls and even whole buildings are turning into media objects. But what kinds of experience are being created by the existence of multiple screens that are simultaneously competing for our attention?

"A media façade answers to the trend that media is everywhere," says Rogier van der Heide, who designed several media façades as a lighting designer at Arup and who is currently the Chief Design Officer at Philips Lighting. "We carry with us technology all the time. It is wearable technology, and we're surrounded by that. And architecture is something that surrounds us too, and it just wants to speak the same language."

"Many of our clients," he continues, "are really interested in the marketing proposition that a media façade design offers. Because the media façade is transformable, it can respond to the urban context, or to the behavior of people, of target groups or customers, and, in that way, it engages the company with its audience, because context changes."

van der Heide and his team at Arup, working with architects of UN Studio, created the media façade for the Galleria fashion mall in Seoul. The building was a failed department store that the owners were seeking to revitalize through design. The designers wrapped the exterior of the building in large ceramic disks that during the day have an iridescent sheen, but when the sun sets, the LEDs are turned on and they become a dynamic colorful skin covering the building. "It is as if the building is dressing up for the evening," van der Heide describes. The Galleria has become a visual icon in Seoul, attracting people who want to have their photograph taken, in a matching outfit, in front of the building.

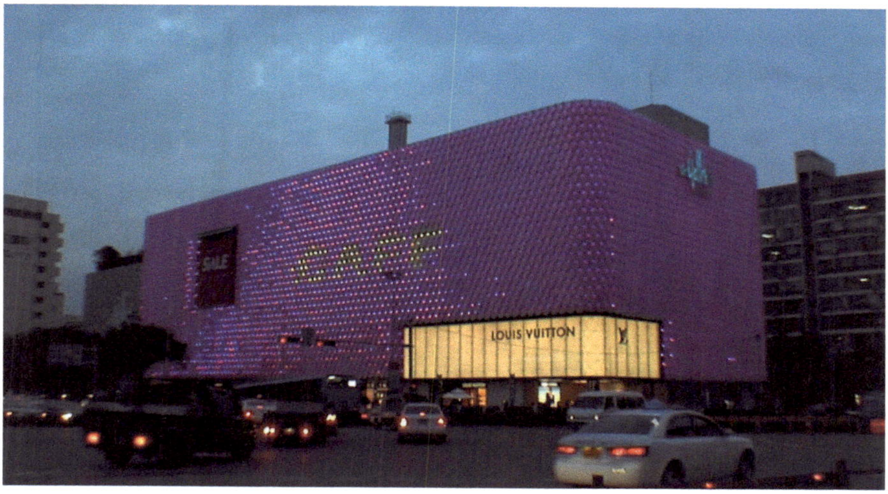

Galleria, Seoul

van der Heide thinks that the investment in media façades is related to the development of the experience economy because they make the shopping experience more stimulating. "We're not satisfied just going shopping," he says. "We want to have a certain context in order to be inspired to buy. And the big real estate developers and the big companies that surround us, they all know very well how to seduce us. And I think media façades are very seductive."

Erkki Huhtamo traces the fascination with media façades to the magic lantern, a nineteenth-century projection technology. Magic lanterns were used by newspapers to project breaking news, such as election results, in public spaces. "In the city centers," relates Huhtamo, "the people operating the magic lantern would quickly scratch the current situation on these (black painted glass) plates and project them for the crowds that have been gathered on the streets. So this anticipated the urge to transmit and receive news in real time. And this was actually the origin of dynamic images in the public space."

The origin of the word "screen" is also interesting to investigate because it has multiple and contradictory meanings. Screens show and reveal, but they also hide and conceal. The screen is an information surface, a window we view, but screens are also surveillance devices, and we actually appear on screens as much as we are looking at them. Closed-circuit surveillance cameras dot the urban landscape, recording our movements and collecting our data. And now, businesses are using this kind of camera data, to see who we are, so that they can appeal to us, as individual consumers.

Sustainability is another important issue associated with these screens, but one that is not often addressed in business meetings and in city planning committees. Dennis Hathaway notes in the film that while digital billboards use energy-efficient LEDs, they still draw a lot of power – 33,000 W for one digital billboard. "But even if it just ran at half of that wattage, it would be equivalent in energy use to 10 houses," he says. Warren Levy, a marketing expert who works extensively on sign campaigns in Times Square, is also concerned about this issue. "As a matter of fact," he relates, "the Walgreen's sign was delayed a few months, because the power requirements were so great that Con Edison had to add a power station in order to support the load on the circuits, just for that screen." But Rogier van der Heide broadens the scope of sustainability beyond energy use and makes the point that a media façade can be a more sustainable option than renovating a retail space or creating a new building. "Because with the media façade," he notes, "you can basically build your building out of software. When a fashion store needs an update, it doesn't need a physical update because you can do so much in the field of software."

If you are a media artist, though, how can you push back against these powerful commercial images to get your work noticed, your message heard? This is something that media artist Jason Eppink thought about as he confronted an increase in advertising signs in New York City in the early 2000s. He created the *Pixelator* as a culture jam of the digital advertising screens that are on many of the subway entrances in New York City (there are about 80 of these screens in the city). The

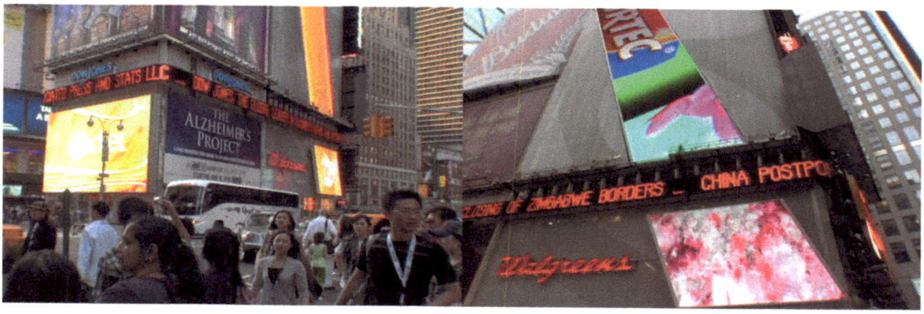

Walgreens screen

Pixelator is a light box, essentially, made out of foam core and diffusion material, which he puts in front of the video billboards.[12] “And what they do,” says Eppink, “is diffuse the video content that Clear Channel provides, and turns them into 45 blinking, color changing squares. So what happens is, I am creating an artwork with the help of these multinational corporations, and they don’t even know about it.”

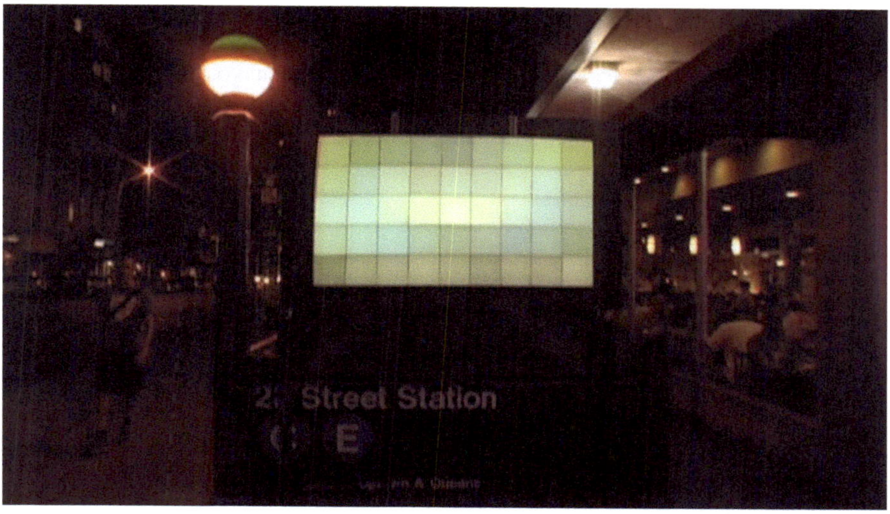

Pixelator project

The increase of screens is also, as some critics have noticed, having a social and cultural impact on public spaces. Professor Zukin comments, “I am constantly amazed by how many people are totally transfixed and living in a virtual world, either overhead or in their palm. You know, what we value about public space is

[12] *Pixelator* website: http://jasoneppink.com/pixelator/

the feeling that we are all in it together. But we can't be in it together if we're also tethered to someone at the other end of cyberspace. And then what happens to the public space that we are physically in? We feel no responsibility to others in that public space and they feel no responsibility for us."

And what would Walter Benjamin think about Times Square today? Professor Zukin thinks that "Walter Benjamin would find the lure of commodities and goods too abstract in this space. Our visual field is overtaken by the signs and symbols of corporations. Not just by products, by the immediacy of products, but by the immediacy of the brands."

Rogier van der Heide is aware that the content he creates for the screens is often better than real. He describes the screen environment as a hyperreal world because what looks like a building during the day can be transformed through software into a commercial at night. He asks, "What is the real self of that building? Is it the media, is it the message, is it the Coca-Cola commercial on that façade? Is that the reality of that structure? Or is it the office that it is during the day? There is no distinction anymore and I believe that reality is definitely not what is out there, you know."

Sharon Zukin provides some final insight in the film by observing that the signs are more exciting than the stores themselves. "If you've ever been inside a Walgreens, it's a pretty ordinary space. These electric signs, they romanticize commerce, and the signs are much more exciting than the actual experience of consuming the product. I mean, you put on a pair of jeans, ok, you put on a pair of jeans, but if you put on a pair of jeans that's moving around on a giant electric sign, then you are consuming the vast experience of this sign."

Urban Lightscape

> Now at the end of my journey, I realize how important public space is to people. It allows us to create a personal narrative that connects us to the places we inhabit. And that ability to shape our environment helps to nourish a democratic society. (Narrator, *Electric Signs*)

The city has become a platform for the media to utilize, to project their messages and information to all the passersby. The media city is a matrix of physical and networked space that expands our experience of the city and alters the dynamics of public space. This urban lightscape is a visible sign of the power of the market economy. The signs shine seductively with messages of modernity and freedom, of free consumer choice, but they eclipse the public aspects of our public spaces.

To better understand this economy of signs, the film visits Prague, the capital of the Czech Republic. Thirty years ago, Prague was behind the "Iron Curtain," and there was very little advertising in the city, as befits a Communist economy that does not celebrate consumer goods. Petr Cikhart, who grew up in Prague in the 1980s and participated in the Velvet Revolution in 1989 that toppled the Communist government, comments on how he felt about the new electric signs. "When the first electric signs came in – and they always kind of symbolized the West and capitalism. And capitalism for us, especially American capitalism, was freedom, and democracy. And after the end of Communism, we thought it was great." Bara Stefanova, who was 17 years old in 1989, says, "I was very excited to see these billboard advertisements because they seemed like, very existentialist to me... because at the time... I was a fan of Michelangelo Antonioni, and he always used that in his films. And so, I was like, oh yeah, finally these things in Prague. So I was excited for the first five years, before I stopped seeing this as something other than an eyesore." Mikulas Novotny, a university student, remarks that his mother told him stories about how dirty and gray the city used to be, but "the city I know now is overloaded with commercials and it's colorful, but maybe it's too colorful, so one stops paying attention to it."

In today's global cities, logos and brandscapes dominate the visual environment, and our public spaces have become highly controlled. But behind the brilliant skylines, a protest movement gathered steam when the economy dove into a deep recession in 2010. In New York City, the Occupy movement established itself in Zuccotti Park, a privately owned public space in the financial district. The protestors sought to reclaim public space and create a different sort of environment, one that was more open, less dominated by corporate interests and messages. The Occupy protest lasted for a few months, before being dispersed by the police, but the protestors asked an important question: who has the right to shape the city?

The film returns to Walter Benjamin for insights. Benjamin, writing in Berlin in the 1920s, critiqued the burgeoning power of mass media as the first electric signs began to shape the image of the modern city. He perceived that visual technology is a form of power, an exploding force that could "burst the world asunder by the dynamite of a 1/10 of a second" (Benjamin 2008).

In today's city space of signs and passersby, what becomes most apparent from my observations is how the luminous beauty of these screens masks the power of visual technology that drives these signs. And that power is used to control our public spaces, dominate our public lives, and influence our private selves.

Bibliography

Abbas A (1997) Hong Kong: culture and the politics of disappearance. University of Minnesota Press, Minneapolis

Banham R (2009) Los Angeles: the architecture of four ecologies. University of California Press, Berkeley

Benjamin W (2003) The Arcades Project: Paris Capital of the 19th Century, Tiedemann R (ed). (trans: Eiland H). Belknap Press of Harvard University Press, Cambridge

Benjamin W (2004) Walter Benjamin: selected writings, volume 1, 1913–1926, Bullock M, Jennings MW (eds). (trans: Eiland H). Belknap Press, Cambridge

Benjamin W (2008) The work of art in the age of mechanical reproduction. Penguin, London

Cohen L (2003) A consumer's republic. Vintage, New York

Gilmartin G (1995) Shaping the city: New York and the municipal art society. Clarkson Potter Publishers. New York

Gudis C (2004) Buyways. Routledge, New York

Harvey D (2004) Conditions of post modernity. Blackwell Publishing, Oxford

Helfand J (2001) Screen: essays on graphic design, new media and visual culture. Princeton Architectural Press, New York
Huhtamo E (2012) Illusions in motion: media archaeology of the moving panorama and related spectacles. MIT Press, Cambridge
Mathews G, Tai-lok Lui (eds) (2001) Consuming Hong Kong. Hong Kong University Press, Hong Kong
McDonogh G, Wong C (2005) Global Hong Kong. Routledge, New York
McQuire S (2008) The media city: media, architecture and urban space. Sage Publications, London
Nye D (1992) Electrifying America. MIT Press, Cambridge
Zukin S (2010) Naked city: the death and life of authentic urban places. Oxford University Press, New York

Chapter 8
Overload/Absence: The Collapse of Space to Surface in Representations of Urban Space

Annette Weintraub

Abstract As public space is rapidly diminished through gentrification and privatization, it is paradoxically "enlivened" by the introduction of large-screen moving images, digital advertising, and other mediazation. We live in an environment of overload/absence, navigating public spaces dense with video monitors and giant advertisements, while beyond the explicit media zones, the streetscape is increasingly narcotic. The hyperactive and featureless landscapes, seemingly opposites, in fact both prioritize the *skin* of architecture: each is the manifestation of an intense preoccupation with surface, demonstrated in the activation of surface by light and moving images on billboards and screens or in the ubiquitous grids of glass and metal that wrap contemporary buildings. This paper will examine the perceptual shift from space to surface in environments of overload and absence and look at how surface has been used to create new representations of urban space.

From Space to Surface

The contrast can be stark: a nexus of avenues so illuminated by signs and video that the uplight can be seen from an airplane at 35,000 ft, while a few streets away the landscape flattens out into corridors of gleaming but uninspired corporate towers anchored by an anonymous and repetitive streetscape of banks and chain stores. Paradoxically, we live in an environment of overload/absence—we navigate public spaces dense with video monitors and giant advertisements, yet beyond the explicit media zones, the streetscape is increasingly narcotic. The hyperactive and featureless environments, while seemingly opposites, in fact both prioritize the *skin* of architecture: each is the manifestation of an intense preoccupation with surface, demonstrated variously in the activation of surface by light and moving images on billboards and screens and in the ubiquitous grids of glass and metal that wrap contemporary buildings and also represented in popular media in images of architecture that emphasize reflection and surface incident over function and

A. Weintraub (✉)
Department of Art, City College of New York, 160 Convent Avenue, 10031 New York, NY, USA
e-mail: annette@annetteweintraub.com; weintraub@ccny.cuny.edu

F.T. Marchese (eds.), *Media Art and the Urban Environment*, Future City 5,
DOI 10.1007/978-3-319-15153-3_8

substance. This contrast of extremes, of overload and absence, is a reflection of an urban space that has effectively been stripped of locality, sense of place, and history, and at the same time it is amplified by media intrusions both visible and virtual. These changes in urban space have dehistoricized the city, obliterating many of the idiosyncratic visual codes that formerly enriched city space, contributed to a sense of local identity, and fostered multiple and complex geospatial identities. In an environment of overload yet visual impoverishment, space collapses into surface.

This primacy of surface and the resulting concealment/devaluation of function create a visual landscape that exhibits many of the characteristics of simulated space and the "non-space" described by Marc Augé in his anthropology of supermodernity. In *Non-places: An Introduction to Supermodernity*, Augé wrote about "space that cannot be defined as relational, or historical, or concerned with identity and is thus devoid of emotion and memory" (Augé 2008). Augé's "empirical non-places," which he describes as "spaces of circulation, consumption and communication," are largely ahistorical and ageographical and lack the organic growth and dynamic chaos that once characterized the city. The introduction to the second edition (2008) of Augé's book expanded this idea to include the impact of globalization and the constant mediation of personal information devices that cause an individual to "live rather oddly in an intellectual, musical or visual environment that is wholly independent of his immediate physical surrounding." The new sense of time and space resulting from this extension of physical and psychological awareness through media has also been described by William J. Mitchell as "the electronic present continuous," a continuum of simultaneous events and electronic inputs that flow into real space and real time (Mitchell 2003).

Urban space characterized by eroding community, history, and memory yet steeped in the "electronic present continuous" is space that collapses into surface. The visually hyperactive landscape of Times Square-type entertainment zones and the featureless districts of endless developer's towers—while seemingly very different experiences of urban space—are perceptually similar: they are superficial (in the sense of a reduction of nuanced and dimensional space to a two-dimensional plane). Giant screens or large grids of mirrored surface have an effect that privileges facade over function and exterior over interior, essentially compressing space. These spaces of overstimulation and zones of absence reduce environment to a skin. As activated surfaces project streams of images at us, the blank mirrored boxes, like black holes, reciprocally reflect one another—present as a reflection in their neighbor's surface of polished glass.

The gradual disappearance of elements of vernacular architecture and signage (typography and unconventional building materials), the design of skyscrapers primarily for the effect as part of the skyline (as opposed to function and integration at street level), the repurposing of fragments of historical buildings as décor, and the impact of 3D rendering styles in the way buildings are conceived and designed are all consequences and characteristics of a dematerialized architecture that prioritizes surface over space. This chapter will consider the shift from space to surface in environments of overload and absence and look at how surface has been used to create new representations of urban space.

The Erosion of Memory

Landscape is history: much of memory is inscribed in place. As familiar places vanish, memory turns into myth, and the commonplace landmarks of even the recent past become objects of nostalgia when they can only be revisited in memory or thorough photography. When change is rapid, and loss of familiar surroundings continual, there is a feeling that the present is eroding the past and disorientation or decentering results.

The destruction of once grand buildings and important public works (the most publicly regretted demolition that of Penn Station) has been extensively recorded in books and photographic archives. As a response to the widespread gentrification of recent years, there is also interest in documenting the more humble structures that make up a neighborhood and provide character and human scale. Endangered and disappearing local landmarks have been itemized and mourned in contemporary catalogs of evicted stores and iconic streetscapes. Popular urban blogs like *Jeremiah's Vanishing New York* (Moss 2014) or James and Karla Murray's website and Flickr feed[1] *Store Front: The Disappearing Face of New York* (Murray and Murray 2009) each record the relentless disappearance of distinctive and eccentric neighborhood icons that were part of the vibrant street life. The Murrays' extensive archive of vernacular storefronts reminds us of what has vanished in just the recent past and iconizes those that remain. To calculate the number of years of New York City history lost in just the last 12 years (the Bloomberg administration), Jeremiah Moss, the curator of *Jeremiah's Vanishing New York*, compiled a "Master List" (Moss 2013) of businesses that have disappeared since 2001 and estimates that 6,926 years of history has been lost due to gentrification and other neighborhood changes. His *Master List* enumerates decades at the same location and the cause of the closure: usually loss of lease due to a rent increase. What he mourns is not just history, but character and community. The storefronts tracked on these blogs and in other visual records were each the product of an individual sensibility evolved over time: visually unique, evocative of each owner's personal history, and reflective of their specific function. It has become a cliché to describe upscale stores as having collections "curated" by the owner. Yet the contents of these original neighborhood stores do seem truly curated in the sense that they represented a holistic sensibility that was consistent in signage, architecture, and store contents. Quirky and eccentric, they distinguished themselves through use of color and unorthodox construction materials, and the flavor of their display typography was reminiscent of a particular historical era. It is this individuality and eccentric sense of variety that is missed. Once commonplace, now increasingly rare, these visually unconventional storefronts have the idiosyncratic originality and quirkiness of folk art.

---

[1]Also see James and Karla Murray's Flickr feed for extensive documentation of typographic signage. http://www.flickr.com/photos/jimandkarlamurray/sets/72157612285425548/

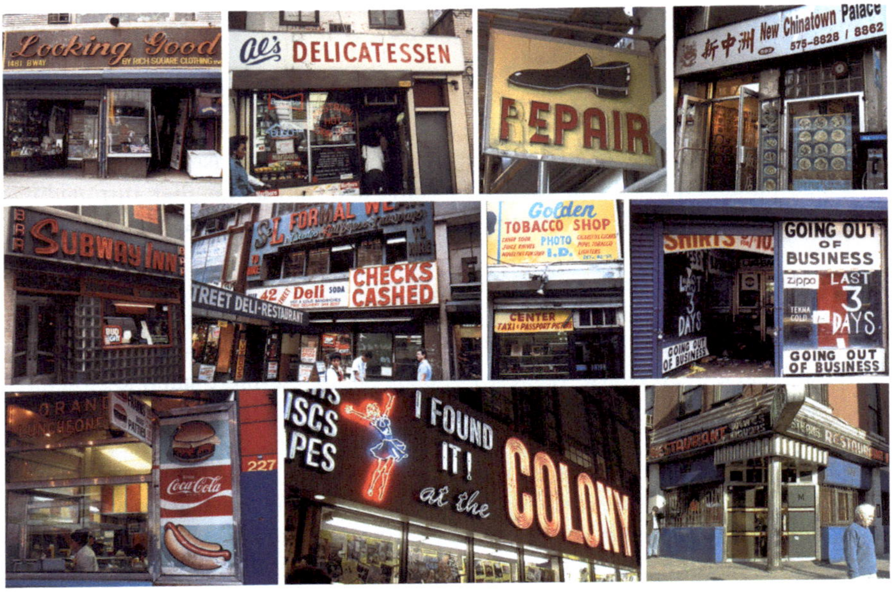

New York City signage, photos 1996–2001, Annette Weintraub

Typography (often in the form of hand-painted lettering—another dying art—or neon signage) was frequently the vehicle of this personalization; hand-fabricated type, even that done by a professional signage artist, had a warmth and sense of time that framed the contents of a shop window free of professional "styling." These facades conveyed the individuality of the owner, quite the opposite of the effect of a generic chain store window. While there is an element of nostalgia and fetishization in the collection and lamentation of these lost neighborhood locations, the variety and personalization of these independent businesses, particularly seen en masse as in the Murray's Flickr feed, makes tangible the loss of a range of distinct visual cultures as the urban landscape changes and individual shops are replaced by national brands. The attraction of these storefronts is in their individuality and diversity, their sharp sense of the local.

These shops are a visual autobiography of place and representative of a communal history that made the street welcoming and surprising. In aggregate, they created a neighborhood, and though highly individual in their design, they forged a collective identity based on individual difference. Their appeal is of an unconscious, uncalculated, and unintended branding of function that contrasts so starkly with the impersonal blandness of the banks and chain stores that have replaced them.

The Architecture of Remoteness

In contrast to the personalized streetscape, recent high-rise high-profile trophy buildings minimize the importance of street level. Designed to be seen at a distance, these buildings are totems that brand the skyline. Ignoring pedestrians and the street-level experience, they emphasize the broad vista over the close-up. Almost invisible on approach, they are monumental and iconic from afar. Set back from the context of neighborhood and pedestrian traffic, they loom large on the skyline as icons. Intentionally remote, they proclaim their aloofness from the fabric of the city even as they use it as a signifier of their importance. This disjunction of near and far seems representative of an attitude toward urban space: creating a presence that situates within the city while remaining apart from it.

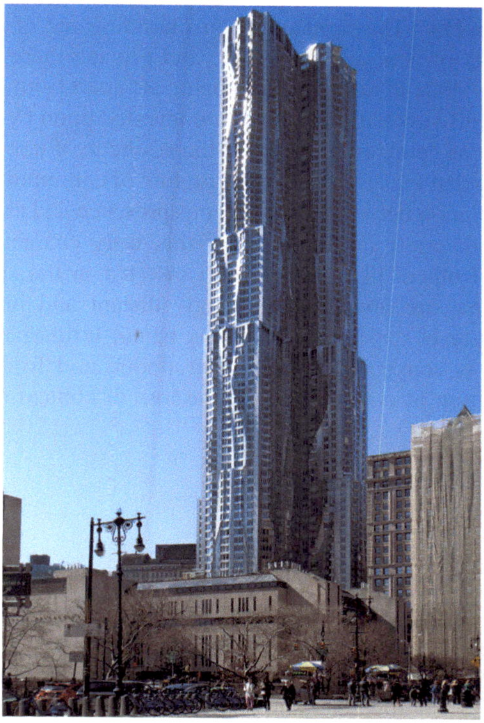

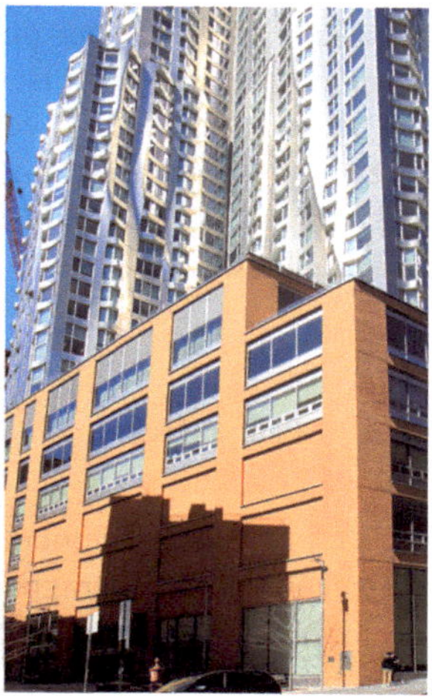

Views of 8 Spruce Street (Photographs by Annette Weintraub, 2014)

An example of the rejection of the street for the vista is the new Frank Gehry building in lower Manhattan at 8 Spruce Street. Billed as the second tallest residential tower in the Western Hemisphere, it is a 76-story undulating silvery column that sits on a squat five-story brick base. There is no visual transition or structural articulation from the spectacular tower to the pedestrian base. From afar, the building is sculpture; at ground level, a big box. On the building website, there are no images of the building at street level. Promotional photographs (mostly aerial) emphasize the tower's sculptural quality by focusing on its crenellated and gleaming surface or show it in an establishing shot of the panoramic Manhattan skyline.

As critics have noted, the upper floors are delicate, the silvery skin elegant and beautifully modulated in the constantly changing light. The tower is visible at great distance from uptown or across the river, seeming to rise from a cluster of lower buildings. At street level the visual effect is altogether different: the base is a generic, five-story brick block, clearly and intentionally unrelated to the tower that rises above. It houses a public school, space for New York Downtown Hospital, and retail space. The base and tower are clearly demarcated: public and private zones clearly marked in bluntly economic terms: bespoke design for the tower, generic box for the public. Nicolai Ouroussoff, in his largely favorable review in the *New York Times* (Ouroussoff 2008), noted the "disparate levels of creative energy invested in the building's public and private spheres." The signature of this building and of others of similar pedigree is this duality of near and far, public and private; there is the contradiction of being in the city while at the same time being set apart from it.

This reversal of the hierarchy of street and vista was commented upon by Augé who saw the cinematic long shot as being emblematic of the aesthetic of non-place and connected it to the distancing that makes us forget the rupture of community and function (Augé 2008). This oscillation between simultaneous presence and absence reflects ambivalence about the diversity and complexity that truly characterizes urban space. Gehry's tower and others like it have been inserted in the skyline like a set of precious jewels that are always tantalizingly distant and in stark contrast to an immediate street-level experience which is banal, utilitarian, and featureless. Cinematic framing is consigned to the upper floors, and from that elevated perspective, the dislocation of community and the loss of context on the ground seem barely of interest.

The Past as Pastiche

Another kind of estrangement—of past and present—occurs when bits of the historic past are saved as fragments removed from their original context. A strategy for development that appeases landmark and zoning boards has been to demolish a historical structure, keeping a decorative element as an "amenity." In this way historical preservation is flattened and transformed into cosmetic decoration. What was once a fully functional structure is turned into framing device to validate what is usually mediocre construction. This is a compromise that disrespects the past and shows a remarkable lack of imagination in the present.

The demolition of St. Ann's church, built in 1847, to construct an NYU dormitory in the East Village of New York City is such an example. This memorable building was largely demolished to make way for an uninspired and cheap-looking tower. All that remains of a lovely Victorian Gothic stone church is a slice of the facade, backed up against a generic brick structure that has accurately been described as "Soviet inspired," and is unrelated in scale, materials, or geometry to its predecessor or to the street. The *AIA Guide to New York City* described the result: "Preserve or demolish? Or preserve and demolish? Here St. Ann's Shrine is gone, excepting bell tower and façade, recast as a folly behind which lurks yet another dorm for NYU .... The exercise if futile: no connection is made, or even attempted, between the old church and the new 26-story hulk, save some pavers laid between. The effect is of a majestic elk, shot and stuffed." (White et al. 2012)

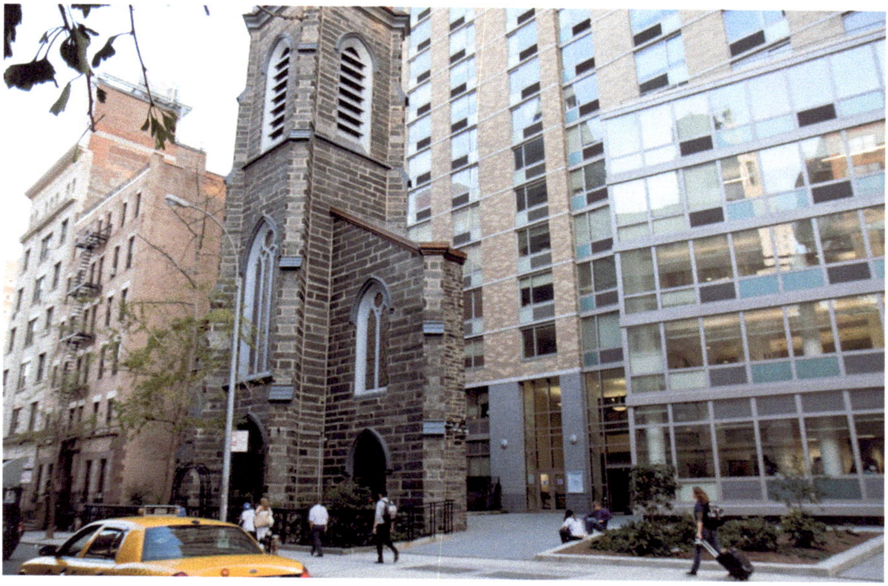

St. Ann's church facade and NYU dorm (Photograph by Annette Weintraub, 2010)

There is an aspect of theme parking in this kind of historical mash-up, which Augé has called "history as entertainment." The amputated fragment is so divorced from original function and context it appears out of place and doubly inauthentic, a failed "preservation" strategy that turns a fragment of a once-distinguished building into an ornamental quotation that makes its prosaic replacement appear even more impoverished. What was once a fully dimensional and inhabited structure has become flattened into a single surface, a stage prop with no function, whose purpose is to point to the generic dorm that looms behind. This type of amputation and grafting is now seen throughout the city, the past as pastiche.

The Romance of Ruins

As the contemporary landscape is stripped of its history, there is a temptation to romanticize the past and exoticize decay. This is a way of confronting absence and loss—the recent interest in urban exploring and documenting architectural decline in our own surroundings serves to reconnect us with a lost past and permits us to experience an authenticity of place that is lacking in our more conventional and predictable urban environment.

Artists' interest in abandoned and crumbling grandeur as subject is typified by Piranesi's eighteenth-century etchings of Roman ruins, although certainly the fascination with the deteriorating remnants of earlier civilizations predates his work. Perhaps there is a universal inclination to voyeuristically observe the remains of a vanished culture or period and speculate on the reasons for its demise while thoroughly enjoying the spectacle. Fascination with the passage of time and the idea of objects as memento mori was a theme of art before seventeenth-century Dutch still life painting established this subject matter as a genre. In the nineteenth century, partially in reaction to the urbanization and the despoiling effects of the Industrial Revolution, Romantic artists invoked the sublime to heroicize the forces of nature, portraying violent extremes of weather and using images of architectural ruin to meditate on the passage of time and evoke nostalgia for a lost pastoral landscape. Romantic images were an antidote to and escape from the grim economic and social realities of industrialization.

The phenomenon of "urban exploration"—illegal forays into abandoned power stations, train tunnels, and sewers—has resulted in greater public awareness of the diminished scale and ambition of our own public works as compared to the more heroic structures of the nineteenth and early to mid-twentieth centuries. In contrast to Piranesi's time, when artists looked back at antiquity to speculate upon the waning of earlier cultures, contemporary interest in ruins is focused on our own *recent* decline and documents the deterioration of our industrial parks, factories, transportation infrastructure, and other public works. Transgressive acts of exploration of inaccessible and dangerous places, when photographed and shared, have led to a wider fascination with the places themselves and created the genre ironically called "ruin porn"—a term that succinctly expresses the complexity of a guilty pleasure. While photographs were initially only evidence of the act, a kind of "I was there" gesture, they have now become the rationale for expedition in themselves. It is an odd by-product of these incursions that a subculture of photographers, video artists, and graffiti and street artists has become the de facto and unlikely public historians and chroniclers of urban loss.

Their images document and valorize the architectural rejects and neglected spaces that are unseen and unappreciated by the general public. One of the original websites devoted to urban exploration was Infiltration.org (*the zine about going places you're not supposed to go*) (http://www.infiltration.org and http://www.infiltration.org/history-timeline.html. Accessed 29 March 2014) which was started by Jeff Chapman, the pseudonymous Ninjalicious (Ninjalicious 2005). His site was a remarkable catalog of exploration locations worldwide, with links to extensive photographic documentation of drains and catacombs, abandoned hotels and hospitals, utility tunnels, and factories; it is still maintained but sadly has not been expanded or dead links purged since Ninjalicious died in 2005. Also, international in focus and extensive in its visual and written critique of current design practice is Mark Minkjan's blog, *The Poetry of Decay—Failed Architecture*, which has become a platform for writing and images that critically examine global urban dystopia (Minkjan 2014). *Failed Architecture* reminds us, in posts featuring numerous contemporary and historical examples of urban planning dysfunction, that dystopia cannot be safely distanced into the past or future but is a component of our own space and time.

Isolated islands of the past, the residual monumental remnants of nineteenth- and early twentieth-century industrial architecture stand in stark contrast to the more conventional contemporary landscape. These plants were built on an imposing scale; they were visual confirmation of a dominant and vibrant industrial machine. Designed to convey optimism, boundless confidence, and the ascendency of capitalistic culture, their design in scale and materials harks back to the public works of ancient times: they were functional and unadorned, yet magnificent as works of engineered space.

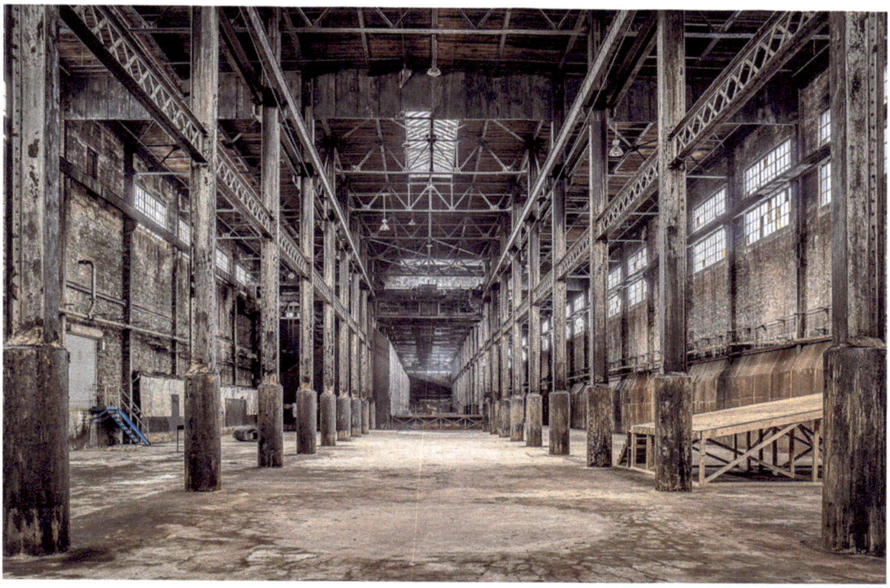

Interior, Domino Sugar Refinery, Brooklyn (Photograph by Paul Raphaelson, 2014)

Photographer Paul Raphaelson's images of the deserted Domino Sugar Refinery (http://www.paulraphaelson.com/domino/. Accessed 14 April 2014) in Brooklyn convey this duality of former grandeur and current desolation. At its peak, Domino refined more than half the sugar used in the United States. Now, although partially landmarked, the site has been purchased by a developer. Plans for development include a large residential tower described by the Municipal Art Society as likely to overwhelm the historic parts of the refinery that will escape outright demolition. Raphaelson's interior images capture both the immense halls and intricate mechanical workings of the plant. The austere open spaces read as an elevation of the utilitarian, while the complex machinery with patinated surfaces of corroded metal, flaking rust, and peeling paint reminds us of the temporality of even the most ambitious undertakings. The Domino photographs have the linearity of drawings and convey the intricacy of complex mechanicals that might appear simply sculptural to us now, but which would have been entirely comprehensible and ordinary to the workers that formerly inhabited the space. Raphaelson's interior photographs emphasize materiality of surface and intricacy of structure in a way that conveys the epic drama of the plant's intended purpose and underlines the poignancy of its current diminished state.

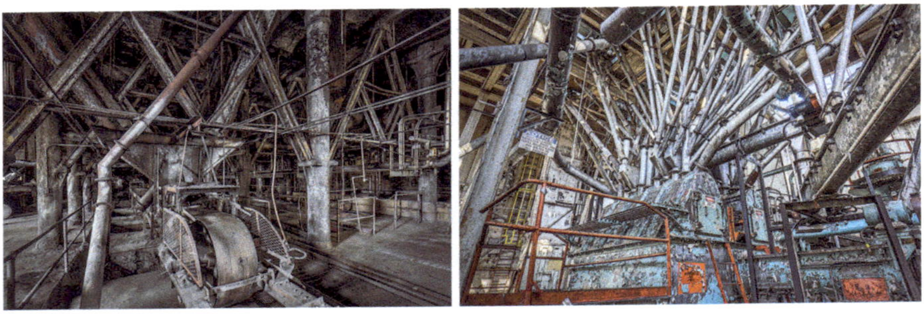

Interiors, Domino Sugar Refinery, Brooklyn (Photographs by Paul Raphaelson, 2014)

Images that aestheticize decay and destruction and record the ravages of time create a visual language that is in dramatic opposition to the polished visual language of commercial architectural photography and computer rendering. This attraction to ruin was described as a "mediating power, between the old and the new, and between nature and culture" by the nineteenth-century art critic John Ruskin. Images of destruction provide a counter narrative to the seduction of real estate development; idealized utopia and its mirror image of urban devastation both contain similar elements of nostalgia, longing, and anxiety about time. Writing about how architecture transmits the illusions of the current dominant ideology, Augé describes a utopian dimension that might be attainable, but which is always slightly out of reach and that "reproduces in reverse the relation with time expressed by the spectacle of ruins." Time thus expressed through the ruins of the past or the barely glimpsed utopias of the future is a way to "grasp in the present a lack that structures the present moment by orienting towards the past or the future" (Augé 2008).

Rendering as Reality

Architectural renderings and the depictions of urban space in video games seem to represent opposed images of the city, utopian vs. dystopian and reality based vs. fictive, yet there are some interesting points of resemblance. Both representations of the urban spaces are modeled, texture mapped, lit, and rendered and use a common set of imaging conventions which have been institutionalized as familiar visual tropes. Yet while the conventions of imaging are similar, they are used with very different intent.

Video games use location in a filmic way, employing a shorthand mix of easily read symbolic visuals including graffiti, yellow taxis, dimly lit industrial streets, and glowing signage. They create a maximally recognizable backdrop with economically few polygons. The images they use as establishing background are so familiar that they are immediately recognizable even as they zoom by at high speed. The strong sense of urban atmosphere that results from these highly symbolic environments moves the narrative forward as much as the game action itself. This is the city as stage set: created in visual shorthand and designed for ritualized action.

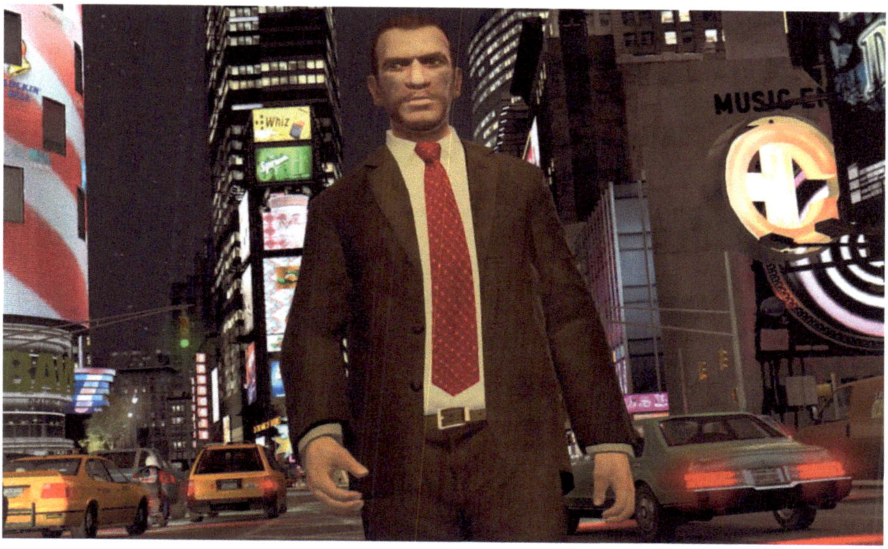

Grand Theft Auto IV's Star Junction. The game's *Liberty City* parodies New York City (Image courtesy of Rockstar Games)

The persona of New York locations in popular films such as *Taxi Driver* or *Escape from New York* is clearly related to the New York of *Grand Theft Auto IV* and *Crysis 2* video games. In a BldgBlog.com post titled *The Inevitability of Prophecy Among Models of New York*, author Jim Rossignol commented that New York doesn't provide merely an interesting backdrop; it creates a strong connection to character. He says, "New York City has become gaming's ideal and idealized urban environment, and it has done so by becoming re-fictionalized and reimagined. The finest example of a city yet given to gaming, that of *Grand Theft Auto IV*, isn't really New York at all, and yet it is more like New York than ever before. It's the essence of New York — a distillation that is only possible thanks to the unique way in which games are able to make the figurative and the abstract resonate with us even more profoundly than the infinite detail of the everyday" (Rossignol 2010). Rossignol correctly locates the sense of presence of the city in GTA IV as coming from its connection to an entire range of films of iconic New York location. Almost every location can be matched with a film memory; this enriches and deepens the play.

While the city in *Grand Theft Auto IV* has been art directed with a painterly and Edward Hopper-like solidity and clarity and depicts a somewhat contemporary New York, *Crysis 2* remakes New York into a future ravaged city that is evanescent and crumbling, poetically evocative of Piranesi. It takes an atmospheric and impressionistic approach, with richly detailed textures and recognizable slices of New York streets in various stages of destruction depicted with an extreme level of detail. There is a deliberate separation of foreground and background with ravaged city landmarks and chunks of skyline rendered in extreme perspective and placed outside of the field of action or looming on the horizon. The use of haze and tinted atmosphere is stylistically connected to artists like Turner or Caspar David Friedrich in the use of extremes of light and shadow to create a feeling of moody and melancholy despair. This game presents a dystopic vision of the city; its visuals aestheticize devastation and thus share some visual characteristics of ruin porn photography.

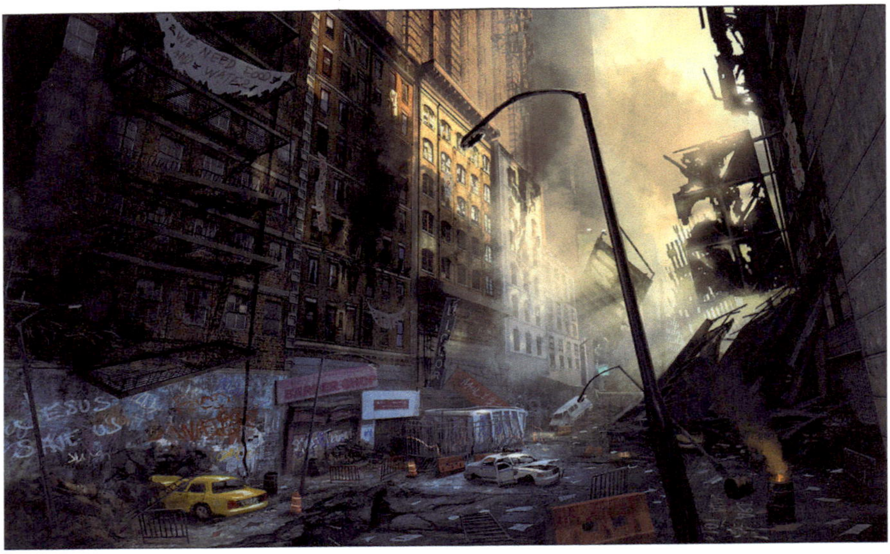

Artwork from *Crysis 2*, 2011 (Image courtesy of Crytek)

The idea that the archetypal idea of the city is more itself in distilled abstraction than in reality is present in many depictions of New York and, I would add, survives as a sustaining mental image despite its visual and cultural transformation of the actual city landscape. Video games represent the city in extremis, as an alternate universe of uncertainty, anxiety, and apprehension located in an approaching dark future of social, political, and ecological deterioration.

It is instructive to contrast how the city is presented in architectural renderings as compared to video games. Architectural rendering is capitalism's "good" twin to the "evil" twin of video games. Where the video game might emphasize claustrophobia and grittiness, the architectural rendering emphasizes expansive and spectacular vistas and unexpected (and improbable) pastoral elements. Well ordered and purposely sterile compared to the visual chaos and deliberate disorder of games, these representations are utopian and present a vision of the city promoting an aesthetic of gleaming surfaces, rational perspectives, and clearly framed viewpoints.

Architect's drawings are themselves hopeful, incorporating wishful thinking and imaginative interpellation along with practical visual information. Their drawings serve not only as a way of viewing a building in situ before construction but as a powerful sales tool that blends the likely and aspirational, emphasizing desire and accessibility. The seductive power of rendering influences architects (consciously or not) to design structures that render well, creating a feedback loop of polished surfaces and burnished materials that promote a particular kind of urban lifestyle. Flipboard's *SkyscraperCity* (Flipboard.com 2014) is a compendium of this rendered aesthetic. Going through the site, it's not always immediately evident if one is looking at a photograph or a rendering since a seamless set of conventions of surface deriving from the world of games and 3D modeling/rendering have become ubiquitous in photography, architecture, and filmic depictions of urban space. This is a visual palette that emphasizes reflected light, strong diagonals, and bold geometries. While these conventions existed in the architectural drawings of other eras, the current tools facilitate hyperreal effects that read as reality surgically enhanced.

Architectural renderings lack the lived-in chaos and human (or alien) presence of video games and also lack their emotional dimension; they emphasize a world of surface perfection that is flawless, unspoiled, and spacious and which presents a marketable model of city life while avoiding any of the more complex characteristics of daily life or reminders of how public space is actually used. The anarchy and confusion of the city and its messiness are contained and sanitized, and spatial representation is condensed to a set of repeatable heroic conventions: renderings that use reflection, exaggerated lighting effects, and intricate pattern to produce blindingly sleek surfaces that subvert solidity; space is downsized to surface.

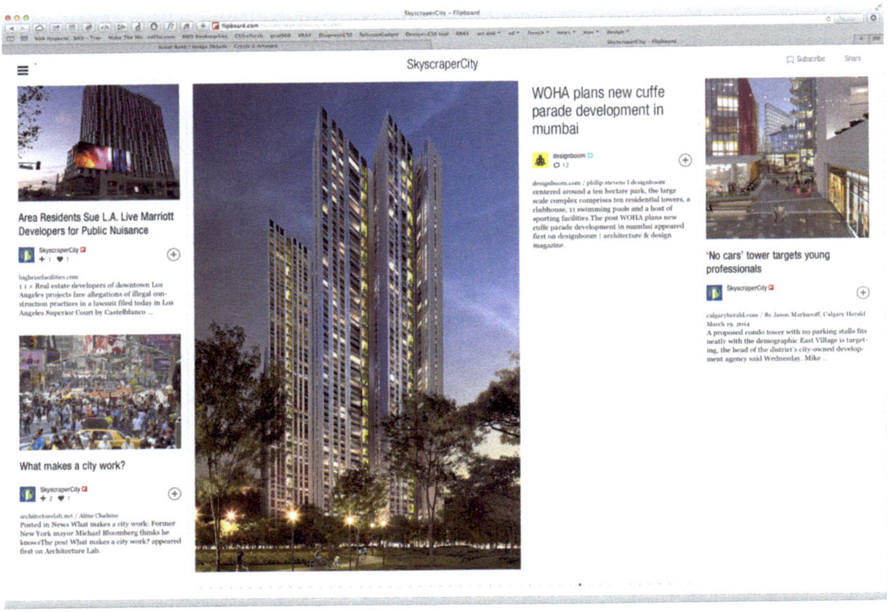

Screen grab from Flipboard's *Skyscraper City*, ND (Image courtesy of Flipboard.com)

Beauty in Non-place

Marc Augé eloquently wrote about the "dilemma of artists doomed to seek beauty in non-place." This search for beauty in non-place is present in the work of many contemporary photographers who record the banal details of shopping center parking lots, bland office interiors, and featureless suburban streets. These once-overlooked subjects have become imagistic subgenres that record and reframe absence. In contrast to photographers depicting the low-key and banal genres of suburban emptiness, Michael Wolf captures non-place by making images of overload and almost unimaginable density. Wolf shoots vast yet completely claustrophobic panoramas of global megacities and makes gigantic prints that use the repetition of window grids and balconies of residential towers in massive urban developments to convey an almost unimaginable density and scale. In his photographic series, *Architecture of Density* and *Transparent City* (Wolf 2014), Wolf creates large-scale images of Hong Kong high-rise towers and uses the impenetrable and abstractly patterned colored grids of these immense and anonymous expanses of buildings to reflect the unimaginable but invisible concentration of humanity. These skyscraper images, shot frontally without land or sky to provide scale, reinforce the sense of divorce from street life and sense of place. In some of Wolf's work, small signs of human presence in the form of laundry hanging from a balcony or even a glimpse of someone looking out a window serve to underline the immense scale and depersonalization of these images. Wolf's images are an abstraction of an extremely concentrated (visually and demographically) urban environment; this abstraction reduces complexity to optical pattern, while paradoxically the immensity of scale (of the images as well as of the subject) renders it ultimately incomprehensible. Wolf's work is an interesting example of how non-place can be represented not as absence but as excess.

The reshaping of the city with attendant changes in social relations and visual codes offers an opportunity for rethinking representations of urban space and exploring emergent narratives. Perhaps an urban landscape in which volume is flattened to skin is more readily understood through pattern and surface, either in the extreme close-up of fragmentary slices of reality or zoomed out into the global perspective of Google Maps.

The Granular Landscape

While restrictions on what can now be photographed in post-2001 public space can be exasperating, much as they are uninformed, it is the increasing blandness and sameness of once richly complex spaces has led me to look at urban space in a different way. As urban landscape becomes more visually indeterminate, the pictorial elements that read as compelling to me are not the visuals so emblematic of the dominant environment of non-place but the small material details in an isolated texture or unusual surface which convey a granular sense of the complexity of the whole. It is the middle distance that reads as banal: the close-up and extreme vista both contain information and reveal underlying structures that can be extrapolated into a visual mapping of the city.

To explore this notion of city as surface, I assembled a collection of images of generic urban visual elements and created a series of animations that connected disparate image fragments using a continually shifting and panning horizontal movement. The movement in this animation corresponds to the experience of optically scanning a space and was intended to place the viewer in the landscape and direct the gaze, somewhat like the mechanism of an establishing shot in a film. This project, *Slide Stories* (Weintraub 2012), is a web-based work that intentionally uses the most insignificant and ordinary detail of the urban landscape to elicit underlying narratives of urban space. These small image fragments, meaningless in isolation, in aggregate combine to obliquely construct sense of place.

Slide Stories is composed of seven moving image sequences, each of which explores a different visual and thematic construction of urban space. It is a poetic exploration of the social and political construction of urban space as well as an investigation of the visual language of architecture. Each scene of *Slide Stories* consists of an extreme horizontal panorama, one or more overlay images, a text, and a track of ambient sound. As the animation loops, the image is reframed, and the path and juxtaposition of text and image change so that the viewer experiences a slightly different conjunction of elements, and the sequencing of images and

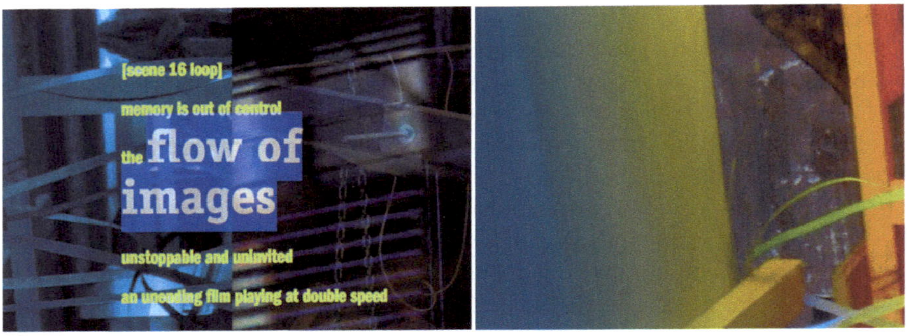

Still from *Slide Stories. Blink, Scene 16: loop,* Annette Weintraub, 2012

text create a narrative. The animations cannot be paused or rewound; the panning movement is deliberately slow and out of phase so that each reiteration reveals a slightly different piece of the whole.

The scenes of *Slide Stories* vary in visual theme and underlying narrative. *Brutalism* explores issues of materiality and space, *The Grid* explores the geometries of familiar places and the patterning of urban space, *Industrial Strength* is a meditation on abandoned spaces, *Nocturnal* is concerned with the transformative effects of artificial light, *Bricks* explores visual repetition and the warmth of organic materials, *Absence* locates memory in the perception of negative space, and *Blink* is about transience and impermanence.

Slide Stories pieces together small granular fragments of city surface to build a sense impression of a complex visual landscape. Using unremarkable and devalued commonplace material detail, *Slide Stories* reconstitutes urban space through surface incident.

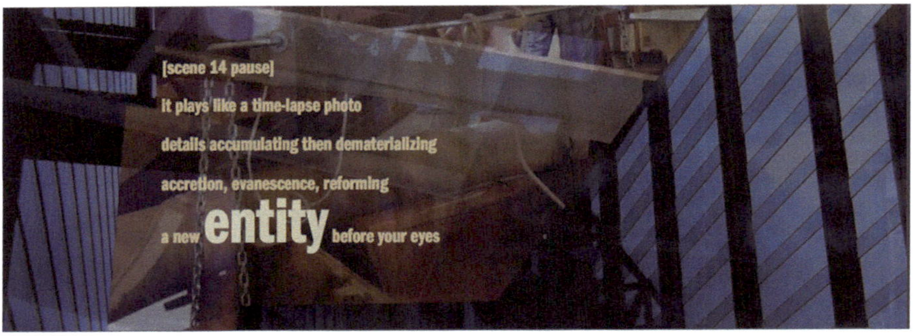

Still from *Slide Stories. Blink, Scene 14: pause*, Annette Weintraub, 2012

Patterns of Organization

Images taken from satellite cameras widen our attention to a global perspective; it is a flattened view of the earth from space and one in which architectural elements become graphic and reductive. The long cinematic shot that Augé described as the way the dominant aesthetic frames megacities is pulled back to frame a global view. This shift in perspective from ground to aerial perspective emphasizes the commonalities of structures; most buildings are mere rectangles when seen from space. By changing perspective so that an image reads not in frontal perspective but in a bird's eye view, Google Maps turns dimensional space into continuous surface—elements such as height and volume are undifferentiated, while color and pattern become critical. The ubiquitous availability of highly detailed satellite maps enables us to look at familiar spaces and structures in a new way: to see overall patterns and extract underlying meanings.

Every Outdoor Basketball Court in Manhattan (2011) is a digital print by Jenny Odell from her series *Satellite Collections* (Odell 2014). These prints extract a set of images of a set of architectural structures of similar function from Google Satellite View and aggregate them in a collection that articulates their incremental similarities and differences. Spliced out of their original location from around the city (or the globe) and removed from original context, they can be read either as patterned abstractions or as icons that convey the very essence of their function. Odell describes taking a view that is not a human one and one that we were never meant to see and finding our common humanity in the repetition and variation of these banal structures.

As Odell states, "The alienation provided by the satellite perspective reveals the things we take for granted to be strange, even absurd. Banal structures and locations can appear fantastical and newly intricate. Directing curiosity toward our own inimitably human landscape, we may find that those things that are most recognizably human are also the most bizarre, the most unlikely, the most fragile." Her "collections," which do not exclusively deal with urban space, point toward a way of imaging landscape that reveals the underlying workings of complex systems. Seen in a coloristic way, Odell's images are painterly geometries, abstractions of landscape archetypes that play with small contrasts of shape and color, but examined as aggregates of specific architectural functions, they reveal much about culture and industry and economy even in the absence of any inhabitants.

Odell's excised images have a toy-like cheerfulness that contradicts the actual function of some of her subjects (water treatment plants and nuclear cooling towers). Detached from their context, the extracted images are presented in a large group, in a juxtaposition of seemingly identical structures whose marginal differences collectively reveal something of their function even as they remain mystifyingly bland and indistinguishable, and charm us with their variety and eccentricity.

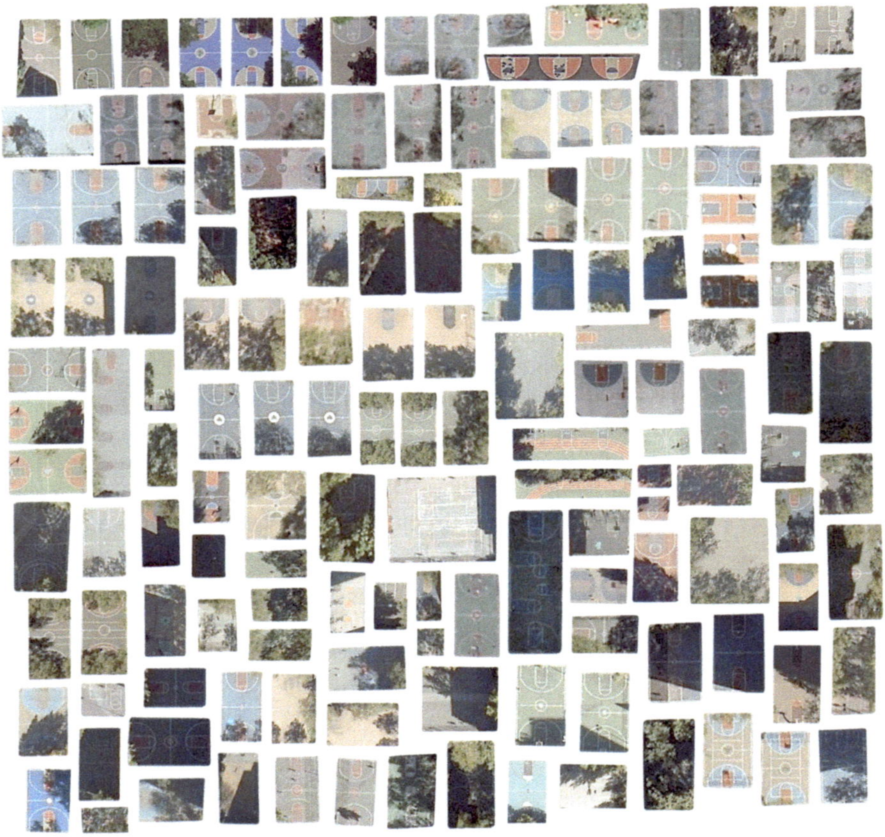

Every Outdoor Basketball Court in Manhattan, Satellite Collections, Jenny Odell, 2011 (Image courtesy of the artist)

Animating Surface

Dynamic movement is as much an essential characteristic of urban space as is the static built environment. The flow of pedestrians and traffic animates urban space as it inhabits it. There are specific and definable urban rhythms: the unique and often frenetic streaming of crowds and traffic on city streets or the flocking and repetitive patterns created by pedestrians in a plaza or open space. These have been recorded as establishing shot in many popular films; they take center stage in Rudy Burckhardt's 1940 film, *The Pursuit of Happiness*, in which he focuses his camera on the feet of pedestrians and obliquely captures the rhythm of the city. The theorist Henri Lefebvre wrote about the importance of rhythm and poetically described the repetitive cycles of movement and patterns of behavior on a busy street in his book *Rhythmanalysis*. "He who walks down the street, over there, is immersed in the

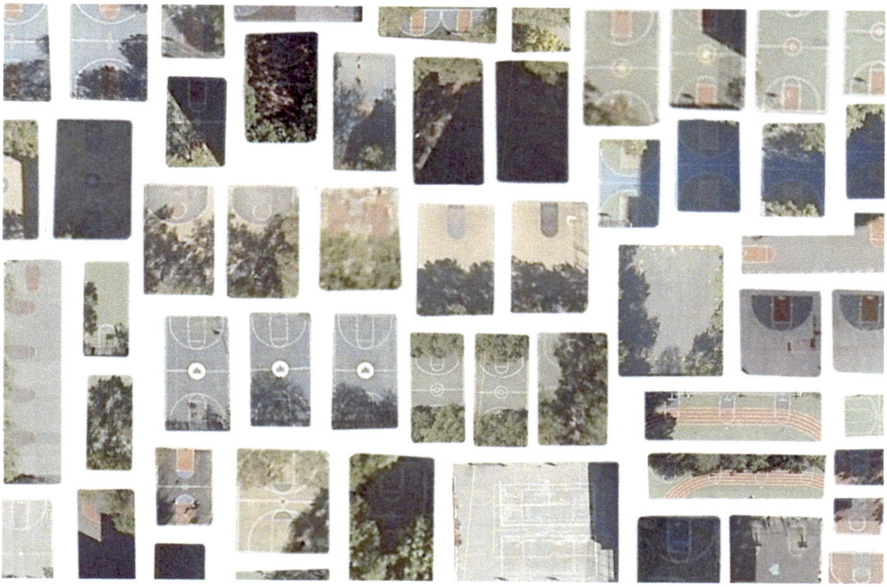

Detail, *Every Outdoor Basketball Court in Manhattan*, Jenny Odell 2011

multiplicity of noises, murmurs, rhythms, (including those of the body, but does he pay attention, except at the moment of crossing the street, when he has to calculate roughly the number of his steps?)... On red, cars at a standstill, the pedestrians cross, feeble murmurings, footsteps, confused voices." (Lefebvre 2004)

It is by capturing the flow and dynamism obliquely, as in Burckhardt's film, that underlying patterns of a particular space can be extracted and examined. Noah Klersfeld's video work, *Times Square (NYPD)* (Klersfeld 2014), breaks up the action of several different video clips into the tessellations of a rotated grid and presents a slice of the city in fragmentary glimpses. He uses a banal architectural element (a section of chain-link fence) as a tiling framework for these fragments. The moving image elements are drawn from one- or multichannel videos, small slices of which are placed in selected negative spaces of the chain link. The audio and video elements activate the tiled surface; the effect is of a constant play of solid and empty, static and dynamic, extended in time.

Klersfeld describes his work as "temporally redistributing the activities taking place on a snowy afternoon in New York City. Focusing on the city's epicenter of tourism and media saturation, the video unifies the vast array of activities by confining everything (people, snow, cars, trucks, buses, digital signage) to a series of identical "viewports". The result is a spatially ambiguous depiction of the urban environment, weaving together multiple timeframes into a singular form."

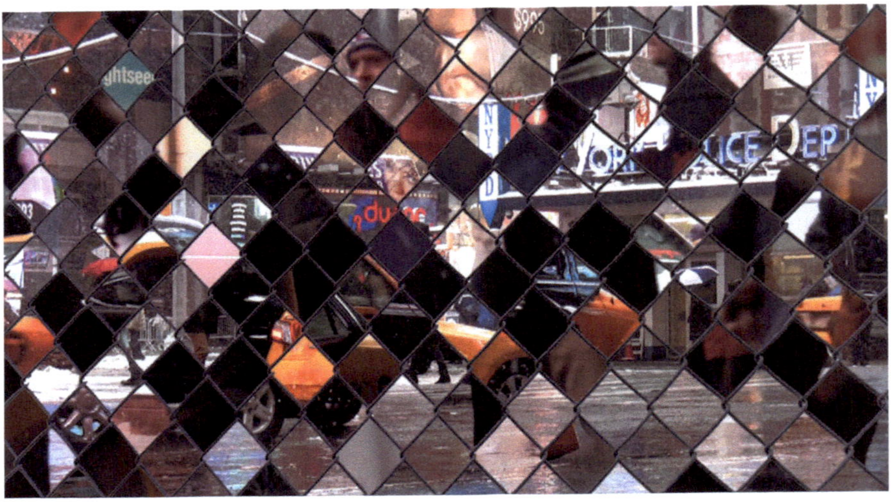

Still from *Times Square (NYPD)*. Video, 5 min. Noah (Klersfeld 2014) (Image courtesy of the artist)

It is the transformation to surface that makes Klersfeld's video resonate a sense of place while showing that place only intermittently and fragmentarily. The framework of spaces in the chain link that he refers to as "viewports," while lying on a single plane, contains elements that are situated at different points in what seems to be shallow space and in different magnifications. The interaction of still image and video seen through this gridded filter captures the tempo and incident of street life, but the elements are so disconnected they provide an indirect and incidental portrait of street activity; they cannot be fully visualized. By inserting slivers of video into an essentially static field, Klersfeld transforms the haphazard movement of the street to pure surface, encoding it so it becomes mysterious—unrecognizable yet familiar—and plays with recognition that hovers barely above the threshold of perception.

Reimaging Indeterminate Place

Perhaps it is not beauty that artists seek in Augé's indeterminate place, but a lens through which to reimagine place in a devisualized landscape. In an urban environment that has been stripped of history and cultural identifiers and that proudly embodies the stylistic conventions of real estate marketing and 3D rendering, sense of place is not only indeterminate but is provisional and transient. The transitional spaces of circulation, consumption, and communication that Augé understood as being emblematic of non-place are no longer merely peripheral or connective; they are the substance of our space, as if the once interstitial spaces had absorbed into themselves the islands of matter that they connected. Artists responding to the phenomena of non-place need not subscribe to the aesthetics of indeterminacy that simply documents and concedes the blankness and alienation. Fragments of visual incident, variation in material surfaces, and analysis and extraction of underlying pattern of the landscape both visual and social can be used to reformulate a vision of urban space that explores its essential dynamism and acknowledges its constant evolution and reinvention, while avoiding the tropes of utopian longing and dystopian dread.

References

Augé M (2008) Non-places: an introduction to supermodernity, 2nd edn. Verso, London

Flipboard.com (2014) Skyscraper city. ND. https://flipboard.com/section/skyscrapercity-bqxHt8. Accessed 8 Mar 2014

Klersfeld N (2014) Times square (NYPD). http://www.noahklersfeld.com/stills.html. Accessed 8 Mar

Lefebvre H (2004) Rhythmanalysis. Continuum, London/New York, pp 28–29

Minkjan M (2014) Failed architecture. http://failedarchitecture.com/the-poetry-of-decay/. Accessed 28 Mar 2014

Mitchell WJ (2003) Me++: the cyborg self and the networked city. MIT Press, Cambridge, MA, pp 104–109

Moss J (2013) Jeremiah's master list of vanishing New York. http://vanishingnewyork.blogspot.com/2013/12/master-list-2001-2013.html. Accessed 28 Mar 2014

Moss J (2014) Jeremiah's Vanishing New York a.k.a. The Book of Lamentations: a bitterly nostalgic look at a city in the process of going extinct. http://vanishingnewyork.blogspot.com/. Accessed 28 Mar 2014

Murray J, Murray KL (2009) Store front-the disappearing face of New York. Ginko Press, Incorporated, Berkeley

Ninjalicious (2005) Access all areas: a user's guide to the art of urban exploration. Infilpress, Toronto

Odell J (2014) http://jennyodell.com. Accessed 8 Mar 2014

Ouroussoff N (2008) Looking skyward in lower Manhattan. New York Times. May 31. p 31. http://www.nytimes.com/2008/05/31/arts/design/31beek.html. Accessed 28 Mar 2014

Rossignol J (2010) Bldgblog.com, The inevitability of prophecy among models of New York. http://bldgblog.blogspot.com/2010/11/inevitability-of-prophecy-among-model.html. Accessed 28 Mar 2014

Weintraub A (2012) Slide stories. http://www.annetteweintraub.com/slidestories/

White N, Willensky E, Leadon F (2012) AIA guide to New York city, 5th edn. New York: Oxford University Press. V117

Wolf M (2014) Architecture of density. http://photomichaelwolf.com/. Accessed 8 Mar 2014

Chapter 9
Design as Topology: U-City

Ulrik Ekman

Abstract This chapter discusses the issue of approaching the design of the ubiquitous city as a matter of topology. The general context here is the design of contemporary global urbanity in the form of u-cities, smart cities, or intelligent cities emerging with the second phase of network societies that increasingly develop mixed reality environments with context-aware out-of-the-box computing as well as the sociocultural and experiential horizon of a virtually and physically mobile citizenry. Design here must meet an ongoing and exceedingly complex interactivity among environmental, technical, social, and personal multiplicities of urban nodes on the move. This chapter focuses on the design of a busy traffic intersection in the South Korean u-city Songdo. Hence, the discussion whether and how Songdo may be approached via design as topology primarily considers the situation, event, and experience in which multiplicities of environmental, technical, and human interactants tend toward gathering and dispersing in a single mixed reality street transport scenario. The need for "intelligent" ad hoc connection, routing, and disconnection of multitudes of humans, technical systems, and environmental entities makes this scenario one of the more crucial design test beds. This article offers a critically comparative discussion of a variety of ontological and epistemological approaches to design as topology, including realist, nominalist, and constructivist efforts in both cultural theory and technical studies. It is demonstrated that design as topology offers significant resources with respect to traits of the u-city such as continuous material and energetic flows, its environmental landscaping of mixed realities, its ongoing virtual and physical infrastructural developments, the folding and unfolding of its architecture, its nodal dynamics, and the relational mobilities at stake. However, this chapter also questions design as topology as an approach when it comes to decisive aspects of urban finitude: citizens' lived experiences, concretization of urban information and communication technology ICT, and sustainability.

U. Ekman (✉)
Department of Arts and Cultural Studies, University of Copenhagen,
Karen Blixensvej 1, 2300 København S, Denmark
e-mail: ekman@hum.ku.dk

F.T. Marchese (eds.), *Media Art and the Urban Environment*, Future City 5,
DOI 10.1007/978-3-319-15153-3_9

Songdo in Transport: Modern Design?

Considering what we today call the Incheon area in northwestern South Korea has been transformed as a natural physical environment into one for human settlement and inhabitation at least since the New Stone Age, we are not really in doubt that this is a rather thoroughly designed phenomenon. This assumption is immediately reaffirmed when we note that almost all traces of its modest modern urbanization as a city housing, around 5,000 people toward the end of the nineteenth century, are now erased, first by the extensive transformations due to the opening of its sea port and then by the considerable changes ensuing from its close neighborhood with Seoul as the national capital.

We must approach Incheon as a designed modern urban area, specifically, as a sprawling suburban development in which almost three million people now live, making it the fourth largest in the world in terms of population. The impression of a purposeful and progressive design activity is perhaps only reinforced by the recognition that the opening of the port led to the modernization of South Korea as a center of the twentieth century industrialization. This is a line of transformative development further pursued in the new millennium by strengthening the three regions of Incheon (Songdo, Cheongna, and Yeongjong; covering a total of nearly 52,000 acres) as a major, tightly structured, and highly functional transportation hub for South Korea and beyond.

Today, as a territory and a *polis* housing a community of millions, Incheon thus appears not least as an ambitious modern urban design plan aimed at having a certain sprawl solve a problem of either complex global, regional, national, and local mobility or ever so many vectors of flow. In terms of human transport, this involves the building, maintenance, operation, and development of Incheon International Airport as a regional air hub and South Korea's primary international airport. In addition, Incheon's seaport now appears as the second largest in the nation and maintains, among other things, primary connections to China. Incheon Bus Terminal holds open numerous inner city lines and also ones linking the city to all other parts of the country. Furthermore, Incheon's well-developed subway system has also become tightly linked with the subway of Seoul, and an airport express train permits transfer between the capital and Incheon Airport in less than 45 min. Last but not least, all parts of Incheon remain interlinked with the rest of the nation via the considerable network of streets, roads, and highways for automobile transport.

Observing all of this, one would largely be adhering to paradigmatic notions of the planning, design, and architecture of a modern urban sprawl and, along with this, some equally paradigmatic notions of modern design. In other words, one recognizes a diagram for Incheon that is scaled, is well delimited, and displays a measurable and progressive development for material, technical, and social or individual housing, leisure, work, and transport. It would not be too difficult to generate for this urban sprawl a two-dimensional geometric symbolic representation using graphs – something like a map, a blueprint, or a sketch. Wemight debate the elegance, clarity, simplicity, and validity of its pattern of entities, its relations, and its suggestive lines of movement, but we would be agreeing on the feasibility and usefulness of making this kind of urban diagram.

Likewise, it would seem obviously relevant to keep with notions of Incheon as a recognizably designed artifact, a planned and purposeful urban functionality and structure. This would bespeak design qua a conception for the completed form of an object or a project, something that can be modeled concretely and for which a limited set of instructions or rules exists. We understand Incheon to have involved activities of planning and devising, as an outcome of such processes and as something whose everyday sociocultural quality of life in a community derives in large part from this (Julier 2008, 4). Here Incheon would be approachable as a design qua a numbered set of urban elements whose order and organizational arrangement we can perceive, mostly without too much trouble (Raizman 2010, 11). In this, one would in all likelihood be operating in accordance with some of the broad but still finite and practically measurable modern ideas of design, which often make implicit reference to cultural anthropology as a discipline. Incheon would be a design in Victor Papanek's sense, for which design is basic to all human activity: "the planning and patterning of any act towards a desired, foreseeable end" (Papanek 1985, 3). Or, Incheon would be a design in John Heskett's sense, a processual actualization and end result of "the human capacity to shape and make our environment in ways without precedence in nature to serve our needs and give meaning to our lives" (Heskett 2002, 7).

However, if not before, then surely at the point when one considers the emergence in and out of Incheon of the u-city of Songdo (the edge city which is the focus of this text), these paradigms of modern design and simple, elegant urban diagramming are no doubt still in play for urban planners, urban designers, architects, and citizens or visitors, but they also become subject to doubt. The stretch or interval of Incheon's spatiotemporal urban evolvement, whose origin, end, and dimensionality we seem unable to determine, should already have alerted us. Perhaps these notions of design and diagram are not entirely adequate but rather misleading, or at least they begin to appear open to a range of questions as to their reach.

It might well be that both Incheon and now Songdo are in transport in ways not to be designed and diagrammed according to such finite delimitation of form. It might be that they are in flow and transformation in ways that do not necessarily bear witness to the linear teleological progress toward a desired human goal allegedly bestowing upon them a definite functional and structural form. Perhaps they, and Songdo in particular, rather involve urban diagramming in kinds of environmental, technical, and social or individual transport, transformation, and emergence which solicit a consideration of topology qua continuous formation and deformation without tear, cut, or break.

Songdo emerges with the development after the mid-1990s of urban planning, urban design, and architecture in the expanded field – in tandem with the unfolding of network societies which draw increasingly on out-of-the-box computing as well as the sociocultural and experiential horizon of a virtually and physically mobile citizenry[1] (Crang and Graham 2007; Ekman 2013; McCullough 2004; Williams and Dourish 2006; Weiser 1991; Weiser and Brown 2008; Weiser et al. 1999). It is intimately related to those contemporary global urbanization processes whose expansion, intensification, and complexity are unresolved key issues for current research in urban studies, urban planning, urban design, and architecture (Banerjee and Loukaitou-Sideris 2010; Bridge and Watson 2010; Haas 2008; Knox 2011; LeGates and Stout 2011; Sassen 2012). More specifically, this edge city partakes Incheon as a "free economic zone" supported politically and economically at the national level, with the declared aim to transform the area into a hub for logistics, international business, leisure, and tourism for the Northeast Asian region. This transformation includes improvement of not least the business environment for foreign-invested enterprises and the living conditions for foreigners. It is aimed to be an area for the installation and development of advanced knowledge-based technologies and a place for eco-friendly urban living.

This 40 billion dollar city development with several international corporations as stakeholders is officially in process from 1994 to 2020, covering a bit more than 13,000 acres, partly on reclaimed land along Incheon's waterfront, that are to house a population of a quarter million people. At the heart of this design process, planned architecturally by the New York office of Kohn Pedersen Fox, one finds the 1,500 acres of reclaimed land where Songdo International Business District (IBD) is built as a ubiquitous city from the ground up. Computational units have been included from the start as part and parcel of the infrastructure. Ubiquitous computing is integrated in all buildings, streets, and offices, linked in a wide area network. Apart from the Trade Tower, the area includes schools, hospitals, apartments, office buildings, and cultural amenities, plus downscaled replicas of architectural hallmarks such as New York City's Central Park and Venice's waterways. Samsung has placed its bio industry here, and four universities have satellite campus areas here as well, including State University of New York (Sony Brook), George Mason University, and University of Utah. The Songdo IBD is being developed as a "sustainable city" – with all major buildings living up to *LEED* requirements, more than 40 % of its area reserved for green space and bicycling lanes, numerous charging stations for electric vehicles included, and a waste collection system that eliminates the need for trash trucks.

Songdo as a u-city is thus obviously related to a growing number of earlier or still ongoing efforts toward designing an exemplary intelligent city or smart city (Droege 1990; Komninos 2008; Shepard 2011). These can be found in Asia (Seoul, Suwon, Singapore, Mitaka), Europe (Amsterdam, Lyon, Birmingham, Glasgow), and the

---

[1] I am referring here broadly to the third wave of computing, after mainframes, after personal computing, and in tandem with the rise of mobile and social media technics. More specifically, I am indicating the developments in urban cultural contexts after Mark Weiser and his Xerox Parc colleagues' early vision in the mid-1990s for a calm human-oriented computing, i.e., the developments during the last 20 years, notably in and around cities in Southeast Asia, Europe, and the USA, of ubiquitous computing, pervasive computing, ambient intelligence, the Internet of Things, and Things That Think.

USA (New York, Dayton, LaGrange). They arrive in various forms – innovative creations of new cities from the ground up, meshes of such new cities with existing urban sprawls, and regenerative efforts internal to old cities. Supposedly, in their planning, design, and architecture, these share an effort to be or become environmentally, technically, and socioculturally "smart" or "intelligent."[2] One would think that this concerns a diagramming of the dynamic coexistence and co-development of sustainable urban environmentality and a mixed reality with human as well as technical context awareness in operation. Here the design of an artificial, dynamic, and intensely networked mixed reality (with a real *and* a virtual environment as well as an augmented virtuality *and* an augmented reality) necessarily assumes that the environment, the city, its cultures, and the social body of its human inhabitants are in co-development with ubiquitous computing[3] (Azuma 1997; Milgram and Kishino 1994). Design of Songdo as a u-city for the second

---

[2]My focus in this text on the design of lived urban cultural experience, technics, and the environment is, of course, a vast reduction of the complexity at stake in contemporary global urbanization processes and their ensuing functionalities and structures. Although this has its own merits and is also dictated by the brevity of an article or chapter, I would insert two remarks. (1) My indication of what a "smart" and "intelligent" city might mean here already deviates on purpose from most other definitions of these terms in existing research discourses, which are moreover usually left rather implicit. I am calling for cities smart and intelligent in their maintenance and development of the quality of the social and individual urban life forms and experiences of their inhabitants and visitors, in their maintenance and development of environmental sustainability, and in their maintenance and development of technics seeing to the membranes between these two and the transports through them in both directions. The complexity of this aside, this is already to signal a departure from the most common de facto approaches to "smartness" and "intelligence" which seem at one in deploying the currently most advanced information and communication technologies as infrastructural and infrafunctional determinants of a city formation, guided by a belief or assumption that this will drive economic growth. That technological determinism and overdetermination by quantifiable economic growth are also primary concerns in the development of Songdo, to the detriment of urban governance, urban culture, social and individual city life, and the environment, is made evident already in the first set of national and international research efforts in the area (Shin 2009; Kim and Kim 2012; Kim 2008). See also Germaine Halegoua's interesting observation that national politics for this have all along been significantly guided by the idea of an urban design experiment generating an exemplary prototype of the u-city which would be native to South Korea, incorporate massive foreign investment, and then be a profitable export model later on (Halegoua 2011). Other critical discussions of the meaning of the "smart" and "intelligent" city, generally supporting my remarks, can be found in Allwinkle and Cruickshank (2011), Aurigi (2006), Böhlen and Frei (2010), Graham (2004) and Hollands (2008). (2) Perhaps one of the strongest and best arguments for pointing to the design of the u-city as a topological issue is to be found by the very noticing of these kinds of predominant reductions of urban complexity and the effects of tendentially privileged weightings of economics and technics. For this makes it clear that another kind of complex, holist, and continuist diagramming might be what is solicited.

[3]I am parenthesizing here the issue of the ontological and epistemological traits of "mixed reality" and "augmentation," along with several existing notions of the human experience of these. I am adopting here shorthand versions of the recognized and widely used definitions from computer science and HCI supplied by Azuma, Milgram, and Kishino. I go into further detail with respect to competing notions in Ekman (2013), 13–18, 44–49. A post-phenomenological approach to embodied experience in and of mixed reality can be found in Hansen. The interested reader can find other treatments of mixed reality in the fields of architecture, design, and construction

epoch of network societies thus presupposes an exceedingly rich and complex interactivity among at least its ecological, technical, sociocultural, and personal multiplicities of nodes.

Only some of these nodes and their vertices are to be approached as physically territorial, while others are virtual or augmented in various ways. Only some of them are partaking the city in a relatively stable and static sense, whereas multitudes of others must be approached as in transformative transport. They are dynamically distributed, mobile, and highly variable in their modalities, times, and places of interactivity. This should suffice to allow one to remark quite generally that a theoretical departure from a number of traditional and more static, objectifying, and purposive modern notions of design is perhaps solicited if one is to begin to do justice to the ways in which Songdo raises questions concerning the diagramming of its multiplicities, their relational dynamics, as well as their potentially ongoing creative or dissipative character as a whole. This right away leaves one with a series of unanswered and pressing questions. Is Songdo a design, one serving a meaningful human purpose by having its environment, functions, structures, and modes of urban life proceed toward a certain aim, or is it an urban complexity in a formative and deformative becoming without any obvious *arche* or *telos*? Is this u-city a designed whole (in process) or a part of a globalized urbanity (to come), perhaps whole as a part in becoming? Is its diagram, if any, a continuous urban planning or rather a discrete urban plan? Are its urban design and architecture limited forms permitting of measure and metric, or are they matters of continuous formation and deformation in a more general topological sense?

Urban Intersection: Cultural Theoretical Topologization

Everyday Cultural Realism

The visiting American professor in math and science studies who has just crossed the canal to arrive for the first time at the exceedingly busy traffic intersection between highway 110 and regional road 77 realizes that he has lost his way. He hardly notices the coastal environment behind where land is still being reclaimed, nor the finished bridge toward the airport, nor the ongoing buildup of Songdo north of the u-city. He is almost oblivious to his fellow pedestrians, of which there are now relatively few, and most of whom seem to be local residents coming from Songdo to go to work in the business district. He does realize that some kind of morning rush hour traffic jam has halted all vehicle transport but does not observe that this is in large part due to errors in the smart traffic control systeminfrastructurally embedded

(e.g., Wang and Schnabel 2008) and in the field of augmented urban spaces (e.g., Aurigi and De Cindio 2008).

here. Arriving at the highway has made it clear to him that he is about to leave Songdo IBD, while he should really be ready for his meeting in the southern part of campus town in about 20 min.

He knows he got off a few subway stops too late, and perhaps it was the combination of this, the busy morning throng at the University of Incheon Station, and the synergetic or planned heterogeneity mix between radial roads and gridded blocks that got him lost. This first-time visitor seems a good deal less concerned with the deeper issues of Songdo's topological design or design as topology than he is with regaining his orientation, finding the right place in Haedoji-Ro, and making his appointment in time. He is quite thankful that he is carrying two simple city diagrams, in the form of a paper tourist map and the map on his smartphone. In less than a minute, recourse to these enables him to see that he just needs to backtrack 100 yards, turn left in the green area down along the canal, and then turn right into the campus town at the next regular road.

It may be quite tempting to reduce the question concerning design as topology in this finite and everyday commonsensical manner. A citizen of Seoul commuting daily by train or car to and from work in the u-city Songdo and a local inhabitant regularly walking to and from the nearest Family Mart for late afternoon shopping will presumably already have incorporated such a finite and simply delimited urban mapping diagram, integrating it into an everyday social and individual cultural habitus to the point of completely taking it for granted in embodied practices. Recalling the undeniable feats of modern scientific cartography and chronology with respect to producing representative diagrams that let us know and live practically and aesthetically with the movement, the space, and the time of the city, one might well wonder whether there is a more genuine design-topological problem with the whole spatiotemporal and experiential development and formation of this u-city. Tourist maps, Google maps, GIS, and GPS plus habitual practice seem to do fine almost all the time for most people. The architects' white paper urban plan for Songdo evidently makes do with two-dimensional representations of environments, neighborhoods, and buildings in three dimensions plus a smaller set of 2D representations of the spatiotemporal flows of the traffic patterns, everyday cultural activities, and the city development. Perhaps this is enough to affirm that Songdo adheres to a well-known diagrammatic finitude.

This could be theoretically supported and strengthened by arguing that the design of Songdo is a diagrammatic experiment with a metric topology for a whole whose transformation concerns a complex but measurable and knowable transport of parts and relations, from an external environment across technics as a membrane or medium to an inner urban environment for social and individual work, leisure, and living – and back, with a difference. Likewise, as Frederik Stjernfelt would argue, simple diagrams, construction precepts, and diagrams equipped with construction precepts (such as city maps, algebraic formulae for architectures and specific urban designs, and graph theoretical urban plans equipped with the requisite algebra for its part and relations) would be approachable with a pragmatic scientific realism. They would work as special urban icons that include some kind of similarity to the city to which they refer (Stjernfelt 2007). The urban planner and designer uses diagrams as icons of sets of rationally ordered and related objects that both suggest a possibility and in some way actually show. As signs of an order in plurality, diagrams are thus a pragmatized variant of a Kantian schematism for the encounter of intuition and concept.[4] If Songdo is to be considered truly diagrammed by a designer in this way, as an order in plurality for a metric topology, the core of this rests with a belief in the ongoing operation of the diagram as a vehicle or formal machine for thought experiment and manipulation that remain in a deductive reasoning process and thus linked to a reach for truth through symbolic logic.

Cultural Ontologization and the Question of Finitude

In one of the cars on highway 110, now stuck in the traffic mayhem a bit from the intersection – on the way from the airport to Songdo IBD – is an urban theorist from Brazil who is here to participate in a conference on the future of smart cities. Parts of his disciplinary training have been in philosophy and math, and his paper is deeply engaged in discussing comparatively the topologies of Mitaka and Songdo with a view to finding certain homologies and invariants in their ongoing city space formations and deformations. Much in keeping with his momentary standstill due to the break in the traffic flow and the seemingly endless looping (or crash) of the smart traffic control system, he readily acknowledges the finitude of his experience and the import of the metric city topology on the macro level of such situations and events. However, his theoretical argument, in the paper to be delivered in another hour and a half, would make of this Euclidean geometry and linearly progressive time a limit case, a side effect (albeit an important and relatively well-known one) of the more general topologizing ongoing at the micro-level. In fact, he will invoke almost an obverse kind of argument compared with that mentioned above, for he plans to start his paper a bit provocatively by quoting very early some of Alain Badiou's remarks to the effect that "mathematics is ontology," "mathematics is the science of being qua being," or "finitude must be rationally denied since thinking is essentially tied to infinity" (Badiou 2005, 4; Badiou 2004, xv). This would right away make the diagram of the u-city an ontological one, a matter of a topology whose link with the infinite must be undergone and performed in a set-theoretical event.

This urban theorist is quite fascinated by Badiou's Neoplatonic reawakening of Cantor's set theory and its axiomatization by Zermelo and Fraenkel and the rather resolute asymmetrical privilege Badiou grants at the core of his work to math as the infinite qua form of being, over and above poiesis, or the infinite qua resource for the power of language. He considers it an efficient opening strategy to have his

[4] For Stjernfelt, the diagram is a special icon, often governed doubly by a symbol, both by the type of rational relations used and by the empirical phenomenon referred to. For a more detailed account of such a diagrammatology, including reinterpretations of Husserlian and, notably, Peircean semiotics, I refer to Stjernfelt's book, notably the introduction and chapters four through six.

paper present the urban situations of Songdo and Mitaka as two pure indifferent multiplicities awaiting a rupture event that will constitute and let take place a truth, with a remainder of an incalculable emergence, for he can see quite some resources here in terms of putting to debate set-theoretical ontology as the way to work on the design as topology of the spaces of Songdo and Mitaka. Badiou's double movement will work well here, first as a scientific opening of the theoretical field of set-theoretical topology and this then accompanied by an artistic-creative articulation, both keeping to a certain fidelity to the event of truth with the subject as its locally active generic procedure. This would go quite far in terms of a theoretical approach demonstrating performatively how urban spatial multiplicities *are* continuously and dynamically creative of their limits and composition, even if Badiou's generically procedural thought of infinity and his respect for Gödel's incompleteness theorem will deny them being whole.

However, it is meant just as an opening, for this urban theorist only agrees with Badiou as regards the key import of topologizing and seeing set theory as a privileged way to do this. In his paper he resists Badiou's pure ontologization of topology because he remains wary of essentialist and metaphysical trappings. He also resists Badiou's attacks on theorization in "the little style" because he remains committed to doing urban theory and science in finitude. Rehearsing these attacks in brief does, however, permit him to make clear that his approach to set-theoretical accounts of the topologies of these two u-cities and their potential homologies and invariants is not to be considered relativist, constructivist, or linguistic but rather "realist" in a certain quasi-Kantian sense. His key thrust is to argue in favor of seeing a traffic intersection of a u-city such as the one he is currently stuck in as a topological design and as a real instantiation of the art of topologizing. Here traffic and Songdo in transport are perhaps a set upon which a topological structure can be defined, not primarily for its contingency, its kind of construction, or the way it may say something normative but rather for its way of being formative, generative, and productive for certain spatially shaped situations, events, and concerns.

He is arguing in favor of always having to topologize – regarding it as a universal method or approach to look at every problem, urban formations, or otherwise. The topological is here akin to a general a priori form under which objects and relations are to be examined in order to reveal their invariant aspects. This is tantamount to conceiving of topology as a generalized transcendental aesthetics, based on a general topological a priori (Mormann 2013, 6). Approaching this u-city and its traffic intersection from this angle will admittedly yield extremely general results as to its spatializations, not strong specific ones. However, his hope is to single out appropriate special classes of topological spaces for Songdo and Mitaka, spaces for which one can indeed prove more specific results. He of course acknowledges metrical space as one such but is quite as interested in investigating metrizable spaces and Hausdorff spaces that might provide more general classes. Ultimately, he aims to examine whether it is possible on that kind of basis to indicate relations between the two u-city topologies: a continuous map proving their nontrivial "equivalence" or higher-level homology. In that case, he would offer a diagram qua a relatively elaborate set-theoretical proof.

He expects that part of the questions he might receive at the conference in Songdo will come from positions defending conceptions of Euclidean space and time against such general topologization of the design of the u-city. He is prepared to acknowledge that from such a perspective, his theorization will almost always appear to remain vague and metaphorical. But he will then also counter that a topologization via point-set theory provides perhaps the most convincing conceptual toolbox to analyze these other more general spaces and their ongoing (de)formations. What is more, he will claim that topology is the best way to pinpoint the finitude of Euclidean knowledge and the best way to delimit a given scientific theory anchored and aimed within that one kind of continuity.

Cultural Nominalism

Passing this way through a presentation of positions on the design as topology of the Songdo u-city in transport runs the obvious danger of ending in a polemic fueled by polarization and something as unproductive as a binary opposition or even a dualism. The formation and deformation of this u-city and the diagram for this are hardly issues of a regular opposition or sharp distinction between Euclidean and topological spatialization, the finitude of realist science and the infinities of a strong set-theoretical ontologization, or between strictly discrete and altogether continuous urban plans and designs. This would not begin to make clear some of the main factors that have such a u-city formation today insisted on raising the question of design as topology. It would also not begin to take into account the evident current broadening of theoretical interest in topology across a number of other disciplines relevant to urban studies (e.g., geography, sociology, science and technology studies, cultural theory, design and architecture, media studies, and aesthetics). This is an interesting development in its own right since topology has led a respectable but relatively specialized life mostly in math and geometry during the last century while being largely parenthesized in science and philosophy, for instance, during massive efforts in studies of logic and language. Specifically, then, the recourse to any such polarization or opposition would be reductive of several other approaches to design as topology, notably more obviously nominalist or constructivist ones.

One should perhaps briefly mention at least four of the tightly coupled factors in contemporary urban development which could be said to solicit this interest in design as topology. The ongoing expansion and intensification of physical urbanization of the environment engenders a mereological question of how to approach the complexity of something like the design of urbanity qua a dynamic individuation of a potentially globalized whole emerging as a spatiotemporal synergy among innumerable interlinking parts each with their own kind of topological individuation. The globalized market and the transactions in some key urban parts of the world economy, such as the Trade Tower in Songdo IBD (Tokyo, New York, London . . .), problematize to quite some extent the spatial distance and separation as well as the temporal interval and delay in exchanges, raising questions of the topological character of these economical transformations in relation to those taking place at other spatial stretches and other temporal paces.

The push toward a kind of ubiquity, pervasiveness, or ambience of networked and increasingly mobile and smart information and communication technology, globally but in cities especially, changes not only the city into a mixed reality but also the way we as urban dwellers are socialized and individuated, take on modes of subjectivation, or live our experiences in mixed realities. Moreover, the increasing global mobility and migration of individuals, social groups, work forces, etc. emphasizes and changes the topology of urban transport as well as notions of belonging to a territory, a *polis*, and a community, all of which are now undergoing physical, virtual, and augmentative transformations at various situative scales and eventual speeds.

All of these contribute to a felt need for theoretical, practical, and concrete analytical resources that may begin to address the urban continuities and interruptions at stake. Topology may seem of obvious interest once one wants or needs to consider such issues as the formation of urban wholes from parts; the formation of parts as wholes, urban connectivity, and disconnection; and the reformation or change of urban dimension, scale, border, belonging, intersection, exclusion, and urban void. Evidently, topology also appears of interest to a variety of current cultural studies, sociology, as well as post-phenomenological and post-Bergsonian notions of urban experience and becoming. Perhaps the urge to topologize the design of the u-city is not just an ontological or metaphysical misunderstanding or a pseudo-problem nor just a relation to the infinite and the ontological to be situated and eventualized in truth. Then one would arguably need ways to address the quite common everyday urban cultural experience in a place like the u-city of Songdo that space is not really altogether Euclidean, time is not one linear progression, and urban individuation and subjectivation enfold multiplicities of physical, virtual, and augmented relations to environments and events, across the world and taking place at different times.

No doubt it is factors such as these which contribute significantly to the increasingly transdisciplinary and broad investment in topology as a resource. The relevance or reach of this is still an unresolved research issue, but it is safe to say that the approaches vary greatly, and the positions cannot be reduced to the ontologization of math, realist scientific theorization, and semiotic or transcendental aesthetic resuscitations of epistemological indexicality. In fact, via more nominalist positions in recent broad cultural theoretical approaches, one would consider the continuities and deformations of Songdo and its u-city traffic intersection a sign of culture “becoming topological” according to a certain moving ratio (or change as a normalized urban condition), part, and parcel of the arrival of a new urban order of spatiotemporal continuity for cultural life along with developments of politics, economics, the social, technology, and other fields (Lury et al. 2012). Allegedly, the articulation of topological ideas generally help dismantling rigid disciplinary boundaries and help toward a realization that continued relationality is of key import, whether we would consider the emergent articulation of the spaces and time of the u-city as a cultural topology or consider cultural topology as a theoretical armature with which to analyze the becoming of such a u-city. Here topology is thought as sociocultural and technical field for the articulation and naming of certain practical abstractions which would make possible the emergence of new relations between ontology and epistemology.[5] The space of the u-city and its traffic intersection would be called "auto-spatializing," in the sense that they are self-organizing, emergent, and productive of new relations between knowing and being.

One interesting example of a nominalist cultural theoretical approach to Songdo u-city as a topological design would be Xin Wei Sha's argument in favor of using the mathematics of topology as a poetic articulation (just the inverse of Badiou in this regard) or a stammering experimental approach to cultural dynamics (Sha 2012). An articulated topological diagram of this u-city traffic intersection is neither a replacement for God, nor an instrument, nor a measure but rather a poetic formulation of its morphogenesis and cultural dynamics, and the diagram would be saying these in ways agnostic to measure, metric, counting, finitude, and formal structure. Topology here contributes those simplest symbolic substances ("topological media") whose composition would be doing a kind of non-reductive and non-anthropocentric justice to the u-city as a continuously dynamic life world: temporalization, boundless morphogenesis, superposability, density, value, etc.

Sha's poetically articulate diagram of this city life world will thus go considerably beyond the intrinsic mathematical formalization of topology as introduced by Cantor, Poincaré, Brouwer, or Hausdorff because it will refer to external fields of articulation as well as notions of lived and shared but decentered experience. Moreover, it will take a rather slanted approach to topology as a reserve of such symbolic substances since this is not so much introduced as the mathematical concepts you may find in the canonical textbooks on topology but rather as essential intuitions, much in the way of an artistically creative reading of Gilles Châtelet's treatment of mathematics (Châtelet 2000). The topological diagram of this traffic intersection would thus be an exact, entangled, and situated description of its experiential formation and dynamics – an intuitively involved and qualitative approximation to the city in transport, one might say. Such a diagram would bespeak the quality of urban life qua the continuity of a nonego-based lived experience in and of the city in movements and transformations prior to number and metric. A diagram of traffic like this would perhaps also be especially open to qualitative change, multiplicity, and complex cultural dynamics (Sha 2012, 242–243).

Brian Rotman's interesting treatment of topology and algebra as potentially useful resources for the characterization of cultural dynamics seems to share much with the efforts of Sha, just as it opens with an epigraph from Châtelet: "Diagrams... for those capable of attention are the moments where being is

[5]As Francis T. Marchese points out in his chapter in this volume, pervasive media technology may permit certain breaks with urban space conceived as a Euclidean grid or container. Urban space is perhaps rather to be considered a by-product of the relational dynamics proper to a multiplicity of contiguously overlapping spaces (physical, technologically augmented, communicational, social, and individual environments).

glimpsed smiling" (Rotman 2012, 247; Châtelet 2000, 10). Rotman provides an excursion through set theorization – not least the Bourbaki initiative to provide the ontological and epistemological vocabulary to formally define and handle all mathematical entities in syntax-based proofs. This in the main serves as a backdrop for an argument in favor of an embodied thought of space and spatializations. For such a material and corporeal account of mathematics, Rotman finds primary resources by going back to Poincaré's fundamental group theory only to move ahead toward more contemporary uses of the concept of categories and their graphical language of arrows. On the face of it, this shares much with Sha's poetic articulation of cultural experience of the life world via topological media.

Both do parenthesize the axiomatics of point-set theory in favor of approaching topology as a question of sensate cognizing of movement in material space (e.g., urban bodies, shapes, and contours in transport), notably, the properties of spaces left unchanged by continuous analogue transformations such as stretching, bending, folding, twisting, etc. Both Sha and Rotman also seem to agree on reversing three main tenets in set theory. Both prefer to grant priority to relations over objects and to grant priority to extrinsic relationality over and above the altogether intrinsic ambitions for set-theoretical onto-epistemology. Thirdly, both grant a language of material embodiment and writing a certain priority over the inherently formalist and abstract logico-syntactical language of set theory. However, the impression of a shared agenda is misleading, on at least two counts.

First, where Sha acknowledges a vast positive potential in the math of topology, Rotman remains much more skeptical with respect to seeing topology as appropriate and useful for cultural theorization. Rotman rightly asks whether the vocabulary of n-dimensional differentiable manifolds and a concept of independent degrees of freedom can be taken from the modeling of the dynamics of purely mechanical systems to generate theorizations of the modes of variation of *cultural* apparatuses (Rotman 2012, 252). In the same vein, one might well wonder whether a vocabulary based on differential equations and calculus and aimed at capturing physical dynamics and change would have more than a rather limited relevance for the *intensive* phenomena and dynamics in the sociocultural domain, perhaps apart from areas such as strictly quantitative sociology and physical geography.

Secondly, where Rotman does find a positive cultural theoretical potential in topology and algebra, it is more modest and displays a divergence from the poetic articulation favored by Sha. Rotman would here be aiming not at topological equalities in any strong set-theoretical sense but rather at finding at best certain dynamic isomorphisms. This is already hinted at by his reintroduction of Poincaré and a categorical language of arrows that would work as an exterior epistemology, letting one know a mathematical object from the outside, in a biosocial and compository transformative movement from species to individual. The isomorphisms at stake for Rotman between a topology of categorical language of mathematics and a topology of social and cultural theoretical language might yield insights in terms of relational and structural invariants for sociocultural multiplicities. But today this is still hard to evaluate, Rotman observes, since algebraic thinking here is not yet widely developed or tried.

Rotman would also not be advocating a poiesis of the u-city in transport through topological media. Rather, he would move through categories and arrows so as to get at the deeper sense of a diagram of the u-city as a mathematical ontogenesis. On this path he finds in Châtelet's work on math not just a poetics but an "embodied rumination" qua a combination of the strands of sensible matter, bodies with agency in movement, and intuitive thought that remains mathematical. Rotman's categories and arrows are to reinstall for topological thought the pivotal role of diagrams which is denied in the abstract, logico-syntactical language of set-theoretical topology. He is thus not so much interested in the axiomatic side of math (qua structures of sets) as in the problematic side (qua responses to problems whose solutions give birth to and characterize these structures). Here the import of Châtelet for Rotman is as one who signals that diagrams coupled with bodily gestures are the very means of mathematical ontogenesis, a primary path toward the becoming of mathematical ideas, objects, and relations. At the traffic intersection in Songdo, Rotman would probably produce a diagram of it by gesturing forth the writing of a weave, figuring a piece of crocheting, or drawing up a knot, hence locating a complex and heterogeneous topology in relation to the sensory modalities and activities of one or more moving human bodies in environmental space. The other way around, confronted by an existing map diagram of this intersection, Rotman would be likely to reverse engineer or deconstruct it by retracing the gestures that have been operationalized into its symbols and then make explicit in this specific physical urban situation and environment what intuitions guided the gestures of writing and what problems they respond to.

But perhaps one of the most extensive major nominalist resources for a topological thought of the design of the u-city is to be found in Michel Serres' transdisciplinary work across cultural theory, technology studies, and the sciences. In particular this would concern both the middle and the later phase of Serres work. I am thinking here of the way in which works from the 1980s such as *Genesis* and *The Five Senses* would permit one to think of this traffic intersection in Songdo as what can be intimated by citizens in situ as a morphogenetic emergence of a complex, mixed, and undecidable environment (Serres 1995, 2009). This would not be a question of a careful diagrammatological mapping of the urban terrain, sociality, and polis but rather a migrant and performative tracing productive of the matter, time, space, and sensation of an urban landscape that cannot be given in advance. Here a traffic intersection would leave a coded and recognizable diagram only as an after burn of what remains a continued processing of open, non-totalizable, and perhaps chaotic systems.

I am also thinking about the figurally inventive ways in which Serres work, notably around the mid-1990s onward, proceeds to think time in nonlinear fashion. This would allow one to think urban traffic time spatially, as dynamic volumes or dimensions in a topological sense. For example, an entire country will continuously change shape according to the train travel times among its cities, train schedules, waiting times, and delays, as explored in the MIT Senseable City Lab project "Trains of Data" partnered in France with the SNCF division at work on the national high-speed trains.[6] Likewise, the specific traffic intersection in Songdo would be temporally moving as a river forking and rolling back on itself, as the kneading of dough, or as a crumpled handkerchief whose seemingly distant and disseminated points in flux can be pulled back and drawn together into contiguity or adjacency (Serres and Latour 1995, 60–61). An urban time of the traffic intersection would concern an implicate order, a complex volume that folds upon itself not just to transform but to gather up and free time. Much like Deleuze generalizes the metaphor of the fold in his studies of Leibniz and Foucault (Deleuze 1988, 1993), Serres finds in folding and implication the infinitesimal germ of form or a sort of topological atom besides the algebraic element (Serres 1994, 49).

The late work by Serres, from the mid-1990s up to *Hominescence* in the new millennium, might be said to offer the reader a particularly evocative projection of the shape generated by the u-city (Serres 2001). It is perhaps here that Serres' work is the closest to delineating the kind of imbrication of localized urban life and tendentially delocalized information materials that we encounter and live with as mixed realities in the second phase of network societies. The traffic intersection in Songdo is, with Serres, perhaps to be thought as a hypertopology qua the time and transport of a network of networks permitting spatiotemporal possibilities for urban mixed realities: one exists out there, without, here and there, outside in, and there beyond[7] (Connor 2004).

Sociocultural Constructivism

Considering the overlapping interests in social and technological agency and processes, science and technology studies and actor-network theory would seem influential approaches that might well open up social constructivist insights respecting the u-city as topological design. In a recent text, Matthew Fuller and Andrew Goffey offer one variant of this, finding topology relevant in the sense that it permits one to think together the processes generating the intensive continua of desiring production with the sociotechnical operations of contemporary ICT infrastructures (Fuller and Goffey 2012). This would lend an approach to Songdo that focused on the ways in which the material infrastructures of ubicomp artifacts give shape to fields of social experience and afford certain kinds of action. Fuller and Goffey are particularly interested in the topology of those codings and decodings of spatial and temporal configurations which may be intimated once one explores digital artifacts and their infrastructures with the technical and the social on a par.

---

[6] See http://senseable.mit.edu/trainsofdata/

[7] For an interesting effort to convey visualizations of such network complexity, see the many examples presented in Manuel Lima's work in design and information visualization, e.g., http://www.visualcomplexity.com/vc/index.cfm?domain=Transportation%20Networks. See also his book on visual complexity (Lima 2011).

This one traffic intersection in Songdo with its landscaped physical environment; its roads, cars, social throng, traffic signals, and displays; and a smart traffic control system partaking of the technical infrastructure of a u-city could thus be approached so as to present a set of operational sociotechnical topological constructs. Here a sociotechnical operation is performative and constructs the forms of topological continuity on which it appears to act. This would allow a consideration of the broader artfulness of technology and a sensibility to the ways in which power is organized continuously with a field of experience. An obvious example of such sociotechnical topological constructs would here be the multiplicity of apps on smartphones moving through this intersection and its Hertzian urban space, constructing technical, medial, and social communicational continua with quite some organization of power relations. Another example would be the surveillance cameras and the embedded networks of sensors and actuators in this information-intensive environment which generate a proliferation of posts in urban relational databases and its knowledge economy with powerful continua of flows through data, work, and libido. In other words, on this path one might become aware that it is one continuous topological movement of urban deformation that we tend to describe as an affinity between a technical network topology for traffic control on the one hand, and, on the other, an affective network topology for social production and desire.

One must acknowledge that urban design as topology today could be said to solicit an extraordinarily wide and internally torn set of approaches at the intersection of cultural theory and the sciences. This including strong defenses of epistemological finitude wards off topological idealization and strong ontologization of the infinities of several orders in set theory. Moreover, it includes advocacy of several kinds of nominalism and constructivism emphasizing topology as a transdisciplinary resource or reserve, whether for epistemologies finitizing the infinitude of urbanization or delimiting its infinite finitude. One could be tempted to diagnose this state of affairs as a matter of an apparent need to try to come to terms with what is intimated as the uncomfortable and uncontrollable contingency of complex urbanization as a relational whole. At the very least, this puts quite some emphasis on a general need for new kinds of design theorization in an expanded and intensive field. Design here relates, enfolds, and unfolds any artificialization, and perhaps one could think of design as an emergent continuity generating what Ezio Manzini calls an "un-known artificial world that we must examine to discover its qualities and laws" (Manzini 1992, *Artefacts* 52).

Urban Intersection: Technical Theoretical Topologization

The call for a look to a holist, relational, and morphogenetic diagramming of the u-city as emergent also means that one is not only dealing with design at the intersection of urban culture and the sciences. It is of import but not the whole story to have a notion of design as topology lead one to ask questions of the coexistence and co-development of topologization with urban planning, urban design, urban architecture, and urban cultural studies.[8] The u-city emerges with ubiquitous computing, and technics must be considered as the (de)formative envelope, curtain, or membrane of exterior and interior urban environment. It is very interesting to note that the question of the technical design as topology of the u-city will lead one to recognize a set of approaches and positions in computer science, hardware, and software development that appears to have the same width and the same tensions as observed above.

A smart traffic intersection design for a u-city such as Songdo must consider the interface with the dynamics and events of the real environment qua a topology of physical landscaping, neighborhoods, streets, buildings, vehicles, and living beings such as humans. It must also consider the interface with the dynamics and events of the virtual environment qua a topology of computational systems, including those of wired urban infrastructure (traffic signals, a variety of smart systems), a multiplicity of wired and wireless networks of sensors and actuators (temperature, wind, humidity, movement, pressure), a multiplicity of wired and wireless embedded mobile systems in vehicles and the like, and multiplicity of more personalized mobile systems (implants and prosthetics, wearable computing, smartphones, tablets, laptops, etc.). In addition, in a mixed reality such an intersection must consider how real and virtual topologies are to mix in augmented reality and augmented virtuality interfaces for a multiplicity of human interactants connecting and disconnecting ad hoc and on the fly.

Technical Ontologization of Topology

By necessity, the urban traffic controllers, the hardware engineers, the software developers, and the interaction designers will consider the construction, maintenance, use, and development of system topologies in ontological and idealized virtual variants. However, they are most likely to diagram these using some kind of graph theory to display invariant structures or patterns in various urban flows, basing this on mathematical models that can be inferred from statistical physics (Blanchard and Volchenkov 2009). More concretely, they will seek an intelligent optimization of these system topologies by integrating operational data froma multitude of

[8]Interestingly, addressing the question of the morphogenesis of the u-city as a mixed reality from a holist angle of approach will lead one to a diagramming of continuities across the existing major disciplinary borders separating the scales and speeds of urban planning, urban design, and architecture. This would be tantamount to a topologization of the u-city that would articulate as one process of deformation the slow speed and long time frame of change for the environment and the urban region (urban planning); the medium speed over more than a 5-year change rate for the city, its neighborhoods, and streets (urban design); and the faster pace and frequent minor changes within a few years supposed to characterize an urban place, an intersection or square, and buildings plus their interiors (architecture).

their near real-time systems, e.g., statistics from traffic flows and congestion, power consumption, public safety events, waste processing, and throughput in transportation systems (Harrison et al. 2010).

Nominalist and Constructivist Technical Topologizations

Variants of nominalist and constructivist work on design as topology typically appear at the middleware-level and in higher-level applications. For instance, middleware for South Korean u-cities draw upon context-aware networks recognizing urban resource requests and citizen activities or intents so as to have an operational and decidedly *semantic* service discovery. At the user end, all parts of the u-city infrastructure and services are supposed be *meaningfully* present and accessible via all networked computational devices regardless of time and location. Security and surveillance systems for the u-city will also draw upon a variety of sensor networks and a massive amount of data to recognize human presence, movement, body posture, racial and ethnic affiliation, facial expression, and affect. On this basis semantic inferences will be made, and past or future histories or narratives of action will be considered. In a sense this both recognizes and constructs human context and contextual meaning in close to real time for events in a given 3D situation qua an urban place. Smart u-city design at a traffic intersection thus also tends to mean construction, installation, and operation of a system that can prevent crime by forecasting context and cope with criminal acts – in the name of a high level of safety for urban life (Cho 2012).

One can easily imagine that the intersection of the highway from the airport and regional road 77 in Songdo suffers from congestion and something far from the formation of a continuous traffic flow due to precisely this kind of demand for semantic inference. Perhaps the smart traffic system is an early generation variant. It probably has plenty of infrastructural sensor and actuator networks and is thus in one sense well equipped for the task of being intelligently context aware. It also probably has good tools in terms of ontologies constructed to deal with the topologies of the physical environment, vehicles, and humans. However, urban management policy and traffic control seem to have been rather ambitious, demanding that the systems integrated not only follow, predict, and adapt to vehicular movement, and provide semantic inferences as to earlier and forthcoming security issues for both vehicles and humans, but also that semantic inferences should be produced with respect to the likely intent of human pedestrians. Assuming that the developers have delivered a relatively good semantic reasoning engine for each of these demands, these will, even when they do work, tend to cause noticeable delays. The idea for this technical urban intersection framework is powerful, but semantic inference demands quite excessive computational power due to its complexity, and this gets almost impossible to handle with any kind of efficient reductionism once vast amounts of data are involved. Pedestrians, taxis, trucks, and all the rest incline toward a topological deformation qua standstill because their passage through a mixed reality is meant to be smart, i.e., both efficiently *and* meaningfully controlled.[9]

Technical Topologization as Control of Contingency

Perhaps one of the key technical designs as topology challenges for this kind of urban intersection is something that sounds enticingly simple but whose actual dynamics and complexity almost always make it remain a utopian ideal: continuous routing of transport in real and virtual environments with both wired and wireless networks that include stationary as well as mobile nodes. The high degree of locative and event-specific contingency and the relatively recent arrival in this context of mobile and wireless computing along with networks of sensors and actuators could well be said to make wireless ad hoc routing a crucial test bed for u-city design as a topology of continuous transport and transformation. If that is so, one could also expect that this aspect of contemporary computer science is rather uniquely significant as an externalization of an urban environmental membrane. Its actualization of proposed technical hardware and software solutions to a design as topology problem allows one to read here not a little in terms of ideas and ideologies for urban culture and environmentality in the second phase of network societies.[10]

In computer science, the hardware and software developers will think of this part of the u-city topology as a problem of network communication. Topology is thus a diagrammatic problem of controlling a dynamic whole of ad hoc communicational connectivity in a heterogeneous mesh network. The u-city design is thus implicitly one of the dynamic and heterogeneous mesh networks, the key being its ongoing topological becoming as a whole of connectivity.[11] More specifically, the routing of this traffic intersection in Songdo is perceived to call for a design of network connectivity that takes into consideration a number of constraints and demands, such as:

---

[9]Naturally, technical designs exist which will reduce this kind of problem significantly, although it tends not to disappear. A distributed reasoning engine ecosystem with some kind of peer-to-peer agent architecture will probably be capable of handling the problem via a divide and conquer strategy that reduces the inference times (Almeida and Lopez-de-Ipina 2012).

[10]For an interesting study of wirelessness along such lines, including a treatment of digital signal processing in wireless chips as a conjunctive envelope, see the recent work of Adrian MacKenzie (2011, 59–86).

[11]The last 15 years of technical urban developments along these lines are accompanied by both analogous and surprisingly different sociocultural developments of normative and qualitative approaches to living in cities that are their virtual and physical environmental connectivity in emergence. This is perhaps most easy to diagnose in terms of studies of cultural developments and the use of mobile and social media for urban world spacing as a media ecology, individual cocooning, and loosely aggregated social participation (hanging out, playing around, or flash mobs). For an initial set of interesting examples of such work, see Choi (2010), Goggin (2008), Ito et al. (2005), Ling (2008), Varnelis (2008) and Wajchman et al. (2008).

- The network is decentralized and highly distributed.
- Components are prone to signal interference and failure.
- Components have finite battery lifetimes and limited energy conservation.
- Components are heterogeneous and mobile.
- The roles played by nodes change dynamically.
- Sensing is done by collaboration among nodes.
- Software configuration of nodes is dynamic.
- Network routing is ad hoc.
- Communication is subject to significant fluctuation.
- Routing of communication is subject to time constraints (security).

Despite the complex and topological character of this challenge concerning continuous network routing undergoing variation, computer science work will typically treat it by several kinds of reductionism. First it will be embedded and placed with the hierarchical layering approach proper to almost all work on networks, and then it will be embedded within work on the most relevant protocol for communication, in this case most likely one of IEEE's 802.11 standards. Within this protocol, the problem will typically be handled as a cross-layer problem but one mostly to be handled at the second level of architectural abstraction, i.e., not so much as a problem of the physical layer as the link layer, i.e., as a problem of topology control and management to be handled prior to network and transport layer processing. At this level, the problem will become one of finding the best adaptive optimization algorithms for securing the best multi-hop routing paths in the topology of a self-organizing network and its varying flow rates. A variety of work is still ongoing in this context, including approaches such as graph-based topology control, hierarchization of networks by dominating sets (node clustering), hierarchization of networks by comparison of dominating sets and local minimum spanning tree, or variants of social network analysis of actors, relations, and attributes of these.

It is hard not to notice here a certain analogy with the predominant modes of operation in cultural theorization confronted with the contingency of emergent u-city design as topology. As in most realist, nominalist, and constructivist work in cultural theory, so also in computer science, kinds of reductionist control are preferred, but it is also acknowledged that this has its limits vis-à-vis urban topologization. One could perhaps say that at least until the last decade, most of the solutions deployed in computer science displayed a preference for something that would draw on quasi-static minimum-cost routing trees.[12] A gradual shifting of weight then takes place, so that later work begins to investigate in earnest a more agile and near contingency kind of approach. Notably, research now passes through work on the existing extensive theorizations of dynamic backpressure routing so as to begin implementing this in practical systems, trying to deal with its known

[12]For a good overview of topology control techniques developed, including homogeneous and nonhomogeneous (location based, direction based, neighbor based), see Santi (2005).

problems (packet looping, link losses, delays, scalability). Once we get this far, the traffic intersection in Songdo is seen to constitute a topology of networked communication transport whose routing should control flow and congestion by way of stochastic network optimization (Neely 2010), very often done via a kind of Lyapunov optimization algorithm.[13] Then routing takes place for each individual communication packet and does not involve any computation of a path from source to destination. Instead, the routing and forwarding decision is made by computing for each outgoing link a backpressure weight that is a function of localized queue and link-state information. Each node in the network is computing the backpressure weight for all its neighbors and uses this as the basis for making independent decisions as to routing (who to send to) and forwarding (whether or not to transmit the package).

Concluding Remarks

Judging from the developments in technical studies and cultural theory, design of the u-city as topology arrives as a more pressing concern in tandem with increased uncertainty with respect to relations among the ontology and the epistemology of urbanity. Doubts as to the being and becoming of the urban and how we live with it knowingly appear to fuel investigations of urban design as topology. At present there is rather strong dissent and no agreement or consensus in research and development, except for the felt need to remark upon the topological. This, however, is already a first signal of agreement that, generally speaking, a theoretical departure from a number of traditional and more static, objectifying, and purposive modern notions of design is solicited if one is to begin to do justice to the ways in which u-cities raise questions of their multiplicities as complex wholes, their dynamics as volumes and dimensions to be delimited at the edge of continuous deformation, as well as their creative or emergent character as synergetic relational composites.

It seems warranted to argue that topology – understood more specifically as morphogenetic, genetically structuring, and transductively individuating (Simondon 1964) – offers significant and productive resources when trying to do justice to such

[13] Lyapunov optimization refers to the use of a Lyapunov function to optimally control a dynamical system. Lyapunov functions are used extensively in control theory to ensure different forms of system stability. The state of a system at a particular time is often described by a multidimensional vector. A Lyapunov function is a nonnegative scalar measure of this multidimensional state. Typically, the function is defined to grow large when the system moves toward undesirable states. System stability is achieved by taking control actions that make the Lyapunov function drift in the negative direction toward zero. Lyapunov drift is central to the study of optimal control in queuing networks. A typical goal is to stabilize all network queues while optimizing some performance objective, such as minimizing average energy or maximizing average throughput. Minimizing the drift of a quadratic Lyapunov function leads to the backpressure routing algorithm for network stability, also called the max-weight algorithm.

traits of the u-city as its material and energetic flows, its environmental landscaping of mixed realities, its ongoing virtual and physical infrastructural developments, the folding and unfolding of its buildings, its nodal dynamics, and the relational mobilities at stake. Thinking u-city via notions of design as topology may thus begin to move toward a notion of such an urban formation as an ensemble with a certain generative unity and wholeness.[14] Topological approaches may also begin to problematize existing design efforts and their theoretical armature as too limited, insufficiently transversal, and too exclusive, as when the design of the u-city remains marked by strict privileging of ghosts of economic and technological determinism.

However, one should perhaps also scrutinize quite critically the tendential generalization or even ontologization in all stronger topological thought. In that vein one would question design as topology as an adequate or fruitful approach to the complexity of a u-city as a kind of artificial environment. For example, design as topology qua a design permitting arbitrary point-to-point transformations that are continuous remains an unjustly general and ontologizing move in many cultural and social respects. The lived interactions and experiences of the human inhabitants of the u-city are perhaps not best affirmed as topological designs in this fashion, given their finite and thus noncontinuous individuations which not least involve certain cuts, tears, breaks, trauma, caesura, and death. Something similar must be said as regards the technics of the u-city. More or less ubiquitous out-of-the-box computing readily admits of a topological design approach, as shown above. It would be a mistake, however, to keep to the promises of the infinite and the continuous and thus overlook the modern scientific reductionism accompanying this, always making it a "topological control" that is to work and function in relation to certain finite ends. Moreover, insofar as a core and paradigmatic role is played here, as in modern computing and cybernetics, by discrete mathematics, the theory, method, and practical deployment of technical topologizations obviously tend toward excluding the smooth variability of real numbers and continuous deformation.

---

[14] As indicated in my reference above, I am arguing in favor of having design as topology also reconsidered via the thought of Gilbert Simondon which is perhaps still too little read in non-Francophile contexts. More specifically, this would think urban design as a matter of a relational ontogenesis that need not be Euclidean but may rather be a self-maintaining of topological conditions for an urban spacing that continuously relates a milieu of interiority to a milieu of exteriority. Insofar as the u-city is alive, its ongoing design can be said to maintain as technics of a membrane that is polarized and asymmetrical. Urban technics will let pass some kinds of bodies in centripetal or centrifugal directions, but it will oppose other kinds of bodies. This is the way in which the urban maintains itself topologically, always repolarizing itself asymmetrically via selectivity. It keeps an interior in relation to an exterior milieu by living at and on itself own limit of self-maintenance, more or less sustainable. The u-city of Songdo is one example of a highly complex urban organism with diverse levels of interiority and exteriority. The structuration of Songdo is not solely a matter of integration and differentiation, but rather also a prior dynamic topology: the continuous transductive instauration of numerous mediations of interiorities and exteriorities. See also Simondon's remarks on information and ontogenesis in the second chapter of the second part of his book on physical and biological individuation (1964, 222–266).

Songdo, not unlike many other "smart" cities, is supposedly to be designed as a "green" city, an urban formation for "eco-friendly" living. Insofar as the solicitation of sustainable urbanization is meant in a stronger and more serious sense, certainly over and above efficient waste disposal and minor signature projects decreasing the urban carbon footprint slightly, design as topology would appear perhaps the strongest theoretical resource in terms of thinking the continuities of interior and exterior environmental multiplicities as a matter of complex and dynamic self-maintenance and sustainability. However, insofar as the ecological environment of the urban remains a finite domain, this is not quite as obviously so. Topology is then at best a resource in terms of providing urban theoretical breakdowns demonstrating how and where continuous sustainability becomes impossible.

So far, u-cities such as Songdo seem to stay on the main path of modern and contemporary urbanization which includes a teleology of progress toward quantifiable economic growth, technical expansion (car and air transport not least), and a better quality of urban life qua increased consumption. Even some of the most interesting and most laudable theorizations of sustainable design of the urban and everyday culture do not break with these teleologies in favor of intrinsically sustainable topological design but at most try to steer via extrinsic approaches toward future design of less unsustainable cities (Manzini and Cullars 1992; Manzini and Jegou 2003; Vezzoli and Manzini 2008). Songdo's transformative becoming – including its destruction of ecologically important tidal flats, its footprint from massive transport systems, and its late capitalist consumer society – mostly confirms the teleological asymmetry of modern urbanization. Inhabited cities tend to appropriate, assimilate, and waste exterior environmental resources (geography, climate, animals, vegetation, etc.), first via technics as an interposed membrane and then via the interior environment of culture and shared pasts. Rigorous topological design of the intrinsically sustainable u-city remains to come. In the best case, inquiring into the design as topology of the u-city Songdo leads to an aporia. The becoming of Songdo makes clear that this passage toward sustainability is impossible, but it remains undecidable how to pass on to another sustainable urbanization even though we must.

References

Allwinkle S, Cruickshank P (2011) Creating smarter cities: an overview. J Urban Technol 18(2):1–16

Almeida A, Lopez-de-Ipina D (2012) A distributed reasoning engine ecosystem for semantic context-management in smart environments. Sensors 12(8):10208–10227

Aurigi A (2006) New technologies, same dilemmas: policy and design issues for the augmented city. J Urban Technol 13(3):5–28

Aurigi A, De Cindio F (2008) Augmented urban spaces: articulating the physical and electronic city. Ashgate, Aldershot

Azuma RT (1997) A survey of augmented reality. Presence: Teleoper Virtual Environ 6(4):355–385

Badiou A (2004) Theoretical writings. Continuum, London

Badiou A (2005) Being and event. Continuum, London

Banerjee T, Loukaitou-Sideris A (eds) (2010) Companion to urban design. Routledge, New York
Blanchard P, Volchenkov D (2009) Mathematical analysis of urban spatial networks. Springer, Berlin
Böhlen M, Frei H (2010) Ambient intelligence in the city. In: Nakashima H, Aghajan HK, Augusto JC (eds) Handbook of ambient intelligence and smart environments. Springer, New York,pp 911–938
Bridge G, Watson S (eds) (2010) The Blackwell city reader, 2nd edn. Wiley-Blackwell, Chichester/Malden
Châtelet G (2000) Figuring space: philosophy, mathematics, and physics. Kluwer, Dordrecht
Cho J-H (2012) A context awareness and prediction support system for efficient management of U-city. J Internet Technol 13(3):509–520
Choi JHJ (2010) The city is connections: Seoul as an urban network. Multimedia Systems 16(1):75–84
Connor S (2004) Topologies: Michel Serres and the shapes of thought, np. http://www.stevenconnor.com/topologies/
Crang M, Graham S (2007) Sentient cities, ambient intelligence and the politics of urban space. Inf Commun Soc 10(6):789–817
Deleuze G (1988) Foucault. University of Minnesota Press, Minneapolis
Deleuze G (1993) The fold: Leibniz and the Baroque. University of Minnesota Press, Minneapolis
Droege P (1990) The intelligent city: Southeast Asian city of the future? MIT Press, Cambridge, MA
Ekman U (ed) (2013) Throughout: art and culture emerging with ubiquitous computing. MIT Press, Cambridge, MA
Fuller M, Goffey A (2012) Digital infrastructures and the machinery of topological abstraction. Theory Cult Soc 29(4–5):311–333
Goggin G (2008) Mobile phone cultures. Routledge, London
Graham S (2004) The cybercities reader. Routledge, London
Haas T (ed) (2008) New urbanism and beyond: designing cities for the future. Rizzoli, New York
Halegoua G (2011) The policy and export of ubiquitous place: investigating South Korean U-cities. In: Foth M, Forlano L, Satchell C, Gibbs M (eds) From social butterfly to engaged citizen: urban informatics, social media, ubiquitous computing, and mobile technology to support citizen engagement. MIT Press, Cambridge, MA, pp 315–333
Harrison C, Eckman B, Hamilton R, Hartswick P, Kalagnanam J, Paraszczak J, Williams P (2010) Foundations for smarter cities. IBM J Res Dev 54(4):1–16
Heskett J (2002) Toothpicks and logos: design in everyday life. Oxford University Press, New York
Hollands RG (2008) Will the real smart city please stand up? City 12(3):303–320
Ito M, Okabe D, Matsuda M (2005) Personal, portable, pedestrian: mobile phones in Japanese life. MIT Press, Cambridge, MA
Julier G (2008) The culture of design, 2nd edn. Sage, Los Angeles
Kim TJ (2008) Planning for knowledge cities in ubiquitous technology spaces. In: Yigitcanlar T, Velibeyoglu K, Baum S (eds) Creative urban regions: harnessing urban technologies to support knowledge city initiatives. IGI Global, Hershey, pp 218–230
Kim CT, Kim MS (2012) Deduction of a U-city strategy by analyzing cases worldwide. Information 15(2):483–495
Knox PL (2011) Cities and design. Routledge, New York
Komninos N (2008) Intelligent cities and globalisation of innovation networks. Routledge, London
LeGates RT, Stout F (2011) The city reader, 5th edn. Routledge, London
Lima M (2011) Visual complexity: mapping patterns of information. Princeton Architectural Press, New York
Ling RS (2008) New tech, new ties: How mobile communication is reshaping social cohesion. MIT Press, Cambridge, MA
Lury C, Parisi L, Terranova T (2012) Introduction: the becoming topological of culture. Theory Cult Soc 29(4–5):3–35

Mackenzie A (2011) Wirelessness: radical empiricism in network cultures. MIT Press, Cambridge, MA
Manzini E (1992) Artefacts: vers une nouvelle écologie de l'environnement artificiel. Ed. du Centre Pompidou, Paris
Manzini E, Cullars J (1992) Prometheus of the everyday: the ecology of the artificial and the designer's responsibility. Des Issue 9(1):5–20
Manzini E, Jegou F (2003) Sustainable everyday: scenarios of urban life. Edizioni Ambienti, Milano
McCullough M (2004) Digital ground: architecture, pervasive computing, and environmental knowing. MIT Press, Cambridge, MA
Milgram P, Kishino F (1994) Taxonomy of mixed reality visual displays. IEICE Trans Inf Syst E77-D(12):1321–1329
Mormann T (2013) Topology as an issue for history of philosophy of science. In: Andersen H, Dieks D, Gonzalez WJ, Uebel T, Wheeler G (eds) New challenges to philosophy of science. Springer, Dordrecht, pp 423–434
Neely MJ (2010) Stochastic network optimization with application to communication and queueing systems. Morgan & Claypool, San Rafael, CA, USA. http://www.morganclaypool.com/doi/abs/10.2200/S00271ED1V01Y201006CNT007
Papanek V (1985) Design for the real world: human ecology and social change, 2nd edn. Thames and Hudson, London
Raizman DS (2010) History of modern design, 2nd edn. Pearson Prentice Hall, Upper Saddle River
Rotman B (2012) Topology, algebra, diagrams. Theory Cult Soc 29(4–5):247–260
Santi P (2005) Topology control in wireless ad hoc and sensor networks. ACM Comput Surv 37(2):164–194
Sassen S (2012) Cities in a world economy, 4th edn. Sage/Pine Forge, Thousand Oaks
Serres M (1994) Atlas. Julliard, Paris
Serres M (1995) Genesis. University of Michigan Press, Ann Arbor
Serres M (2001) Hominescence. Le Pommier, Paris
Serres M (2009) The five senses. Continuum, London
Serres M, Latour B (1995) Conversations on science, culture, and time. University of Michigan Press, Ann Arbor
Sha XW (2012) Topology and morphogenesis. Theory Cult Soc 29(4–5):220–246
Shepard M (2011) Sentient city: ubiquitous computing, architecture, and the future of urban space. MIT Press, Cambridge, MA
Shin DH (2009) Ubiquitous city: urban technologies, urban infrastructure and urban informatics. J Inf Sci 35(5):515–526
Simondon G (1964) L'individu et sa genèse physico-biologique. Presses universitaires de France, Paris
Stjernfelt F (2007) Diagrammatology: an investigation on the borderlines of phenomenology, ontology, and semiotics. Springer, Dordrecht
Varnelis K (ed) (2008) Networked publics. MIT Press, Cambridge, MA
Vezzoli C, Manzini E (2008) Design for environmental sustainability. Springer, Berlin/London
Wajchman J, Bittman M, Jones P (2008) The impact of the mobile phone on work/Life Balance. http://www.oii.ox.ac.uk/research/Report_on_Mobiles_and_Work_Life_Balance.pdf
Wang X, Schnabel MA (eds) (2008) Mixed reality in architecture, design, and construction. Springer, New York
Weiser M (1991) The computer for the 21st century. Sci Am 265(3):94–104
Weiser M, Brown JS (2008) Designing calm technology. http://sandbox.xerox.com/hypertext/weiser/calmtech/calmtech.htm
Weiser M, Gold R, Brown JS (1999) The origins of ubiquitous computing research at PARC in the late 1980s. IBM Syst J 38(4):693–696
Williams A, Dourish P (2006) Imagining the city: the cultural dimensions of urban computing. Computer 39(9):38–44

Chapter 10
The Emergent City: 2004–2012

Stanza

Abstract The Emergent City comprises visual artworks informed by critical analysis of city spaces. I am interested in the patterns we leave behind as well as real-time networked events that can be reimagined and sourced for information. Themes include the urban landscape, surveillance culture, privacy and alienation in the city. This chapter explores my work using connected city spaces, sensors and data visualisations of the city.

Introduction

I am an artist who exploits the changing dynamics of city life as a source for creativity. I have created meaningful artistic metaphors utilising new technologies and have integrated new media artworks into the public domain as part of my ongoing research into the visualisation of city space. In essence, I research data as a medium for creativity and show how meaningful experiences of our cities may result.

From a technological perspective, I research sensors, motes, new display technologies and interactive architectures. This research includes investigations into concepts for the relationship of mobile computing within urban space and the built environment. There are three strands to my working process: collecting data, visualising data and then displaying this data.

My art is about the patterns and hidden values within systems and networks that can be disclosed through artistic practice. Interlocking environments and variables in these systems create other lenses and differences in given situations. The outputs

Stanza (✉)
Independent Artist, London, UK
e-mail: stanza@sublime.net; http://www.stanza.co.uk

F.T. Marchese (eds.), *Media Art and the Urban Environment*, Future City 5,
DOI 10.1007/978-3-319-15153-3_10

from networked interfaces and online visualisations can be realised as real-time dynamic artworks, as diverse installations and as real objects made out of new display materials back in physical space.

I focus on the things that change, the flow and the data that describes our experience of the city as space. Data from all sides of systems that can be mediated by all, with varying visualisations communicated over the Internet and represented on different display systems.

My work has covered experiments in these areas and traced a shift in practice from modernist approaches of asset gathering (linear construction) to arrangements of data sets in fixed lists or databases (interactivity) to new approaches of mining information across networks in real time: generative and real-time systems, culling data from CCTV networks and making visualisations of cities from wireless sensor networks.

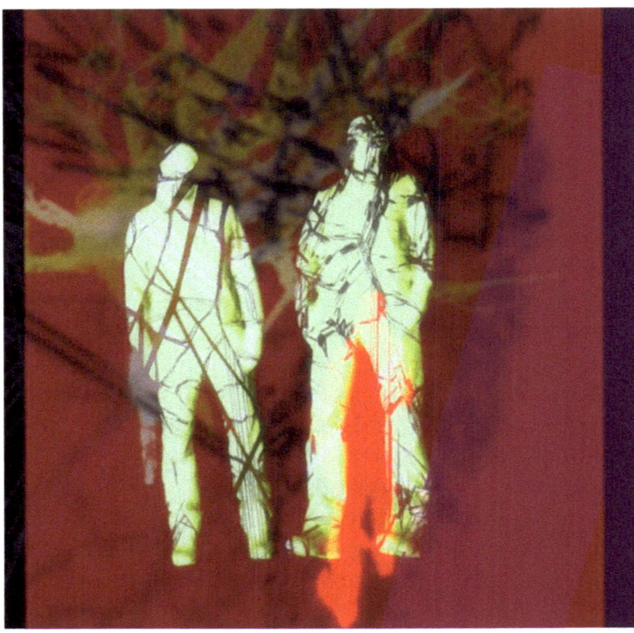

The Data Body. 2013. Software system of real-time transport data

Resulting artworks represent the real-time conditions of the city. Works like *Sensity* create real-time interpretations of social spaces that inform the world (online) and hopefully create new meaningful experiences, allowing critical reflection on the real-time city and the social political undercurrent embedded in the search for the real-time city.

Urban Tapestry. 2005. Output from software system Urban Generation using live real-time CCTV images (Artwork on aluminium 2 m by 2.2 m)

From Fixed Assets to Open Systems and Media Visualisations

The newness of the Internet is still a viable claim. The Internet's ability to draw assets and data to create postmodern allegory allows all sorts of media visualisations. The Internet presents myriad ways to create ephemeral art across networks using all sorts of data information and media. The Internet provides gateways and access for all, creating artists and authors, as choice makers and as decision makers. One can create these gateways oneself to allow others access to one's information and data enabling the creation of organic, generative, ever changing artworks and experiences, which is what I do.

The development of my work has shifted from asset gathering and media collection as artworks into a new studio space online. In earlier works, I used video cameras and sound recorders to gather assets to make artworks that would be presented in various forms. These were mostly used for distribution as editions, or later online, or displayed as outputs as art in an art gallery.

Even in earlier interactive pieces circa 1995–2002, I used assets and remixing techniques, reworking these assets in a system, usually mouse controlled, all displaying some level of user control and a generative visual aspect. That is, in these interactive works, I was manipulating or allowing the user some control over my content.

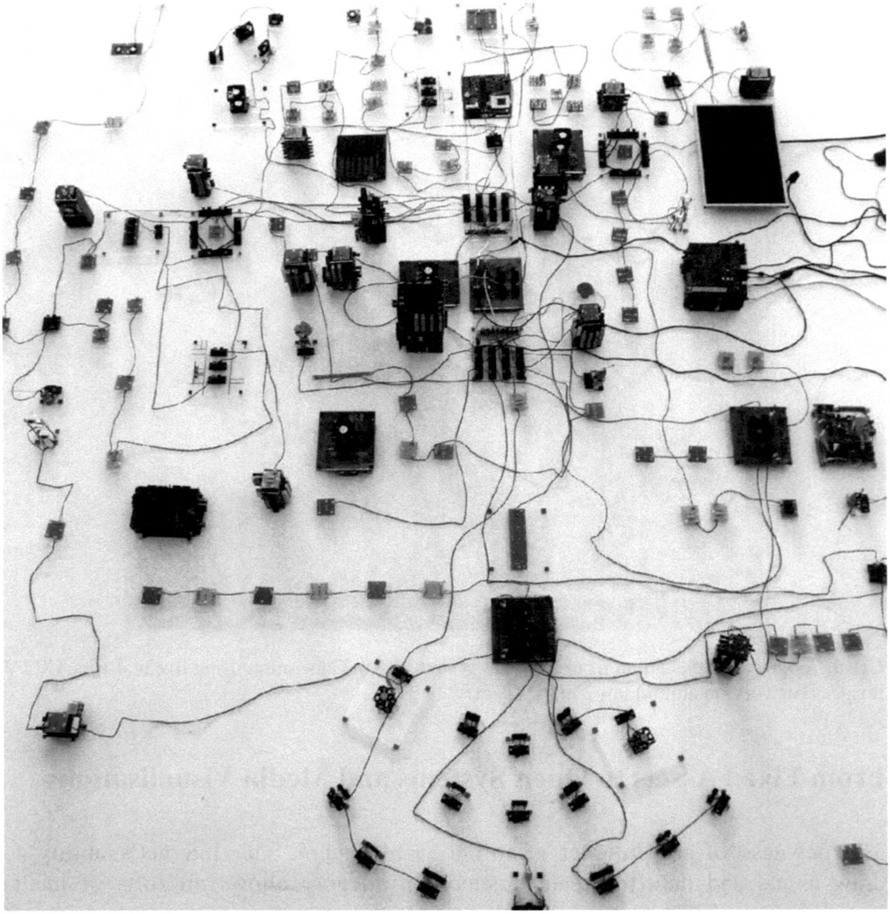

Capacities. 2010. Installation using environmental sensors

These artworks have moved from fixed assets to interactive systems to open generative systems. In 2002, I started to develop less fixed systems that culled data and media from other sources. These mash-ups or interactive collage systems include *Subfusion*, *CCITYV*, etc. In these systems, there are no fixed tangible lists of assets (i.e. they are not databases) they are drawn or harvested via software from (online) spaces.

I also moved away from real-world studio practice to an online studio space for experimentation, output of ideas and finished artworks. From artist as author to system as author, the question is who does the output being to?

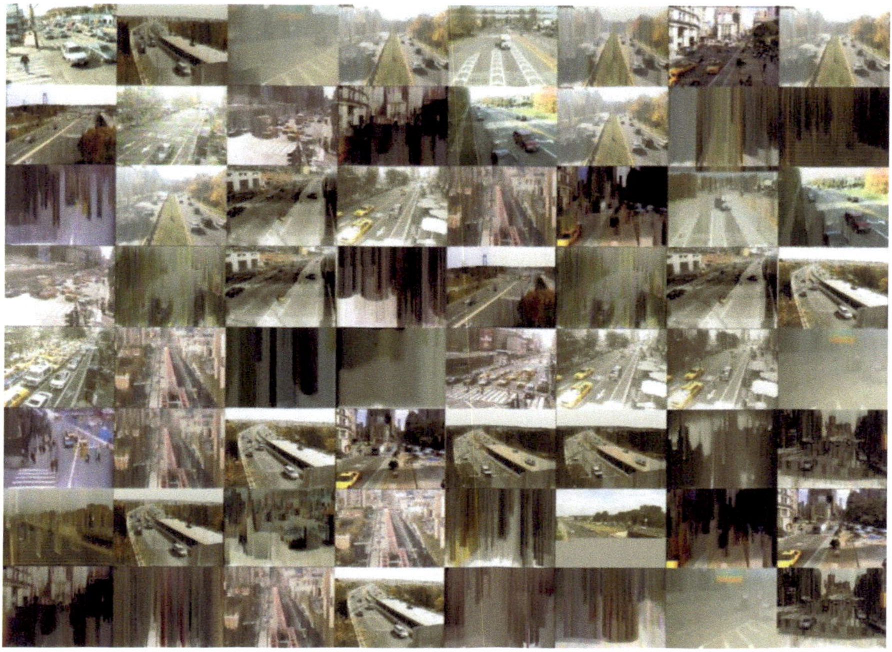

Meltdown. 2004. Unique C print on aluminium 120 cm by 100 cm

The City of Data

The Emergent City has become a body of work connected by a central theme. A city is a web of interconnected networks. In essence, the city fabric is a giant multi-user, multi-data sphere. The city is made up of traffic patterns, pedestrian patterns and bird flocking patterns. Patterns can be seen in the architecture, in buildings, and in the architectural fabric of the urban design network. All of these spheres can be represented by media and therefore by data within the digital realm. And all of this data can be interpreted and mediated. It becomes a matter of choice. Collections of data can be stored to be retrieved later. The mobile data infrastructure becomes a data source so powerful, so interwoven, that its scale can only be imagined as metaphor. The size and scope of such an archive, of such rich-mediated data experience, can support many projects. As such, it can be interpreted via a variety of interfaces.

Cities offer the opportunity for unique types of data gathering experiences via a variety of sources. My objective has been to 'mediate' data into conceptual and cultural artefacts. With this perspective, there are many unimagined threads of data and connections that describe our world that can be explored through wireless mobile networks within which we can create artistic interpretations.

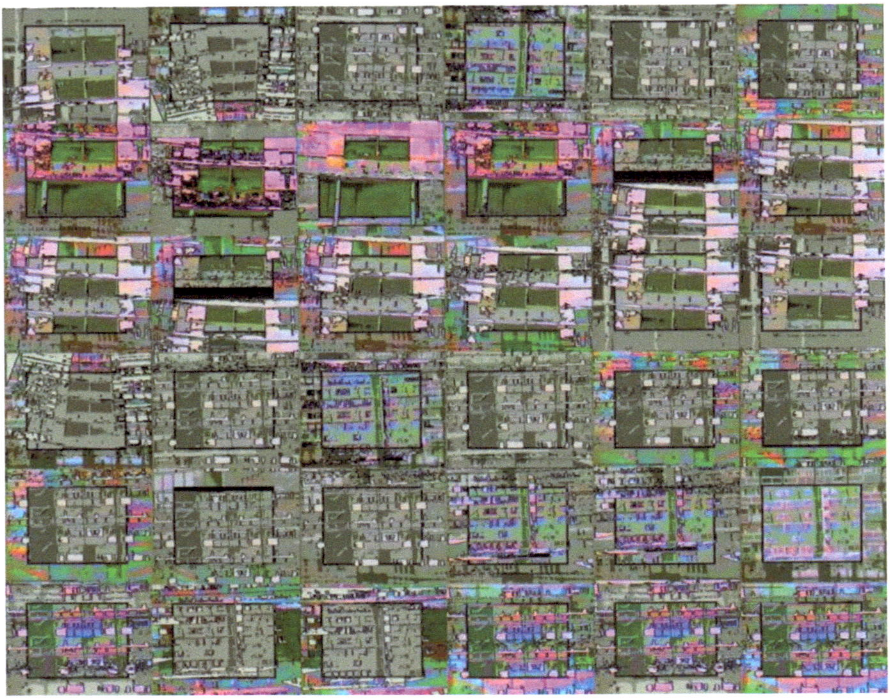

Situations of Panic Noise. 1986. Video still from *Artitexture* series. 120 cm by 100 cm

Various types of data can be reimagined. This includes pollution data recorded via sensors in the street to create audio files. Weather and forecast data, acquired via weather station equipment, can be used to create ambient soundscapes and morphing visualisations as the wind shifts direction or the rain increases. Noise monitor levels, and noise maps, create a symphony of true urban sounds that can be used to make sound-reactive sculptures.

Under this umbrella title of *The Emergent City* project, I made a number of artworks, installations, sonifications and visualisations between 2004 and 2012 that have moved beyond the process of research, beyond what I term asset gathering, that are formed into software and installation prototypes.

Parallel Realities

An example is the artwork *Urban Generation; trying to imagine the world from everyone else's perspective, all at once* (2002–2005). Multiple CCTV cameras are accessed randomly in real time to make this urban tapestry. What you see is an

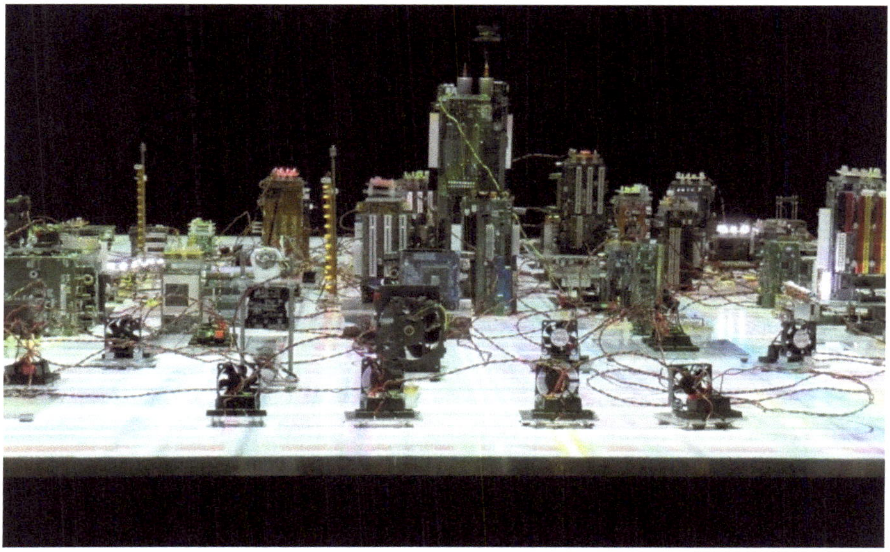

The Emergent City. A Life from Complexity to the City of Bits. 2012

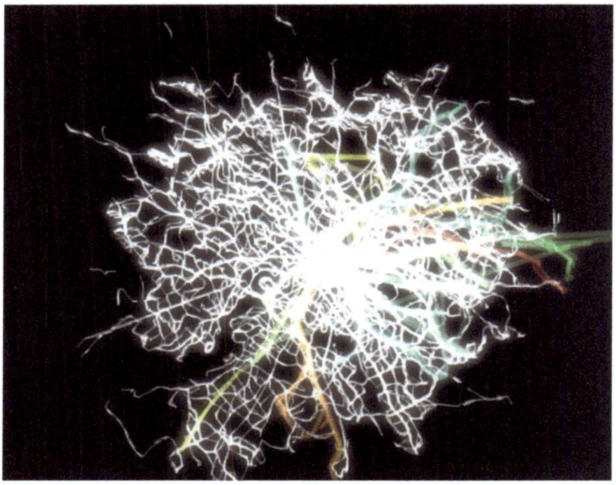

Syncronicity. 2013. Software system using real-time tube and bus data

evolving, generative artwork. These images are taken from London, and they happen as you see them, in real time.[1]

This online artwork represents many realities that exist in city space. The observed real-time surveillance society is reworked into a series of grids. This

[1]The installation versions of this work can be presented in art galleries using projectors or plasma displays.

presents London to a global online audience. The data that you see is protected by the data protection act. Here it is remixed into what you see, which is an online artwork that looks like a filmic experience, but it is not a film. It is a real-time experience of the city from multiple perspectives I call a 'parallel reality'.

The online version now runs as a series of 12 real-time perspectives of the Emergent City experience. This 'filmic system' is constantly evolving and will never be the same again because the images are not recorded. Each screen is a live real-time image from a camera in the city of London. The artwork seeks to explore the rhizomatic multinodal networked experience. *Urban Generation* draws on images across the networked city; the artwork creates a unique interpretation of a multi-point perspective that exists in time always in the present.

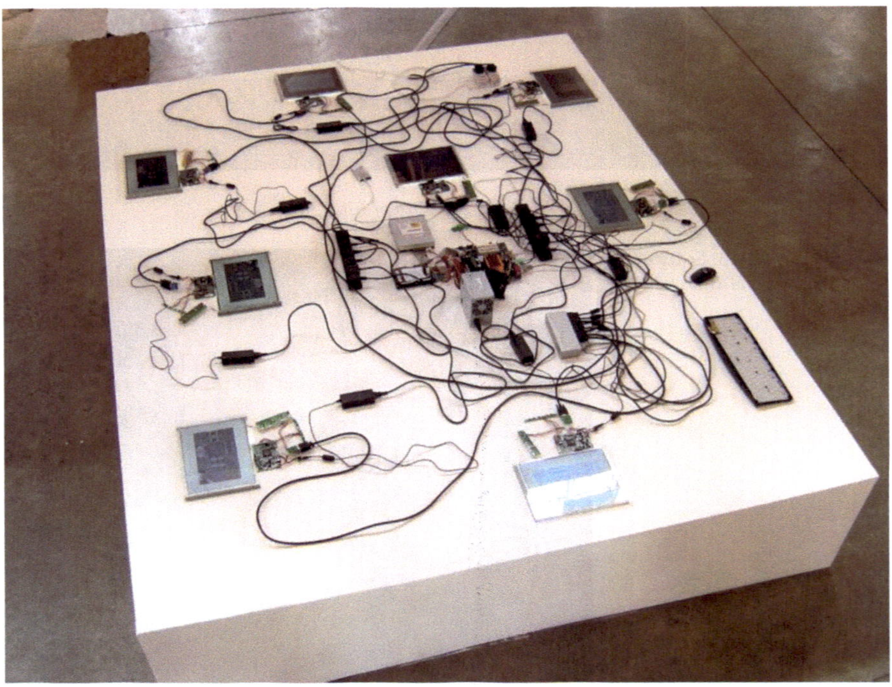

Urban Generation. 2005. Installation and software system using live real-time CCTV images

Data Cities and Control Spaces

The city has millions of CCTV cameras. In essence, the city is the biggest TV station in existence. Millions of hours' worth of data are recorded every day by these cameras on city TV. One can take the sounds and images off live web streams and re-represent them thus creating new interpretations of the city in the process.

The city already has a recorded source of data; CCTV is everywhere. Using data from CCTV, you can bring the outside in. Selected feeds are collected from around the world in real time. These real-time images are fed into a software system where a series of specialised channels rework these images. The channels are always on, and always changing, a constant view of the world changing and evolving around the clock. This artwork uses specially created software and technology to randomly find images in real time from anywhere in the network, in this case anywhere in the world.

The increase of technology infrastructure in the daily existence of a city means that technology will be, more than ever, everywhere in our environment. Mobile data mining will be part of the fabric of the landscape. We will be carrying these data in pods, phones and ID cards. Everything is, or will be, tracked: CCTV, car sensors, tracking inside our phones, ID card movement and tracking in the guise of antiterror activity.

The patterns we make, the forces we weave, are all being networked into retrievable data structures that can be reimagined and sourced for information. These patterns all disclose new ways of seeing the world. The value of information will be a new currency as power changes. The central issue that will develop will be the privilege and access to these data sources. Uses of this information and data should allow rich new interpretations on the way our world is built, used and designed.

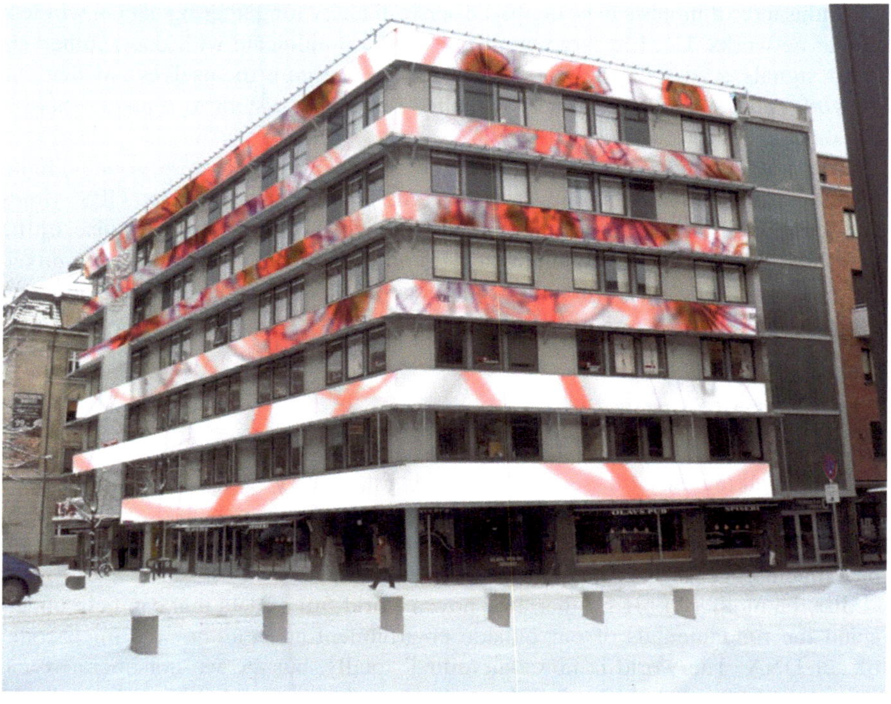

Facade Norway. 2010. Live data visualisation connecting the city using sensors

So we need to imagine the city on a different scale. The possibility is to extend our imagination and enable that perception of the city as a dynamic network. We can now put systems in place that can re-employ our perception and thus create new understanding of how this behaviour unfolds. There are patterns, they are connected, and the systems that evolve can be simulated and acted upon.

We can influence the process and the system, and we can also create variables in this system that allow understanding of the by-products of the system, the data and the resulting information.

The Art of Environmental Data

Visualisations and Sonifications of the Real-Time City

In 2004, I started to layer the city with sensors for the *Sensity* projects. Dozens of sensors access the 'environmental data' and make it public. I wanted to claim this space as a public domain and to create a series of social sculptures affected in real time by the changes in the city.

The overall aim was to make smart networks that have data open to all and not closed off spy surveillance-oriented systems. These networks can be thought of as open social sculptures that inform the world and create new meaningful experiences.

Thousands of motes can be deployed across the city for gathering data in wireless sensor networks. Used in large numbers, they communicate with one another via radio signals across the network. They can reconfigure themselves, so that the network stays stable. The data is funnelled through a system to a point where it can then be interpreted.

The motes monitor the environment for changes in temperature, sounds, light, position, acceleration, vibration, stress, weight, pressure, humidity and GPS. Motes and sensor boards monitor the micro incidents of change in the city: the noise, traffic flows and people flows. The interactions of all these data are controlled via mixed up interfaces that can reform and recontextualise experiences in real time as social sculpture.

The Control Space

Imagine walking out of the door and knowing that every single action, movement, sound, micro movement, pulse and thread of information is being tracked, monitored, stored, analysed, interpreted and logged. The world we will live in seems to be a much bigger brother than first realised.

It's the mother of big brother. It is now a world full of data that can help understand the fundamentals of our outside environment and monitor the microcodes of our DNA. The world is now 'surveilled' totally, but are we not liberated and empowered by data! When will the technology become more than gimmick and start to actually work for us? This is where these projects and artworks begin. In

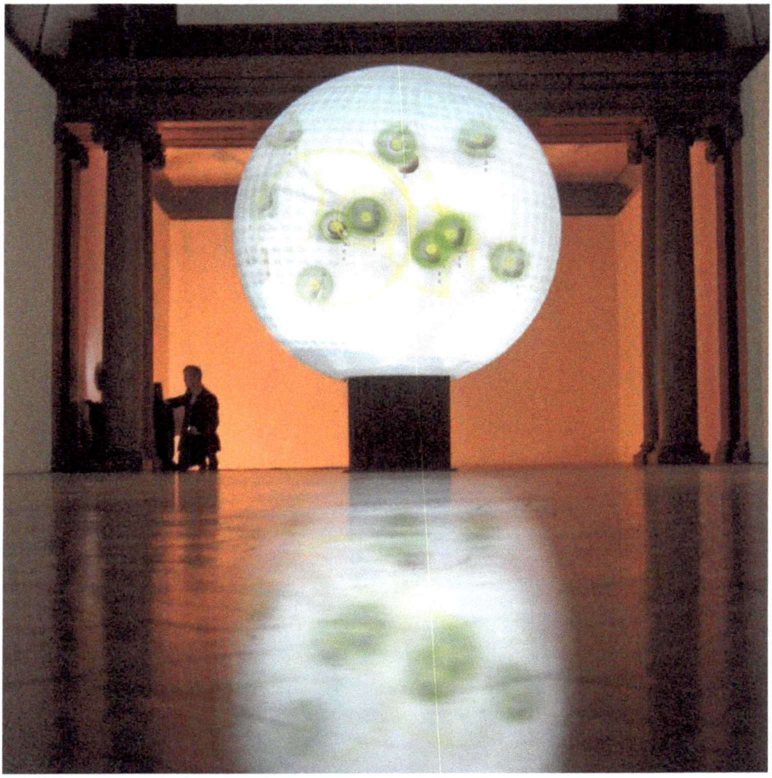

Sensity. 2006. Live data visualisation connecting the city using sensors

addition, the artwork alludes to a more socially engaged practice, based on critical reflection of notions of privacy, surveillance space and control space, speculating on the interactive city and meaning of real-time space.

Towards the Emergent City 2006–2012

The *Sensity* artworks were made from the data that is collected from urban environment locations. The networks of sensors collect data, which is then published online. The sensors interpret the micro-data of the interactive city. The output from the sensors displays the 'emotional' state of the city online, and the information is used to create installations and sculptural artefacts. I believe them to be in effect emergent social sculptures visualising the 'emotional state' of the city.[2]

[2]The sensor network can be moved from urban to rural setting and different types of visualisation can be made depending on the environment.

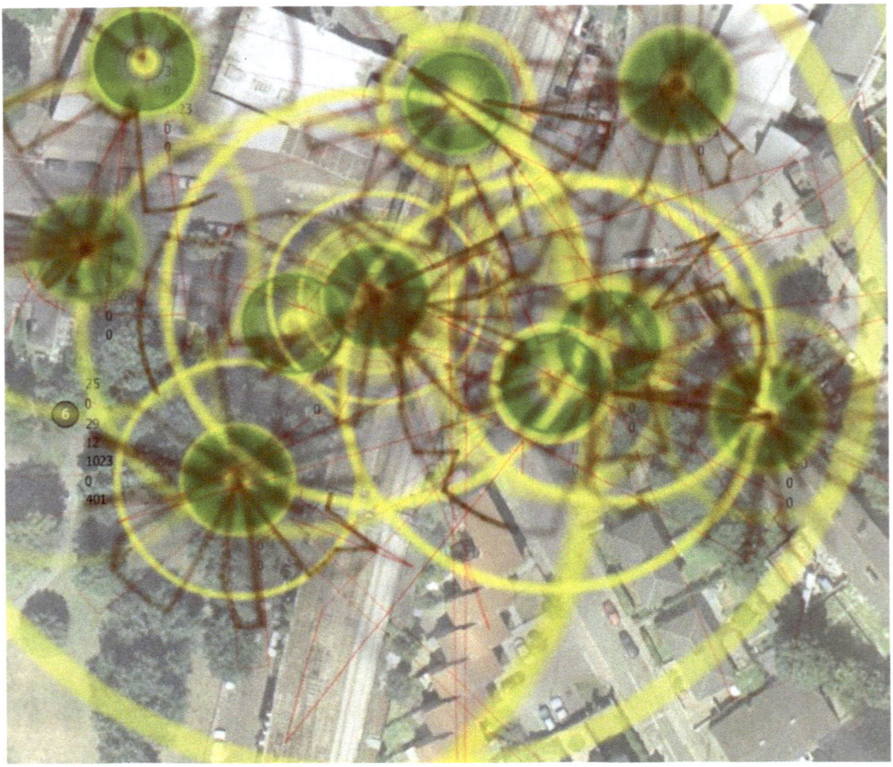

London Sensity. 2004. Software system and live data visualisation

Sensity is also a highly technical project that can output vast amounts of information about the fabric of our cities. By embedding the sensors in this way, we can re-engage with the urban fabric and weave new artistic metaphors within city space. Custom-made software enables these sensors to communicate with one another in a network over a proxy server in real time. The data is also available for others and can be used to create visualisations in the open-source environment that is online.

Representations of these data sets allow unique understanding of the urban environment from this real-time perspective. The interactions of all this data, controlled via interfaces, can reform and recontextualise experiences in real time. *Sensity* becomes a holistic city system. The sense city is a city of accumulated incidents of love, abuse and death, the micro incidents of changes in the weather, the noise traffic flows and the people flows.

Sensity leverages the real-time data city and represents it online showing the life of the system and the emerging changing behaviours of the space.

The Data Is the Medium

In artworks such as *Sensity*, *Facade*, *House*, *Sonicity*, *Capacities*, *Body*, etc., I connect networks of real-time information flows. The shared data space can overlap, and there is a new space: the space in between that only two nodes share. I have merged collected data from various cities and created an aestheticisation of the shared city space.

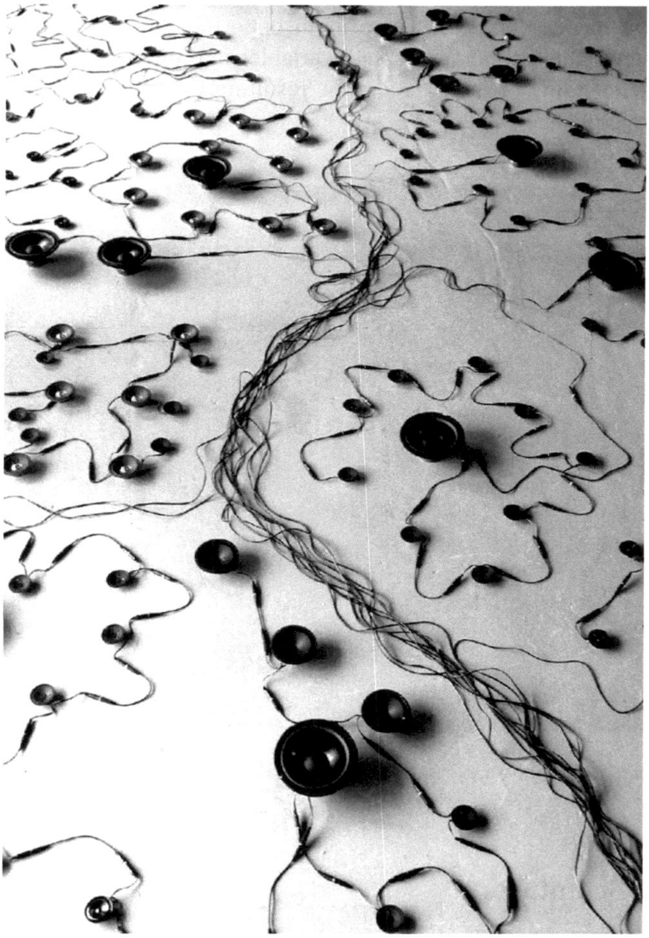

Sonicity: Songs of Atoms Time and Space. 2010. Software system and installation, live data sonification

Marcus Foth states[3] 'Alive with movement and excitement, cities transmit a rapid flow of exchange facilitated by a meshwork of infrastructure connections. In this environment, the Internet has advanced to become the prime communication medium'.

I believe there is a new social space that exists between these independent networks. Future cities will be merged into real-time connected up data cities. A connection of networks of real-time information flows. The results created lead to mashed-up cities and real-time performative city experiences.

These systems re-employ our perception creating new understanding of how this mixed city behaviour unfolds. There is an opportunity to influence this process and the system, and we can also create variables into the networks that will allow greater understanding of the data and the resulting information. Data has become the medium of the age.

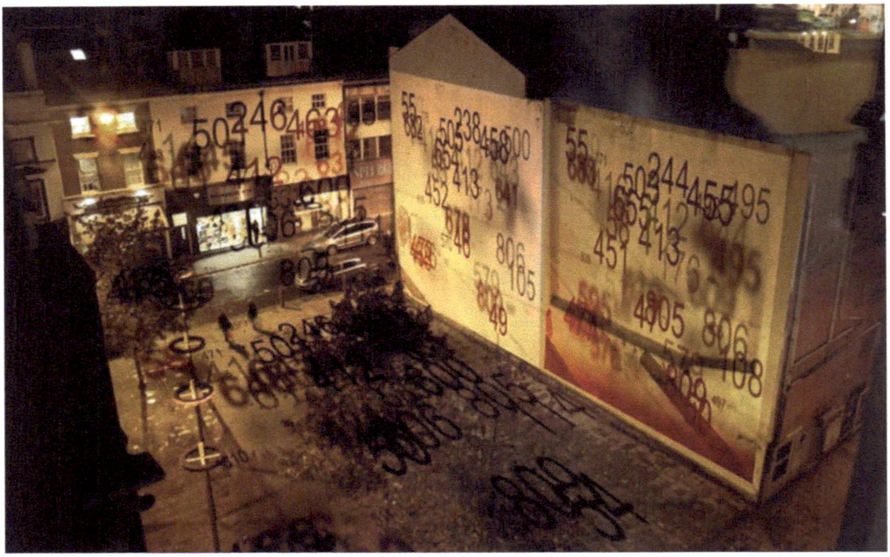

data data data. 2010. Software system and live data visualisation connecting city space using sensors

A City of Sculptures

In one of my experiments *Capacities* (2010), I made a new city of 'sculptures' representing real-time spaces and data environments. The work investigated the loop from the real to the virtual and back to the real space. This notion of playing or

[3]Foth (2009).

manipulating with a malleable form (data) is made possible as each stream, each node, each sensor or even the entire network can be communicated with using XML online gateways.[4,5] The project was instigated by setting up its own wireless sensor networks across London to collect environmental data which was then published online in real time for an extended period of time. The output from the sensors displays an interpretation of the real-time city online, while that same information was evaluated and then revisualised in the creation of numerous artworks, back in the public domain. The resulting artworks represent the real-time conditions of the city.

The artistic aim was to create new meaningful experiences allowing critical reflection on the real-time city and the social political undercurrent embedded in the search for the real-time city. This allows for a greater community of interpreters and beneficiaries to see, and to come to their own understandings arising from, these data about our socially networked environment.

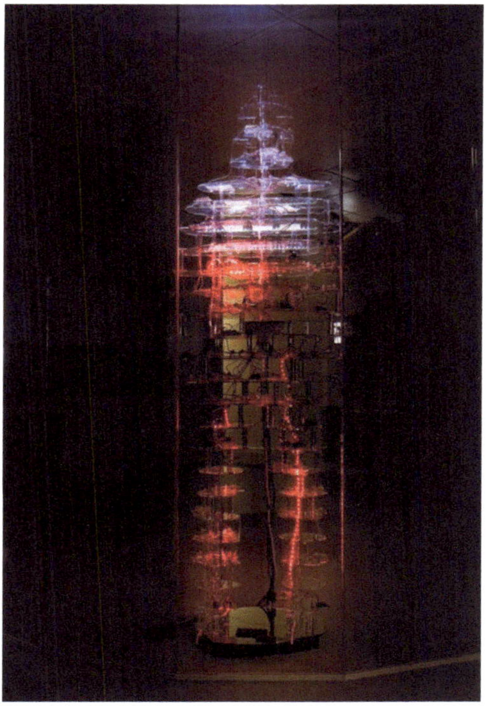

Body 0100001001101111011001000111 1001. 2013. Sculpture which responds to the emergent properties of the environment

[4]http://igor.gold.ac.uk/motes/mts310/motestatusxml_html.php

[5]http://igor.gold.ac.uk/motes/mts310/motestatus.php

The Third Space: Future Avatar Cities

I proposed in a recent interview for the Internet of Things Council[6] that future cities will be merged into real-time connected up data cities. Not just one space but a connection of networks and real-time information flows. The results created will lead to mashed-up cities and real-time performative city experiences.

I am now interested in how this shared data space can overlap, creating a new space in between, which only two nodes share, a future avatar city. The aim here is to give tangible form to this new space, the space where the cities overlap, presenting an alternative urban virtual environment. In the prototype, the audience sees the data and will be able to mix data from cities; they can even mix the data to make music with it. This novel approach allows a critical reflection on the real-time city.

'The Third Space' speculates that social sensing might lead to a new social space and, eventually, to new business enterprises which can be exploited. What is possible is that significant breakthroughs in knowledge about the shared data experience can be achieved through user-based interface online, on mobiles, media facades and other platforms (multi-distribution will spawn microbusiness).

In the last few years, several systems have been deployed to monitor city environments. Mostly they exist as isolated networks. My plan is to ask them to link up. This takes a step towards the 'City of Bits' that Bill Mitchell talks about.[7]

Future cities will be merged into real-time connected data cities. A connection of networks of real-time information flows as demonstrated in trials for my earlier AHRC-funded project (*Sensity* 2006–2009). 'The Third Space' focuses on how data flows can overlap in livestreams, demonstrating the possibilities for novel artistic experiences and technical outputs, including sonifications, visualisations and sculptural objects.

'The Third Space' creates outputs that directly show how networks can be connected and data harvested for creative uses. In the last few years, several cities have started to use mote sensors to create data spaces as test beds including Newcastle and soon Santander. 'The Third Space' also impacts on connected environments, smart places, smart cities and wireless sensor networks.

By using the standardised XML data streams that were developed previously to make the data/city available in the public domain, the aim is to improve this availability and connect up multiple data spaces (cities) to show the impact on our experience of the city and thus create an ecosystem prototype. The data environment becomes a virtual data map of the real events. What is demonstrated is that the collected data can be remade or reconstituted to be real again enabling physical objects to interpret the virtualised city data. The analogue is made digital, and the digital can be reformed into a variety of output devices.

---

[6]http://tinyurl.com/3trotzq

[7]Mitchell (1996).

One has to experiment with the technology and understand how to get under the bonnet, and then new outcomes will shed light on how data flows overlap in livestreams demonstrating what are the possibilities for novel artistic experiences and outputs as well as new business models which will benefit from sonifications, visualisations and sculptural objects, as well as apps and feedback into other networks and integrated systems.

Problems Inside the Emergent City

Can we use new technologies to imagine a world where we are liberated and empowered, where finally all of the technology becomes more than gimmick and starts to actually work for us, or are these technologies going to control us, separate us, divide us and create more borders? Will the securitisation of city space create digital borders that monitor our movement and charge us for our own micro movements inside the system? The conceptual ambition of the project is to answer these questions.

The result will be used to test the main hypothesis which I believe to be a new social space that exists in between these independent networks. What happens when future cities are merged into real-time connected data cities? The results created will lead to mashed-up cities and real-time performative city experiences. For example, how can we merge collected data from various real-time cities to visualise this new space, the space where the cities overlap? This could allude to a new architectural and urban virtual space.

This might also allow for a greater community of interpreters and beneficiaries to see, and to come to their own understandings arising from, this data about our socially networked environment.

Social Spaces

Underpinning this research are a whole series of potential problems about observation, surveillance and the ethics of the control space. By researching current surveillance systems, tracking software and pattern recognition software, one can come to an understanding about the social and ethical implications of such technologies both in artworks and public domain space and to speculate where these technologies could lead us in the future. By building my own art systems and tools using custom-made software, I try to raise further questions about the ethics of the control space and surveillance space.

Alongside this technology research, I have focused on the aesthetic development of system outputs by investigating different ways to represent the data and different ways to question the meaning of the system.

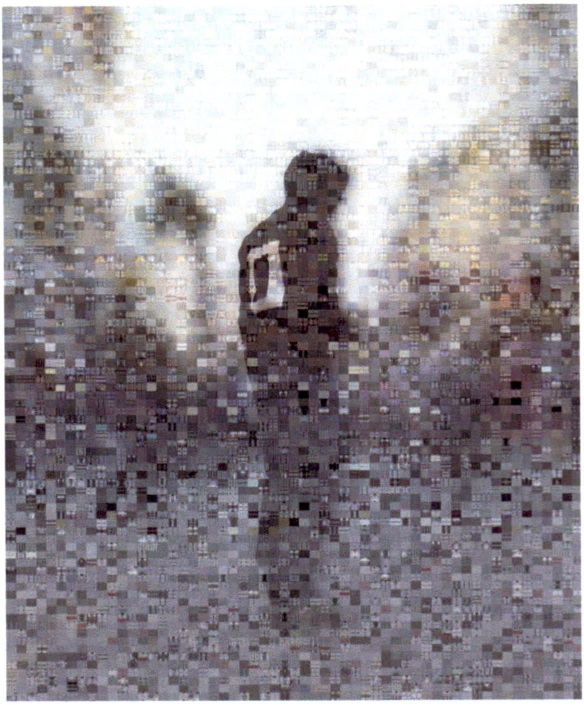

Self-Portrait. Inside CCTV. 2009. Unique C print on aluminium, 200 by 220 cm

Cultural Contexts

How we understand and value information is of great importance. It seems reasonable to suggest that visual metaphors might simplify our understanding of data in space. Adopting visual and poetic metaphors for gathered data enables a perspective which I call a 'parallel reality', a multi-point perspective. I endeavour to make this artwork eternally current allowing a real-time multiple perspective of an identified space.

The Art of Gathering Environmental Data

Selected projects and artworks I have made since 2004 that demonstrate the art of gathering environmental data. These works came into being because of a Nesta Dreamtime award and an AHRC creative fellowship grant. Most of these artworks were made between 2004 and 2012.

The Emergent City. A Life from Complexity to the City of Bits

http://stanza.co.uk/emergentcity_show/index.html

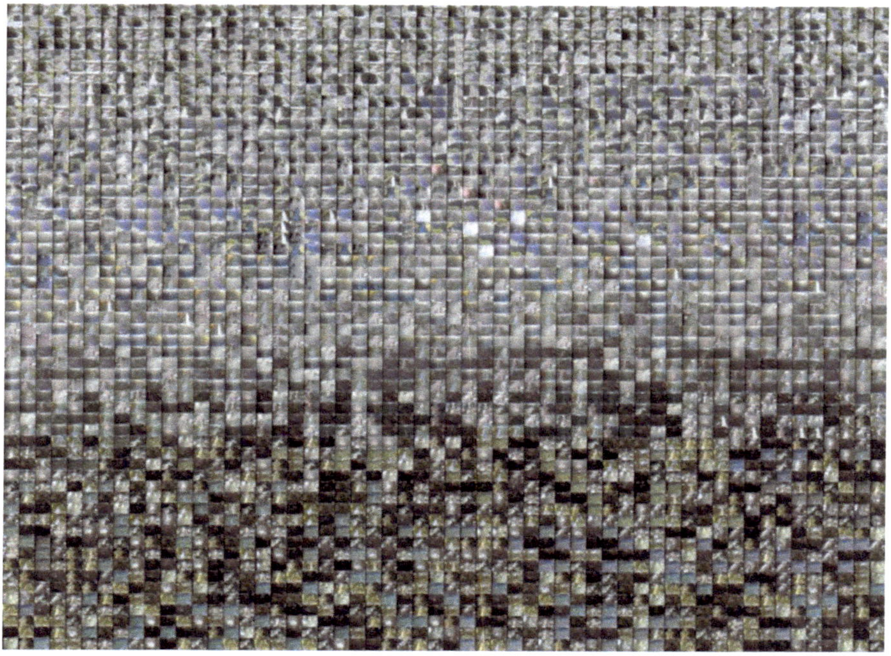

Parallel Reality. Madrid 2006. Output from online software system using real-time CCTV images (Artwork on aluminium 120 cm by 100 cm)

The artwork captures the changes over time in the environment (city) and represents the changing life and complexity of space as an emergent artwork. The artwork explores new ways of thinking about life, emergence and interaction within public space. The project uses environmental monitoring technologies and security-based technologies, to question audiences' experiences of real-time events and create visualisations of life as it unfolds.

Body 010000100110111101100100011111001

http://stanza.co.uk/body/index.html

Body is a sculpture which responds to the emergent properties of the environment in South London where the artist's wireless sensor network is situated. It represents the changing life and complexity of urban space as a dynamic, kinetic artwork.

Sensity

http://www.stanza.co.uk/sensity/index.htmly

Sensity artworks are made from the data that is collected across the urban and environment infrastructure. The sensors interpret the micro-data of the interactive city. The output from the sensors displays the 'emotional' state of the city online in real time, and the information is also used to create offline installations and sculptural artworks.

Data Cities

http://www.soundcities.com/data.php

These data maps show live environmental data from a 40-mote wireless sensor network that can be deployed anywhere. They monitor light, temperature, humidity and noise.

Intelligent Sheep: Baa Ram Ewe . . . to Your Clan Be True

http://www.stanza.co.uk/sheep/index.html

This is an interactive sound performance and concert. This artwork uses local environmental data collected using ad hoc wireless networked devices for environmental monitoring, which are attached to the sheep. In this case, the dozen sheep collect and send data about the environment and respond to the space as a collective as they move about.

Faith in God We Trust

http://www.stanza.co.uk/ingodwetrust/index.html

Faith is an artwork made using data harvested from sensors scattered over the cathedral. The sensors respond to changes in the environment they are located in this case the Liverpool Cathedral. The data is turned into a sound stream, this sound stream represents God's presence, and you can listen to this sound, the sound of God.

House

http://www.stanza.co.uk/house/index.html

House is a dynamic public sculpture viewable over the Internet. House describes the space, a real Victorian terraced house, in this case, that the artist lives in. House is a live embodiment of change and renewal. In 'House', the private interior has been made public. Sensor data unfolds and discloses the inherent properties of the space, creating an online artwork.

Tree

http://www.stanza.co.uk/tree/index.html

A tree that makes music and sings a song about the environment. The first version of *Tree* used 40 networked multisensors. The sensors are hidden all over a tree, broadcasting sensor data (light, temperature, humidity, noise and GPS location). The data is translated to music. The results produce a singing networked tree which can be heard in the park.

A World of New Possibilities

http://www.stanza.co.uk/possibilities/index.html

The landscape becomes virtual, dynamic and encoded. The artwork discloses the underlying data that we see that is changing all the time in front of us.

Gallery

http://www.stanza.co.uk/gallery/index.html

The gallery becomes the artwork formed by the emergent real-time data in the space. The gallery laid bare as a work of art. *Gallery* proposes that the data is art. The art is a real-time flow of the things around us that allow our senses to invoke understanding. The gallery space becomes the art described by the shifts in light, temperature and noises in the space over time.

data data data

http://www.stanza.co.uk/data/index.html

data data data is a live real-time data visualisation made using sensors which are scattered over the building. The sensors respond to changes in gallery space, i.e. the environment of the building. The changing data is turned into this visual event and projected outside across the city, in this case Liverpool. This artwork is networked, in real time, and it takes data from a wireless sensor network that is placed in the real space.

Façade

http://www.stanza.co.uk/facade/index.html

The façade is a live dynamic interface, an artwork that changes its behaviour as a result of the changing conditions in the environment. This works by sensing the city and the environment to make art. The results become representations of the real-time spaces and environment of Trondheim in Norway. The output from the sensors displays the real-time environmental and emotional state of the city online in real time, and the information will be used on the façade and online interface to control it.

Capacities

http://www.stanza.co.uk/capacities/index.html

In *Capacities*, the whole gallery space becomes one large artwork made from real-time city information and data. The aesthetic and feel of the space looks like an electronic city. The city is made of units, grids, repetition and building blocks. In the gallery city called 'Capacities', the leads, the wires and the cables are incorporated into the artwork to look like a city map. Capacities looks 'designed' like a piece of urban design, a city surveyed and controlled. The whole space becomes a map to wander through.

Sonicity: Songs of Atoms Time and Space

http://www.stanza.co.uk/sonicity/index.html

This artwork is a responsive installation, a sonification of the real space and environment. The sounds you hear are the sounds of the changing environment, i.e. the changes of noise, light and temperature of the space are turned into a real-time sound stream using dozens of wireless sensors presented as an installation on 170 speakers. My system monitors the space (the building) and the environment (the city) and captures live real-time data (light, temperature, noise, humidity, position) to create an ambient sonification, an acoustic responsive environment, literally the sound of the micro incidents of change that occur over time.

References

Foth M (2009) Handbook of research on urban informatics: the practice and promise of the real-time city. Information Science Reference, IGI Global, Hershey

Mitchell WJ (1996) City of bits: space, place, and the infobahn. MIT Press, Cambridge, MA

Chapter 11
The Art of Urban Engagement

Francis T. Marchese

Abstract Digital media artists have exploited pervasive technologies to explore urban ecology. Through their interventions, they challenge a city's inhabitants to consider and experience their local communities in alternate ways that may lead to new approaches by citizens for sustaining and enhancing their living conditions. This chapter reviews a breadth of digital-based work including graffiti, cartography, and security.

Introduction

The urban environment poses daily challenges for many inhabitants relating to inequality, spatial density, social conflict, and disorganization that are symptomatic of cities. This is a particularly challenging problem for aging cities that cannot easily build their way out of these problems through traditional modes of urban renewal. Yet the pervasiveness of Wi-Fi and cellular technology, the Internet, and social and locative media permits a city's population to transcend the current state of the urban physical grid, providing means for individuals and groups to adapt, personalize, and merge the urban and virtual landscapes. Indeed, the concept of the U-city, or ubiquitous city, is founded on pervasivecomputing resources that are made to appear everywhere. In such a city, ubiquitous computing is seamlessly integrated within the urban environment so that information, physical, and social systems are merged, and virtually every device and service is linked to an information network through wireless networking and RFID tags and sensors.[1]

[1]Currently, Korea leads the effort in building these cities, with fifteen currently planned. Songdo, Incheon, is the largest, presently nearing completion. Built on 1,500 acres of land reclaimed from the Yellow Sea off Incheon, about 35 miles from the South's capital Seoul, Songdo has been designed as a sustainable city with more than 40 % of its area reserved for green space, including

F.T. Marchese (✉)
Department of Computer Science, Pace University, New York, NY 10038, USA
e-mail: fmarchese@pace.edu; ftmarchese@gmail.com

F.T. Marchese (eds.), *Media Art and the Urban Environment*, Future City 5,
DOI 10.1007/978-3-319-15153-3_11

The power of pervasive media technology provides a means for breaking the bond to the Euclidean grid, allowing the redefinition of urban space. Here, physical urban space is considered not to be a single container, in which humans and their physical artifacts are situated, but rather a space that is the by-product of complex social interrelationships and dynamics that generate a multiplicity of contiguously overlapping spaces which concomitantly influence each other (Lefebvre 1991). By adding communication technology to the mix, it becomes possible to short-circuit traditional passageways through the urban grid and interconnect nonoverlapping urban spaces through virtual digital wormholes which reconfigure urban spatial relationships. For a city resident, these digitally augmented urban environments offer alternate approaches to spatial awareness, engagement, personalization, social collaboration, and wayfaring.

Wayfaring is the means through which urban residents encounter their environment. Indeed, the flâneur, the French word for "stroller," or an idle man-about-town, has been employed as the foundation for analysis of urban engagement (Shaya 2004). The flâneur, a well-educated man who wandered the streets of late-nineteenth-century Paris, was characterized as a gastronome, connoisseur, idler, and artist. The concept of flâneur, first explored in the poetry of Baudelaire (2010), received formal philosophical grounding in Walter Benjamin's analysis of Baudelaire's Paris (Benjamin 1999). In Benjamin's ideal, "the stroller is constantly invaded by new streams of experience and develops new perceptions as he moves through the urban landscape and crowds" (Featherstone 1998). But Benjamin came to see the flâneur as emblematic of capitalism and the alienation of the city (Benjamin 1997). As Shaya (2004) has noted, the shop windows of the new department stores and arcades of nineteenth-century Paris distracted the flâneur's perception, turning his strolling into mere window shopping. Today, that distraction persists with shops and shopping districts clamoring for a contemporary flâneur's attention.

The flâneur's sense of exploration is the foundation for contemporary artistic practices of engaging the urban environment. A noteworthy by-product of these artistic interventions has been the transformation of contemporary cartography from a scientific discipline that endeavors to introduce objective ways of knowing and understanding one's urban setting into subjective, personalized mappings that empower the city's populace (Crampton and Krygier 2005; D'Ignazio 2009).

a 100-acre park, 16 miles of bicycling lanes, charging stations for electric vehicles, and a waste collection system that eliminates the need for trash trucks. It is the first city in the world to have all major buildings meet or exceed LEED (Leadership in Energy and Environmental Design) requirements. By its completion date in 2015, the city is planned to contain 80,000 apartments, 50 million square feet of office, and 10 million square feet of retail space. Computers have been built into the houses, streets, and offices as part of a wide area network. Ironically, in 2003, the organization Birds Korea called for a halt of the reclamation project due to concerns at the loss of important tidal flats. Prior to reclamation, the Songdo tidal flats had supported several threatened water bird species and provided a staging ground for migratory waders as they traveled between the Northern and Southern hemispheres.

Many of the artworks discussed in this chapter fall under D'Ignazio's (2009) list of cartographic mediations: "psychogeography," "locative media," "experimental geography," "site-specific art," "new genre public art," "critical cartography," and "critical spatial practices." However, we are not precisely interested in how art has contributed to the redefinition of cartography as a science; rather we are exploring how artistic interventions have attempted to reengage or re-enchant the urban stroller. Clearly, the artwork examined here either is physically bound to the urban grid or explicitly utilizes cartographic technology. Thus, the sensations of cartography are never far from this artwork.

Urban Awareness

The density of visual stimulation that is intrinsically part of a city's nature tests its residents' abilities to extract meaningful information from noise. Urban settings such as Tokyo's Ginza and New York City's Times Square are notorious examples where pedestrians are visually bombarded by blinding, densely packed digital signage in relentless outpourings of information, with each stream demanding a few seconds of visual attention. In such settings, persuading people to stop, look, and think about their surroundings can be a challenge. One approach is to treat an urban dweller as voyeur. In *Omnivisu* (2004),[2] the Berlin art and design group TheGreenEyl invites people to peer into a box in which a camera films each eye. The live video feeds are projected onto two windows of an office tower, transforming it into the face of the capital with the eyes of its inhabitants. This artwork is installed at a key location in Berlin's history – the Oberbaum Bridge that demarcated the border between East and West Germany and the grounds of the former Narva bulb factory, which represents the failure of the East German economy and renewal (e.g., a new glass tower). *Omnivisu* was deployed at this locale to raise questions about the role of the city's inhabitants in the developments that shape Berlin today.

Stopping, looking, engaging, and thinking are the root of Karolina Sobecka urban artworks, created to reconnect people with public space and explore the way they interact with the built environment. To this end, she has produced interactive storefront installations that engage pedestrians. In *Sniff*[3] a pedestrian encounters a 3D computer graphic (CG)-rendered dog that adapts it behavior in response to the movements and gestures of the spectator (Fig. 11.1). *Sniff* was designed by Sobecka to explore associations created by an individual's presence in an environment. According to her, the dog's conduct externalizes our practices of assessment, evaluation, and testing, procedures that we perform each time anything new enters our scope of experience.

[2]http://www.thegreeneyl.com/omnivisu

[3]http://www.gravitytrap.com/artwork/sniff

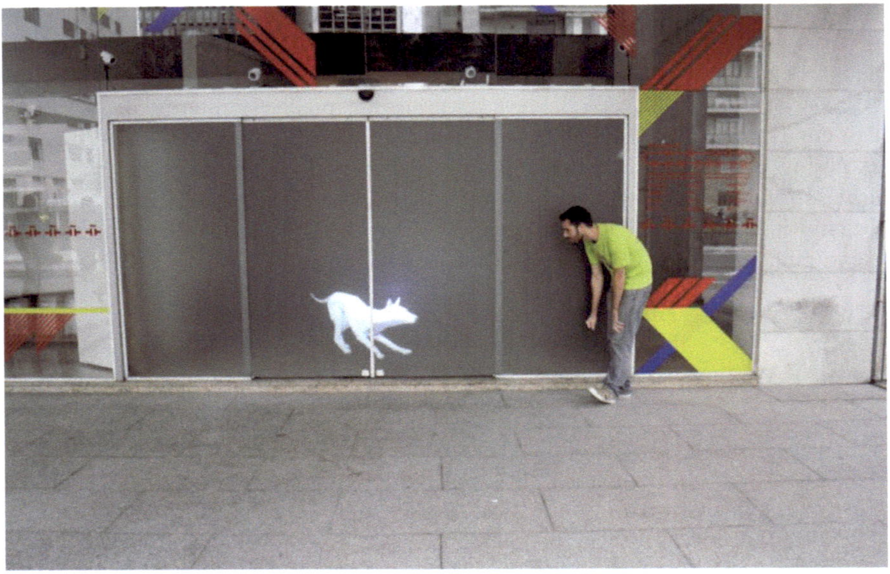

Fig. 11.1 *Sniff*, interactive storefront installation engages human with CGI dog (Photo: Karolina Sobecka, used with permission)

Sobecka expanded upon this work with *It's You*,[4] an interactive storefront installation that examines issues of public gatherings, the processes of spectatorship, and how an individual's actions are affected by the crowd. In it, CG human figures obscure something from a pedestrian's view (Fig. 11.2). When an individual moves into a position behind the CG crowd, as if to look over their collective shoulders, the group steps aside to permit viewing of what they appear to be observing. The pedestrian now becomes part of the group, a willing participant in the social dynamics of spectatorship to the point of literally obscuring the view from other passersby. If the pedestrian lingers sufficiently long, the CG characters redirect their attention to him with an anticipation of role reversal – the viewer now becomes the performer and CG crowd the spectators. If the CG crowd finds something the pedestrian does entertaining, it will react by applauding the performance. For Sobecka, the audience's attention reengages the pedestrian with the daily street scene. To participate, the pedestrian must instill the CG simulation with the specifics of his or her individual circumstance, thus prompting reflection. Finally, *It's You* calls into question an individual's anonymity in public spaces that may or may not be occupied at any moment in time.

Vehicles, construction machinery, sirens, building mechanicals, and more all contribute to the aural cacophony that comprises the urban soundscape, the acoustic

[4]http://www.gravitytrap.com/artwork/its-you

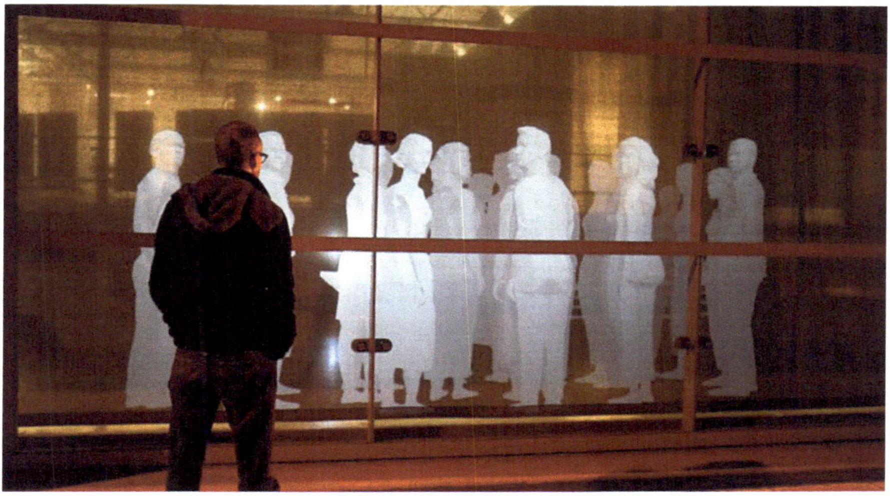

Fig. 11.2 *It's You* transforms spectators into street performers (Photo: Karolina Sobecka, used with permission)

properties of cities that help people relate to their diverse spaces (Southworth 1969). In general, a soundscape is "the collection of biological, geophysical and anthropogenic sounds that emanate from a landscape and which vary over space and time reflecting important ecosystem processes and human activities" (Pijanowski et al. 2011). Filtering out noise has become an autonomic act for city dwellers that, by habitually doing so, may have lost awareness of novel or beneficial aural-sensual aspects of their urban environment.

Reengaging urbanites with their sonic environment has been the target of many media artists and designers. In *Infrasonic Soundscape* (2001),[5] Hidekazu Minami's created a visualization of New York City's (NYC) soundscape to provoke people into realizing the importance of the infrasonic sounds and their relationship to urban space. He selectively sampled from a series of distinct locations a collection of low-frequency sounds that are not normally perceived by residents and integrated them into an interactive sonic geographical browser that takes on the form of a radar screen, allowing users to search the city for sounds and play them. Each sound source is a unique object carefully selected for its atypical soundscape, such as the former World Trade Center vibrating in the wind, massive crowd footsteps on Grand Central Terminal's marble floor, and water lapping against a wooden pier on the Hudson River. In a similar vein, the artist China Blue and her team (2007) documented the Eiffel Tower's ambient acoustics and vibrations. Employing both in-ear binaural and custom-designed seismic microphones, recordings captured the ambient soundscape to document the sound of the wind against the tower's steel

[5]http://www.soundtoys.net/toys/infrasonic-soundscape

surfaces, the intermittent rumbles of its elevators, and the vibrations from the 30,000 daily visitors as they moved from the ground to its summit.[6] She expanded upon this work with *Aqua Alta* (High Water) (2008),[7] a work inspired by the effects of global warming on Venice, its canal environment, and the seasonal flooding of its piazzas, campos, and calles. This immersive sound installation points to Venice's dependence on and the threat by water, submerging the listener in the tidal sounds of water as it splashes on buildings and stairways and underwater sounds of boat movement and sea life. In all, the artwork's rhythms represent the dangers Venice faces from global-warming-induced changes to its aquatic environment.

The works of Hidekazu Minami and China Blue disengage their soundscape recordings from the environment, isolating presentation from the aural chaos of the street. Other artists and designers have engaged urban dwellers in situ. In *Recycled Soundscape*[8] (2004) (Franinovic and Visell 2007), an existing urban soundscape is transformed by an apparatus that is part public sound sculpture and part acoustic-ecological remixer (Fig. 11.3). Its installation in a public space engages onlookers to transform elements of the existing din into a new soundscape by using three large sculptural interfaces which capture and perform sounds that are obtained from its surroundings. A tall red instrument called the Beludire allows listening via headphones and recording sounds from its surroundings with an embedded parabolic microphone. Upon capture, sounds are echoed from loudspeakers integrated within the other two instruments, called Sonic Bowls, and archived as part of the acoustic memory of the location. This aural memory may be evoked and performed by spinning the sonic bowls to select and generate sounds that vary based on the manner of bowl rotation. When the bowls are used in concert, the generated soundscape varies depending on the mutual state of both bowls and comprises a range of moods, based on rhythm, ambient sounds, and memories. The artwork engages residents in a variety of ways. Many individuals spent a significant amount of time exploring the instrument's intricacies and its recordings. Others created performances by arranging for recordings to be captured on cue. And the sonic bowls were used in tandem as part of invented games. In all, the apparatus became a focal point for contemplation and social engagement providing a means for urban dwellers, at least for a short time, to seize control of their sonic ecology.

In contrast to *Recycled Soundscape*, Altavilla and Tanaka's (2012) *The Quiet Walk* (2011) is an interactive mobile artwork that guides pedestrians in their quest for the quietest place in the city. The system has three parts: a smartphone which displays text messages that guide the user; a web server that continuously collects GPS and acoustical data from the smartphone, processes it, and returns movement instructions to the phone; and a computer in an exhibition space that displays a visualization of the acoustical map based on the acquired sonic data processed by the server. The artwork's goal is to explore the relationship between public and

[6]http://www.chinablueart.com/EiffelTower.htm

[7]http://www.chinablueart.com/AquaAlta.html

[8]http://www.zero-th.org/RecycledSound.html

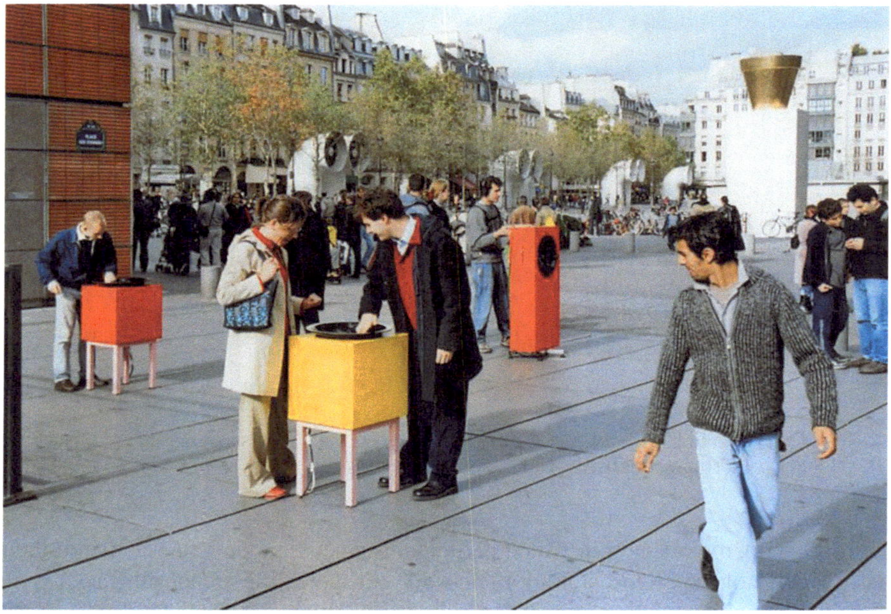

Fig. 11.3 *Recycled Soundscape* by Franinovic and Visell installed at the Centre Pompidou, Paris (Photo: Yon Visell and Karmen Franinovic 2004, used with permission)

private spaces by characterizing daily urban acoustics and by marking map locations to help users shape their personal sonic memories of places encountered. Beyond that, participants found that the artwork expanded their sense of place through the artwork's auditory enhancement of even simple sonic events such as footsteps, birds flying overhead, or buses passing at a distance. Indeed, the strength of this artwork, as with *Recycled Soundscape*, is that it offers urban residents an opportunity to focus on their surroundings and hence a greater awareness of the shape of their local ecology.

Awareness of one's urban setting demands an emotional commitment. Christian Nold has documented this as part of his *Bio Mapping/Emotion Mapping* project (2004–2012).[9] *Bio Mapping* consists of a workshop methodology; a wearable device of his invention that records galvanic skin response (GSR), a simple indicator of emotional arousal; geographical location measured via GPS; and data storage. Workshops were organized for residents to explore their local area while wearing the device. Upon return from their walks, GSR and GPS data were visualized together as an area chart in Google Earth. The varying heights of the track indicate physiological changes that relate to the wearer's degree of excitement. Figure 11.4 shows a 3D map from the 11th arrondissement in Paris. Similar maps were projected

[9]http://www.biomapping.net/

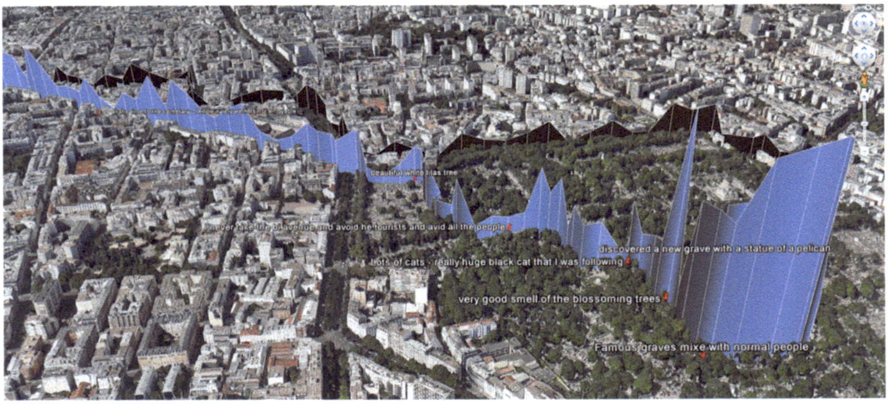

Fig. 11.4 3D emotion map of the 11th arrondissement in Paris showing the results of an individual's walk (Image: Christian Nold, used with permission)

during an open workshop in which participants discussed the data relative to their recollection of experiences along their walk. At workshop's end, all tracks were spatially annotated with comments and integrated into a geographical database that defined the collective emotion map of the area. According to Nold, what was most interesting was "the way people's descriptions tended to blur intimate feelings, events in the physical environment as well as their personal opinions about the area." For him, this "presents a new vision of space, which is relational and material and where geographical space is the common anchoring point for different people's propositions."

Nold's mapping project provides a means for realizing an individual's or group's emotional awareness of urban ecology through the use of a single sensor. Brian House adds more sensors in his work *Forty-Eight to Sixteen* (2012) and broadens an individual's sense of embodiment.[10] House documented one occurrence of his daily 7.5-mile bicycle commute from Brooklyn to Manhattan as a real-time 35-min video that captured his complete door-to-door ride. The video's musical soundtrack was orchestrated based on a number of his biometric measures. His heart rate and the cadence of his pedaling were captured by means of a Garmin Forerunner GPS watch; his breathing was measured with a custom iPhone audio processing app; and the video of his ride was recorded with a chest-mounted Contour camera. House orchestrated the piece so it was precisely timed to the video, and had the cellist Topu Lyo interpret the score. *Forty-Eight to Sixteen* shares many of the aspirations of previously discussed sonic works about the city and its rhythms, including how urban dwellers orchestrate and embody temporal memory.

[10]http://brianhouse.net/works/forty_eight_to_sixteen/

Place Ma(r)king

As long as there have been cities, there has been a need for residents to mark their space. Whether generated as acts of vandalism, social activism, civic pride, or high art, the physical annotation of urban spaces continues to attract taggers, posters, muralists, and now media artists. When considered as acts of vandalism, these works are typically termed *graffiti*, in which writings or drawings have been scribbled, scratched, or sprayed illicitly on a wall or other surface, often in a public place, usually by a nonartist.[11] Equally, the term *street art* may be applied to the forms of visual art created in public spaces, possibly unauthorized, executed independent of traditional art venues. Terminology aside, all these works represent interventions in public spaces in response to the rigidity of the urban grid and the sociopolitical infrastructure that sustains it. Hence, anticipated consequences of either scrawling on or painting public surfaces have been the transformation of public space into private, the construction of distinctive environments, and the promotion of ethnic unity and diversity (Phillips 1996).

The ephemeral nature of graffiti, i.e., the expectation that any work may be eliminated or superseded at some future time, plays into the utilization of digital methods for its generation and conservation. By virtualizing the process of graffiti/street art creation, it becomes possible to intervene in an urban environment, engage its residents, and move on without leaving a trace. Graffiti Research Lab's (GRL) *L.A.S.E.R. Tag* is one instance.[12] A laptop connected to a camera tracks a high-powered green laser pointer across the face of a building, generating graphics based on the laser's position, which in turn is then projected back onto the building with a high-power video projector (Fig. 11.5). *L.A.S.E.R. Tag* produces graffiti in its purest sense, just a tag or simple comment. One strength of GRL's art is that by publishing the documentation required for replicating the system, including all software, it empowers anyone who wishes to become a laser tagger.[13]

The VR/Urban group's *SMSlingshot* (2008)[14] approaches graffiti drawing by employing a digitally enabled slingshot and SMS (short message service) messaging (Fischer and Hornecker 2012). A user types a message on a small keyboard embedded within a specially designed slingshot created by the group, aims the slingshot at a point on the wall that has been selected for video projection, pulls firmly on the slingshot's flexible ribbon, and releases. The text is sent from the slingshot to the computer controlling the projector and appears within a colored splat projected on the wall.

The *MobiSpray* system (Scheible and Ojala 2009) replaces slingshot with a smartphone as a controller to digitally spray paint any surface. Users can select

---

[11]"Graffiti." Oxford Dictionaries. Retrieved 2014-05-05.

[12]http://www.graffitiresearchlab.com/blog/projects/laser-tag/#video

[13]http://www.muonics.net/blog/index.php?postid=15

[14]http://www.vrurban.org/smslingshot.html

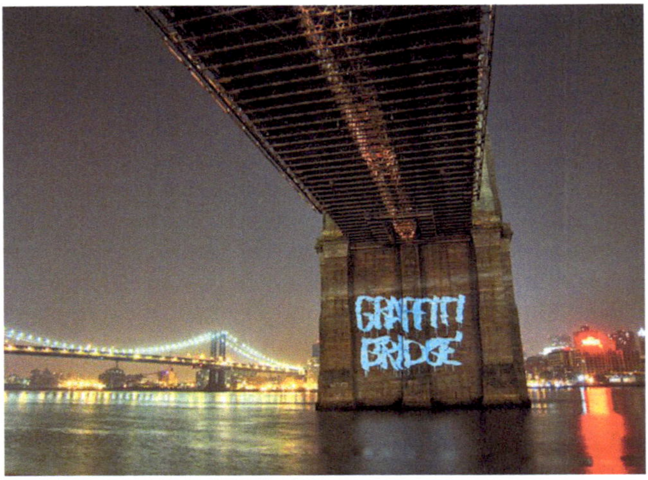

Fig. 11.5 An example of Graffiti Research Lab's *L.A.S.E.R. Tag* on the Brooklyn Bridge, New York City (Image retrieved from http://www.flickr.com/photos/urban_data/tags/lasertag/) (Image supplied by Graffiti Research Lab under Creative Commons License)

nozzle size, color, and intensity using the phone's keypad and add images as well. *MobiSpray* works collaboratively too, allowing up to four individuals to use it simultaneously.

Performance and spectacle have become an increasingly important part of street art. Media artists Ygor Marotta and Cecilia Soloaga, known as Vjsuave,[15] project their animations, paint in real-time, and perform, sometimes on the move, from their platforms of customized tricycles (Fig. 11.6). They use *Tagtool* for the iPad[16] for their live painting, an app developed by OMAi to create on-the-fly animations by individuals or groups in concert. What their artwork demonstrates is that software like *Tagtool* can empower any individual or gathering to intervene in an urban setting to enhance its sense of place.

URBANSCREEN,[17] a collective of eight artists from different disciplines representing architecture, music, stage design, and media art, constructs and choreographs large-scale, site-specific projection installations in urban spaces. Their most notable are the animation of the Sydney Opera House's sail-shaped roofs (2012)[18] and the tiles of façade of the Hamburger Kunsthalle (555 KUBIK, 2009) (Fig. 11.7).[19] The key to URBANSCREEN's work is that instead of hiding the

[15]http://www.vjsuave.com/

[16]http://www.omai.at/

[17]http://www.urbanscreen.com/

[18]http://www.urbanscreen.com/usc/1124

[19]http://www.urbanscreen.com/usc/41

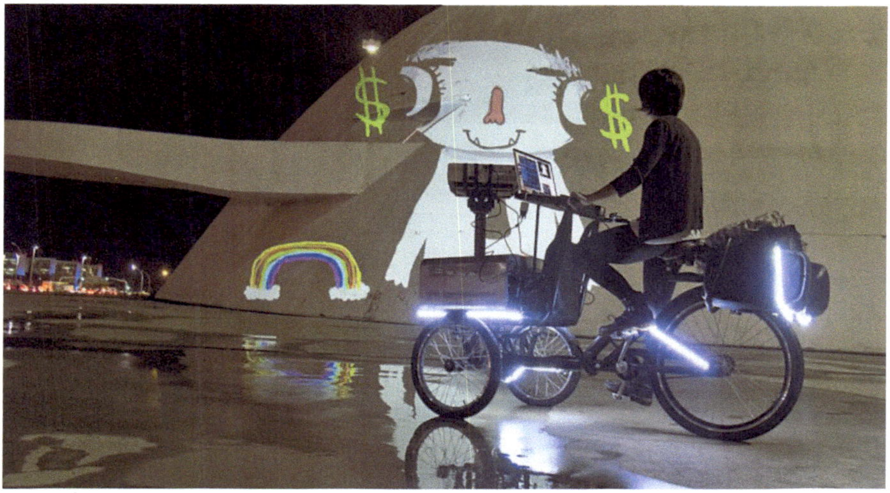

Fig. 11.6 A member of the group Vjsuave on a customized tricycle (Photo: Vjsuave, used with permission)

architecture under layers of graffiti, it is designed to emphasize a building's architectural or environmental features in diverse ways so as to present it or its space to the city's residents in a new light.

Digital graffiti provides synchronous place marking and making experiences for artists and viewers. Once an experience has concluded, the environment returns to its pre-intervention state. Collective memory fades without photos or videos of the event, a situation also true for traditional graffiti. A number of approaches have been taken by artists to preserve urban sites and experiences. Cassidy Curtis created *Graffiti Archaeology*,[20] a project devoted to the study of graffiti-covered walls as they change over time. It captures the process of constant change, making it visible through an interactive time-lapse collage of photographs of certain walls taken over a span of months or years. The photos that make up *Graffiti Archaeology* are of San Francisco, New York, Los Angeles, and other cities. Gathered from diverse sources, including Cassidy Curtis's own collection, other photographers, and various graffiti sites on the web, these images reflect a time span from the late 1990s to the present. When this project began, it was a phenomenon unique to the era of digital photography and the Internet: structured, networked, grassroots assemblage, where the assembly and juxtaposition of scattered fragments can create new kinds of insight.

Re + Public and Heavy Projects take an augmented reality approach to conservation. In *Bowery Wall* (NYC, June 2012),[21] BC "Heavy" Biermann used augmented

[20] http://grafarc.org/

[21] http://www.republiclab.com/projects

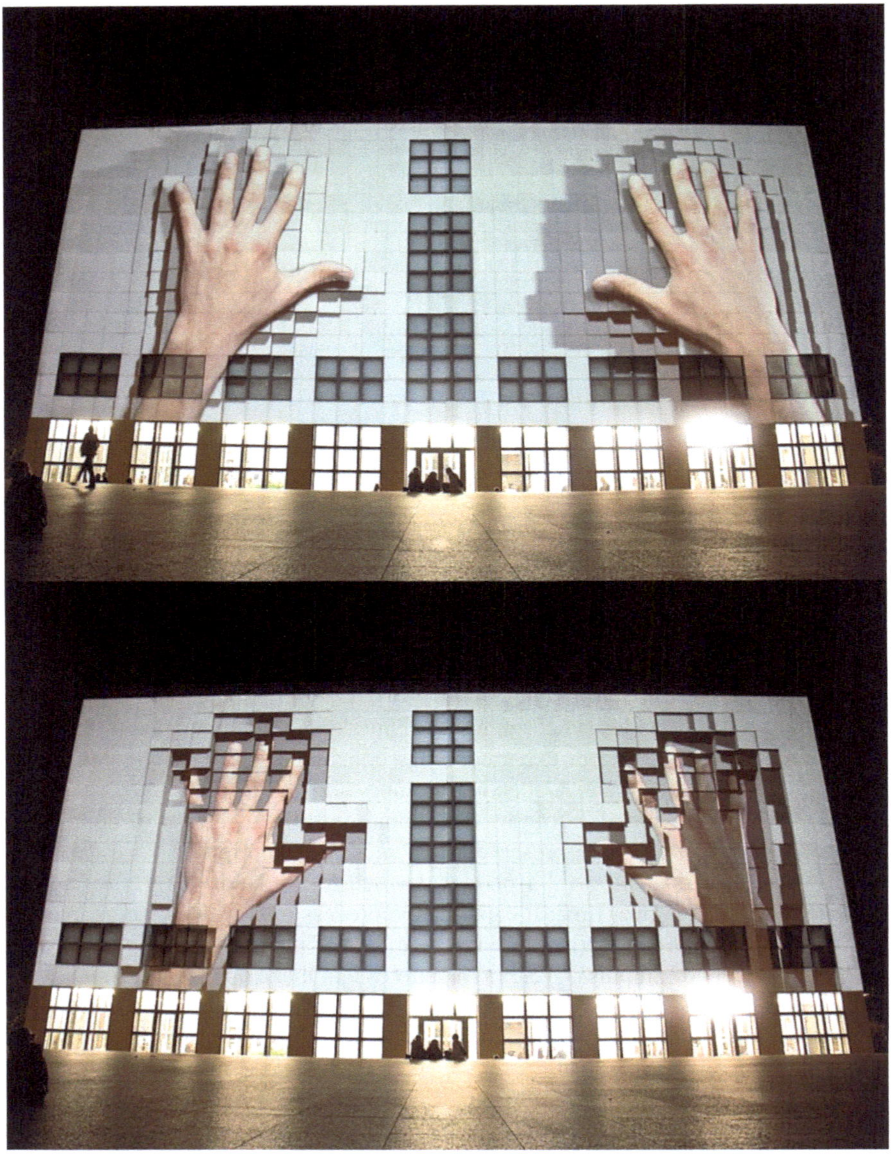

Fig. 11.7 URBANSCREEN's animation of the Hamburger Kunsthalle's tiled wall (Photo: URBANSCREEN, used with permission)

reality to create a virtual history of a well-known mural site in New York City at the corner of Houston St. and The Bowery. In 1982, Keith Haring and his companion Juan Dubose illicitly painted the first mural there, which was eventually tagged and

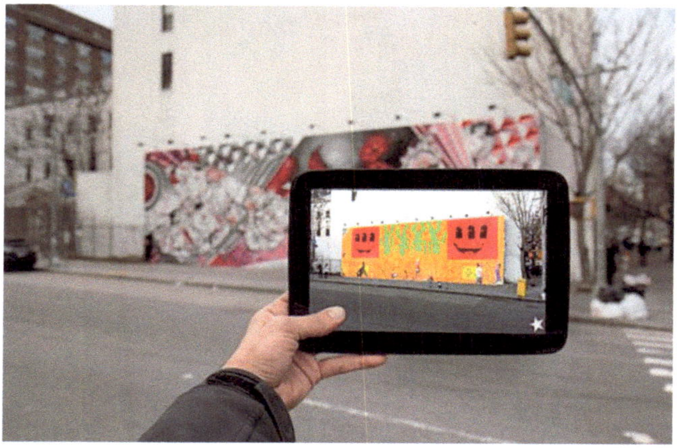

Fig. 11.8 Re + Public and Heavy Project's *Bowery Wall*, an augmented reality project, showing Keith Haring's original graffiti on with an iPad pointed at the wall (Photo: Re + Public (The Heavy Projects + PAC), used with permission)

bombed with ads. In 2008, the wall was restored as a memorial to Haring with the aid of the wall's owner Tony Goldman and curator Jeffrey Deitch at a cost near $30,000 (Koppel 2008). Since then, the Goldman projects have collaborated with artists to regularly reimage the wall, thus creating a natural sequential timeline of murals. Users of Biermann's app can take a virtual tour of the wall and view each piece as it originally appeared just by pointing a smartphone or tablet at the mural (Fig. 11.8). This in situ approach makes reviewing an environment's history an immersive experience that is sharable and integrable with urban wayfaring.

The marking and personalization of space is one of graffiti's strengths. Another is the ability of traditional graffiti to support dialogue among residents by providing surfaces for successive markers to annotate previous work. A number of media artists have augmented the physical annotation of the urban environment with digital annotation as well.

Christopher Allen, Michael Counts, Brian House, and Jesse Shapins created *Yellow Arrow* (2004),[22] a street art project that began on Manhattan's Lower East Side and eventually expanded to 467 cities in 35 countries. Project participants placed uniquely coded yellow arrow stickers at locations or on objects to draw attention to interesting urban artifacts or social or cultural landmarks of note. Once an arrow was set in place, participants sent a text message related to the artifact or site via SMS from a mobile phone to the *Yellow Arrow* phone number beginning with the arrow's unique code (Fig. 11.9). When another individual encountered this *Yellow Arrow*, a call to its particular number would not only retrieve the message but

[22]http://brianhouse.net/works/yellow_arrow/

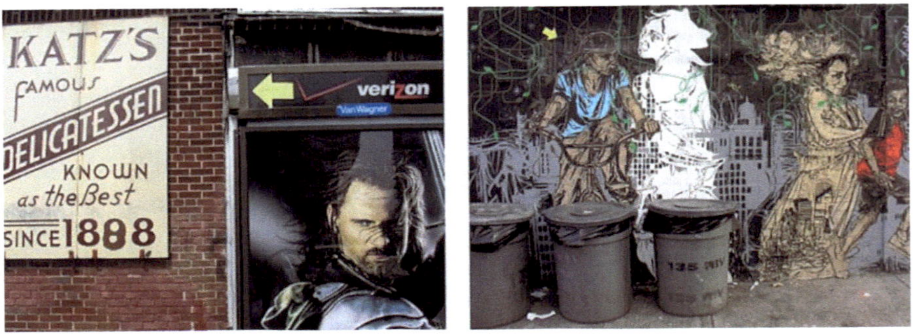

Fig. 11.9 Sample images of *Yellow Arrow* placements. *Left image*: image of sign for Katz's Deli, *Yellow Arrow* jgta4a; location: 205 East Houston Street, Manhattan, NY; photo by sirHC (April 5, 2008). *Right Image*: image of artwork created by Swoon, *Yellow Arrow* p55j; location: 135 Rivington Street, Manhattan, NY; photo by new612 (April 5, 2008) (Reproduced under Creative Commons License)

allowed the caller to reply to its originator. The yellowarrow.org website[23] extended this exchange of information by allowing participants to annotate their arrows with photos and maps. *Yellow Arrow* not only predates but presages Google Maps, Flickr, MySpace, and Facebook in the use of locative media to enable personal annotation of urban space. It afforded a gentle means for formation of social groups among its members through asynchronous communication and collaboration that would have been far more difficult to do synchronously (i.e., face to face) in an urban environment. It also acted as a repository for shared thoughts, secrets, and histories that enriched the urban ecology.

Yellow Arrow promotes a wayfaring approach to exploring one's environment, a practice reminiscent of American hobo culture that was prominent during the late nineteenth through mid-twentieth centuries, where migratory workers or homeless vagabonds traveled the country in search of work. As they visited towns and cities, they learned about the best and worst these urban environments had to offer them and generated a visual code which they used to scrawl on surfaces to document and communicate their knowledge and concerns to fellow transients. Their iconography was created to signal a breadth of information from noting a dangerous individual or neighborhood to where work may be available. In 2011, Golan Levin and Asa Foster updated this annotation scheme for digital nomads wandering the urban landscape with their *QR_STENCILER* software and *QR_HOBO_CODES* coding system.[24] Their QR coding system[25] represented the hobo urban messaging scheme (e.g., "turn right here," "dangerous dog," "food for work") and added 97 more that

[23]The project was suspended in 2006 and the website archived to Flickr (https://www.flickr.com/photos/yellowarrow/collections/).

[24]http://fffff.at/qr-stenciler-and-qr-hobo-codes/

[25]A QR code (quick response code) is a form of two-dimensional barcode used widely to communicate URLs and other short text sequences through camera-based smartphones.

reflected contemporary situations such as "insecure Wi-Fi," "hidden cameras," and "vegans beware." *QR_STENCILER* is a program which generates stencils for spray-painting the codes on surfaces. In all, the *QR_STENCILER* software was developed as a disruptive technology, offering city residents a new way to mark up their environment.

The placement of spatial tags creates a conceptual space that differs from the urban grid. If these annotated locations are taken as a whole, then the connections among related comments remap the network of physical nodes defined by the city plan and its architecture into a mesh of interconnected conceptual nodes that represents the urban dwellers' perception of the city space. This deconstruction or recasting of the cartography fashions a geographic relativism out of a map's absolutism thereby supporting a personalized geography. Such a transformation of the map can be seen in *Street with a View* (2008),[26] the first artistic intervention in Google Street View. Artists Robin Hewlett and Ben Kinsley invited the Google Inc. Street View team and over 100 residents of Pittsburgh to collaborate on a series of *tableau vivants* along Sampsonia Way, a small street on the city's north side. Staging scenes ranging from a parade (Fig. 11.10) and a marathon to a garage band practice to a seventeenth-century sword fight, Street View technicians captured 360° photographs of the street and integrated the images into the Street View mapping platform. In these Street View-captured montages, the artists and local community succeeded in recasting themselves and their environment by emphasizing map locations that were important to the community at large. In so doing, they offered an alternative, personal narrative to the typical indiscriminant Street View visualization.

While GoogleStreet View follows the grid, Jeremy Wood enhances it. Through his wanderings, Wood uses GPS technology to mark out new urban trails as cartographic annotations, incorporating both images[27] and text along his paths (Lauriault and Wood 2009). In his work *Meridians* (2006),[28] he walked a total of 458.6 miles (737.89 km) through London over a 3-month period beginning in January 2005 to create a 44.2-mile (71.12 km) long quote from Herman Melville's Moby-Moby Dick: "It is not down in any map; true places never are" (Fig. 11.11). By traversing parks, golf courses, parking lots, playing fields, cemeteries, and streets, Wood refashions urban spaces through his digital tracings to compel us to think beyond the grid in conceptualizing our local environment. Beyond his tracings, Wood empowers others to do the same through his workshops.[29] Children and adults alike can engage in GPS drawing and mapping to have their perceptions of scale, spatial navigation skills, and local awareness challenged.

---

[26] http://www.streetwithaview.com/

[27] http://www.gpsdrawing.com/gallery.html

[28] http://www.gpsdrawing.com/gallery/land/meridians.html

[29] http://www.gpsdrawing.com/workshops.html

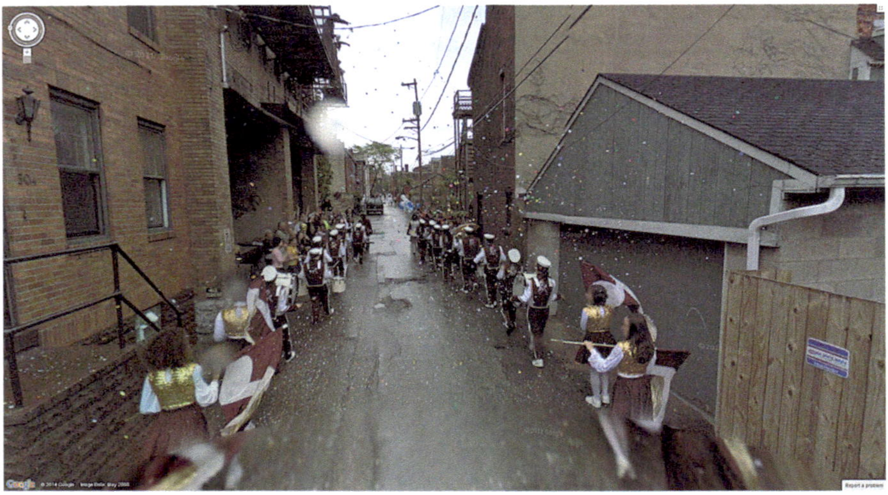

Fig. 11.10 View of parade tableau vivant within *Street with a View* (Image used with permission)

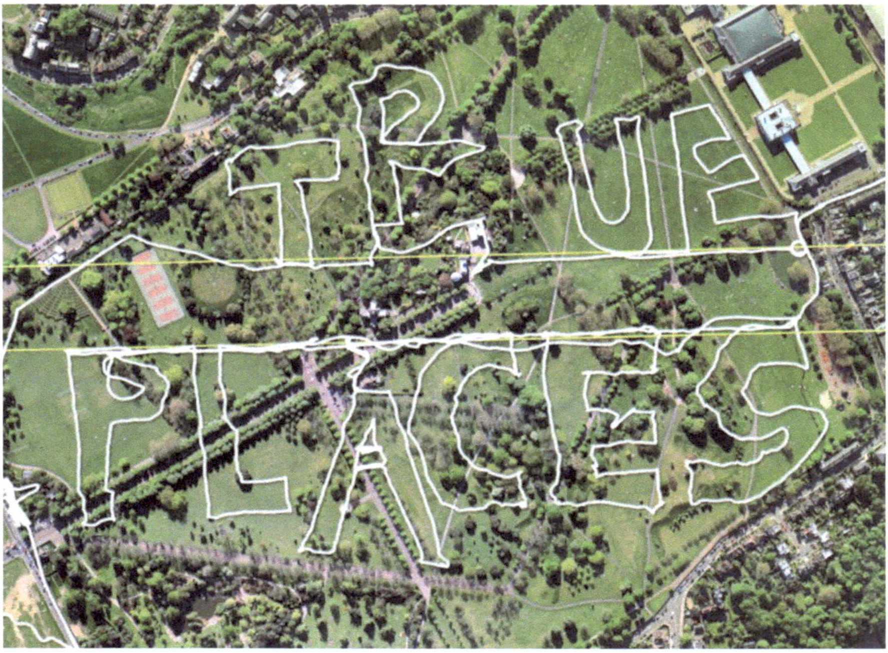

Fig. 11.11 Portion of Jeremy Wood's GPS tracing from his *Meridians* project (Image: Jeremy Wood, used with permission)

Privacy and (In)Security

The exceptional growth in video surveillance installations by private entities and municipalities has raised serious privacy concerns despite civic-minded arguments for their support of public safety. In a recent report by the British Security Industry Authority (BSIA), it was estimated there are up to 5.9 million closed-circuit television (CCTV) cameras in the United Kingdom, including 750,000 in schools and hospitals, one for every 11 people in the United Kingdom (Barrett 2013). The situation is similar in the United States. In downtown Chicago, the camera density is such that virtually every segment of public walkways is under constant high-tech video surveillance (Schwartz 2013). These cameras possess capabilities that far exceed the powers of human observation, including automatic tracking of cars and magnification of small objects at great distances. In response to the 9/11 attacks, New York City has increased the number, density, and sophistication of the CCTV system. The city's move to the use of "smart cameras" allows aggregation of data from 911 alerts, arrest records, mapped crime patterns, surveillance cameras, and radiation detectors, along with artificially intelligent tools such as facial-recognition software (Jeffries 2013). Yet the regulation, transparency (Schwartz 2013), and efficacy (BBC 2009) of these systems remain problematic with their threat to personal privacy continuing to increase unabated.

Clearly, attempting to stroll about a city without being videoed presents a major challenge to wayfaring. As a countermeasure, the Institute for Applied Autonomy (IAA), an activist group (Schienke 2002), developed the mapping application *i-SEE* (2001)[30] which allows an individual to plot "a path of least surveillance" around CCTV cameras to reach a destination. The web-based application only required the user to mark starting and end points, and the software does the rest. But given the distribution and density of CCTV cameras, many routes take the pedestrian around the periphery of Manhattan Island. Today's higher CCTV camera density would essentially eliminate the possibility of finding a completely unsurveilled path. Yet the ability to visualize the CCTV environment provides residents with the ability to assess the degree of either security or privacy they will encounter as they traverse the city's streets.

Not all CCTVs are connected to wired networks. Many private wireless networks exist, broadcasting videos at 2.4 GHz on the ISM radio band, an unregulated portion of the frequency spectrum used for wireless Internet, cordless phones, Bluetooth, and wireless surveillance cameras (e.g., baby monitors). In effect, the collective video output from all these devices forms an ad hoc media network that blankets public and private spaces including homes, offices, and shops, broadcasting the personal narratives of many of the city's inhabitants. Two artists have hacked into this band in separate projects to expose the boundaries between personal, cultural,

[30]http://www.appliedautonomy.com/isee.html

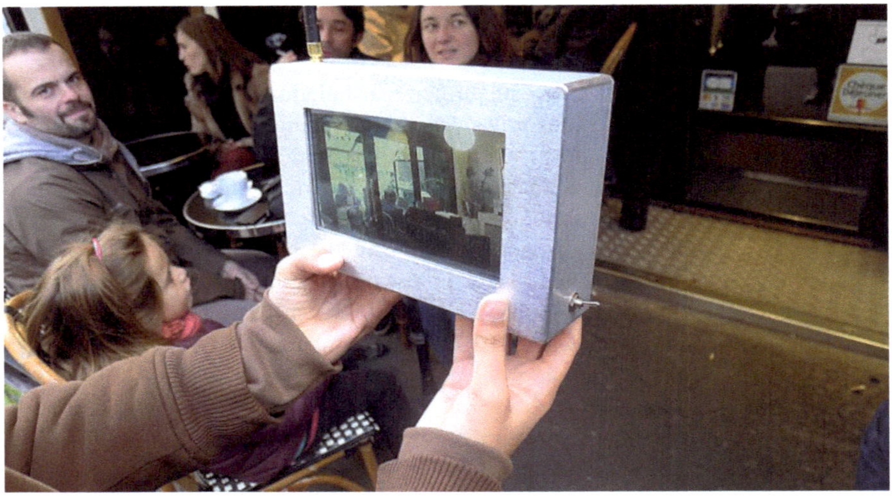

Fig. 11.12 View of a Paris café's wireless CCTV stream (Photo: Benjamin Gaulon, used with permission)

and social spaces. In *Life: A User's Manual* (2003–2006),[31] Michelle Teran creates a series of performances as a nomadic, female character touring the city's streets carrying a video scanner, inviting the public along on her walk. Intercepted live surveillance streams transmit portions of personal narrative from public and private spaces onto Teran's mobile display housed in a shopping cart or suitcase. Viewers are asked to consider the relationships between the physical spaces of the urban setting and the fragmented mediated spaces of the streamed videos. In *2.4 GHz from Surveillance to Broadcast* (2012),[32] Benjamin Gaulon collects and records video footage with a wireless video receiver from various locations in Dublin and Paris, pinning video snippets to their respective locations in Google Maps. He has installed a wireless receiver at various street locations to reveal the presence of cameras and to emphasize that anyone can receive these video signals. Finally, he conducts workshops in CCTV hacking, inviting participants to explore the CCTV wireless networks of their neighborhood urban spaces by searching for 2.4 GHz surveillance video signals. Figure 11.12 shows a wireless receiver in the hands of an urban explorer sitting at a Parisian café displaying a video from the café's wireless CCTV stream.

If one knows the location of surveillance technologies, then it is possible to take countermeasures against them. The artist/technologist Adam Harvey and clothing designer Johanna Bloomfield have created anti-drone detection garments called *Stealth Wear* comprised of hoodies, parkas, and scarfs made from metallized

[31] http://www.ubermatic.org/life/

[32] http://www.recyclism.com/twopointfour.php

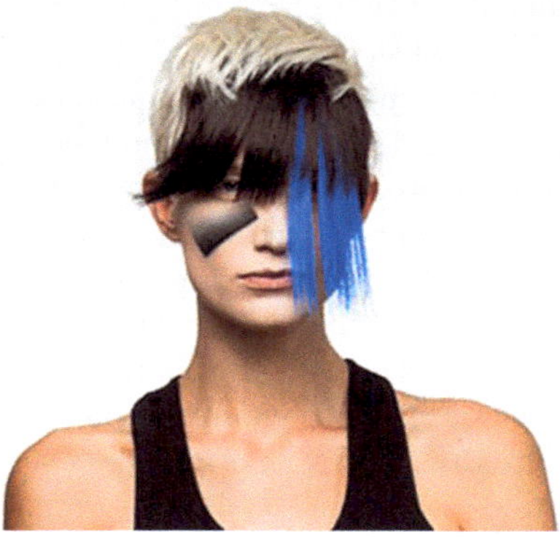

Fig. 11.13 Sample facial camouflage created as part of the *CV Dazzle* project (Image used with permission)

fabric to shield wearers from drones' thermal imaging technology; a copper wallet insert that shields credit cards from RFID skimming; and a phone pouch called *OFF Pocket* that utilizes metal fabric to shield cellphones from cellular, wireless, and GPS signals. Harvey's interest in anti-surveillance measures originated with *CV Dazzle*,[33] an attempt to thwart face-detection algorithms by combining avant-garde hairstyling and makeup to disrupt facial continuity.[34] Since facial-recognition algorithms seek out fundamental design features such as symmetry and tonal contours, Harvey theorized that radically changing hair and cosmetic patterns should fool these algorithms. Figure 11.13 shows an example of Harvey's radical makeup and hairstyles. His artwork inspired Feng and Balakrishnan (2013) who published a tool that could be used to help design facial camouflage that was able to fool facial-recognition algorithms from 82 to 100 % of the time. As they observed, this is just the beginning of what could be fruitful research in helping ensure an individual's privacy within a street setting in an aesthetic way.

Most security-related artistic interventions place law enforcement and populace in opposition. What if the artist and constabulary cooperated? Jill Magid has done that in her work *Evidence Locker* (2005). In 2004, Magid spent 31 days in Liverpool working with Citywatch (Merseyside Police and Liverpool City Council), whose function is citywide video surveillance. Magid staged performances filmed by the police using the public surveillance cameras in the city center. In her video *Trust*, that is part of the *Evidence Locker* project,[35] Magid relies on communication with police officers operating surveillance cameras to guide her through the city as she walked the streets with her eyes shut. Wearing a red trench coat for easy recognition,

[33] http://cvdazzle.com/

[34] *CV Dazzle* is derived from CV, an acronym for computer vision, and Dazzle, a type of World War I naval camouflage which employed cubist-inspired designs to break the visual continuity of a battleship and conceal its orientation and size.

[35] http://www.evidencelocker.net/story.php

she called the officers on duty identifying her detailed location. The officer then guided her through a path by telling her in which direction she should move through a crowded street. In order to obtain access to her video footage, Magid needed to submit 31 subject access request forms – the legal document necessary to outline to the police details of how and when an "incident" occurred. She chose to complete these forms as though they were love letters, expressing her thoughts and feelings. These "letters" form the diary *One Cycle of Memory in the City of L* – an intimate portrait of the relationship between herself, the police, and the city.

The question remains: just to what extent is law enforcement able to recognize it citizenry? The work of Heather Dewey-Hagborg may give some indication. Humans leave traces as they move through the urban environment. Cigarette butts, chewing gum, discarded soft drink containers, and hair are some examples. Each of these artifacts contains genetic information that may be analyzed to extract personal, intimate information not even known to an individual. If done without one's knowledge or consent, it becomes genetic surveillance. Dewey-Hagborg did just that in her artwork *Stranger Visions* by collecting DNA samples from hair, chewing gum, and cigarette butts found in New York City public spaces such as restrooms, subway turnstiles and cars, and on streets, subjecting them to forensic phenotyping in which DNA from forensic samples are translated into a description of a suspect's gender, physical appearance, geographic origin, and possible behavioral attributes. She takes the results of this analysis and feeds this information into a custom computer program she wrote which takes the values coding for physical genetic traits and parameterizes a 3D model of a face to represent them. For example, gender, ethnic ancestry, eye and hair color, freckles, skin tonality, and facial features such as nose width and distance between eyes are adjusted. Upon completion, the model is exported to a full-color 3D printer. The resultant sculpture (c.f. Fig. 11.14), according to Dewey-Hagborg, has a "family resemblance" to the person in which it would be expected that individuals with similar traits and ancestry may look more like a cousin than the person themselves. Yet this sculpture may be far more realistic and true to the "original" individual's appearance than a forensic sketch. It should be remembered that this artwork is a provocation, designed to spur a cultural dialogue about genetic surveillance, privacy issues, and law enforcement. On the one hand, many individuals attempt to secure access to their digital data from the outside world, but these same individuals leave trails of hair, nails, skin, and saliva, without considering its information content and ease of accessibility. In the end, it may be that the trail of biological detritus that we unwillingly and unwittingly leave behind as we traverse our neighborhoods will ensure that we will always be recognized – for better or worse.

Final Thoughts

The works of the artists discussed here were created to stimulate city residents' sense of environmental consciousness by offering new approaches to envisioning their urban spaces and alternative ways of thinking about how to engage their

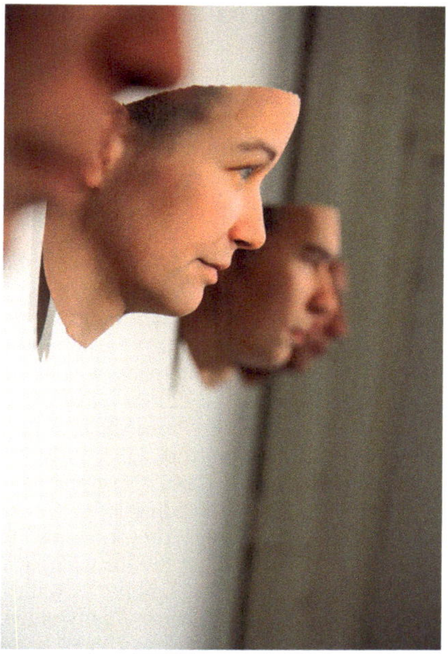

Fig. 11.14 3D printer sculpture-based DNA analysis (Photo: Heather Dewey-Hagborg, used with permission)

neighborhoods. Beyond awareness, many city residents feel a sense of responsibility for community interests and take action accordingly. Whether social or political, this commitment may be viewed as representing an act of ownership, that is, "a sense of belonging to a collective place, commitment to a collective issue, and willingness to share a private resource with the collective in order to allow other citizens to act, without infringing on other people's right of ownership" (de Lange and de Waal 2013). Indeed, these artists have demonstrated their social commitment through their attempts at stimulating positive change in the urban environment through their interventions and educational activities. Whether working at a grassroots level or directly engaging sociopolitical organizations, the creativity and timeliness of media artists are recognized immediately by individuals and institutions alike. As such, these artists possess an agility and ease of urban engagement that no political institution can achieve.

References

Altavilla A, Tanaka A (2012) The quiet walk: sonic memories and mobile cartography. In: Proceedings of the 9th sound and music computing conference, pp 157–162. http://smcnetwork.org/node/1650

Barrett D (2013) One surveillance camera for every 11 people in Britain, says CCTV survey. The Telegraph, 10 July 2013. http://www.telegraph.co.uk/technology/10172298/One-surveillance-camera-for-every-11-people-in-Britain-says-CCTV-survey.html

Baudelaire C-P (2010) The painter of modern life. Penguin, London

BBC News (2009) 1,000 cameras 'solve one crime.' BBC News Online, 24 Aug 2009. http://news.bbc.co.uk/2/hi/uk_news/england/8219022.stm
Benjamin W (1997) Charles Baudelaire: a lyric poet in the era of high capitalism (No. 7). Verso Books
Benjamin W (1999) The arcades project. Harvard University Press. Cambridge, Mass
Crampton JW, Krygier J (2005) An introduction to critical cartography. ACME: Int E-j Crit Geogr 4(1):11–33
D'Ignazio C (2009) Art and cartography. International encyclopedia of human geography, vol 1. Elsevier, Oxford, pp 190–206
de Lange M, de Waal M (2013) Owning the city: new media and citizen engagement in urban design. First Monday 18(11)
Featherstone M (1998) The flâneur, the city and virtual public life. Urban Stud 35(5–6):909–925
Feng R, Prabhakaran B (2013) Facilitating fashion camouflage art. In: Proceedings of the 21st ACM international conference on multimedia, ACM, New York, USA, pp 793–802
Fischer PT, Hornecker E (2012) Urban HCI: spatial aspects in the design of shared encounters for media facades. In: Proceedings of the SIGCHI conference on human factors in computing systems, ACM, New York, USA, pp 307–316
Franinovic K, Visell Y (2007) New musical interfaces in context: sonic interaction design in the urban setting. In: Proceedings of the 7th international conference on new interfaces for musical expression, ACM, New York, USA, pp 191–196
Jeffries A (2013) Privacy is 'off the table' in a 'post-9/11 world,' says New York City police chief. The Verge, 26 Apr 2013. http://www.theverge.com/2013/4/26/4270876/privacy-off-the-table-nyc-police-commissioner-ray-kelly-argues-for-more-cameras
Koppel N (2008) Little angel was here: a Keith Haring collaborator makes his mark. New York Times, 5 Aug 2008. http://www.nytimes.com/2008/08/06/arts/design/06hari.html
Lauriault TP, Wood J (2009) GPS tracings–personal cartographies. Cartogr J 46(4):360–365
Lefebvre H (1991) The production of space, vol 30. Blackwell, Oxford
Phillips SA (1996) Graffiti. Dict Art 13:271
Pijanowski BC, Villanueva-Rivera LJ, Dumyahn SL, Farina A, Krause BL, Napoletano BM, Gage SH, Pieretti N (2011) Soundscape ecology: the science of sound in the landscape. Bioscience 61(3):203–216
Scheible J, Ojala T (2009) MobiSpray: mobile phone as virtual spray can for painting BIG anytime anywhere on anything. Leonardo 42(4):332–341
Schienke EW (2002) On the outside looking out: an interview with the Institute for Applied Autonomy (IAA). Surveill Soc 1(1):102–119
Schwartz A (2013) Chicago's video surveillance cameras: a pervasive and poorly regulated threat to our privacy. Northwest J Tech Intell Prop 11(2):47–60
Shaya G (2004) The Flaneur, the Badaud, and the making of a mass public in France, circa 1860–1910. Am Hist Rev 109(1):41–77
Southworth M (1969) The sonic environment of cities. Environ Behav 1:49–70

Chapter 12
Our Place on That Wall: Community Online Art Projects

Vaughn Whitney Garland

Abstract Through the application of Internet technology, contemporary artists, along with their collaborators and spectators, have the potential to create, build, engage, and exhibit new works of art and propel new concepts for the production and practice of art making. This chapter examines how new online community collaborations transform urban sites into spaces for new art and surveys the nature of community online art projects, as works orchestrated by artists employing the interconnected and participatory nature of the Internet.

Introduction

At its essence, the Internet is a place for storing memories, as well as a place for community connectivity. On the Internet, we learn, explore, and create by searching through the vast virtual space that encompasses public collection and conversation. With the help of computer programs, processes, and software, we may now see the world around us through the discovery and sifting of massive waves of data. Since these massive archives continue to accumulate all kinds of material about our daily lives, many users worry that it might be difficult to maintain individuality and unique human attributes or that certain freedoms may be lost. This may be the case for some types of online data collections. But, the Internet – where vast networks, computer programs, and digital technologies give rise to overlapping and shared data collection – also has the potential to help archive things about communities that may not have been possible otherwise.

Internet archives that employ collaborative participation in order to record community engagement have the potential to foster new personal libraries and activities and to represent aspects of life that may have been overlooked, left out, or lost. In these archives, an increasing number of artists have found valuable material through which to explore growing notions of the online collection. Additionally, these artists perceive new digital technologies as they relate to daily life, in which

V.W. Garland (✉)
Independent Curator and Multimedia Artist, Richmond, 312 West 25th Street, 23225 Richmond, VA, USA
e-mail: garlandvw@gmail.com; http://www.vaughngarland.com/

F.T. Marchese (eds.), *Media Art and the Urban Environment*, Future City 5,
DOI 10.1007/978-3-319-15153-3_12

users engage through their digital devices. It is through these new communal art projects that a more complex representation of a community entwined with technology is possible. In the struggle to define the individual among the digital collection, we may now view engagements within our online communities as an attempt to document the individual's narrative within the group. By storing personal records, images, recordings, comments, and interests, we use the Internet as a place to collect and to amass but also to define who we are, what our interests are, and how others understand our communities.

Storing digital files and identities allows Internet users the ability to isolate personal space within a system that presents itself as immortal – making our online identities and interactions interminable. This space for storage gives rise to further developments through which ideas of remediation are taking new shape and where the material of the past is made new because it is now digital and located online. By investigating how users engage with their virtual environments while, and at the same time, are influenced by the relationships of their physical groups and communities, artists may examine the ways in which societies employ technology and how that technology propels new ways of thinking about collectivism. Internet-based technologies, such as mobile devices, may be an important aspect of this investigation because they now are so ingrained into daily life and utilized in numerous activities. Furthermore, it is through more inquiries into the use of Internet applications like mobile Internet technologies that we may begin to interpret how members of a group engage while making new types of art.

As individuals, companies, and governments build a wide range of online databases, artists have realized that these collections offer insights into and challenges to conceptions of participation, interconnection, and remediation. Internet-based works of art, which I will term community online art projects or COAPs, are new attempts by current multimedia artists to engage, share, and sometimes alter the relationship between art and technology. Central to COAPs is the drive to participate and collaborate with others in order to construct a community and archive online or to connect back to a physical location or group. It is the online archive that gives these art projects a place to be developed and exhibited and to become the record or "object" of the shared creative experience.

The Internet's complex system of smart technology crossed with the immediately accessible and massive networks offers artists innumerable ways to encourage participation and collaboration. Due to the increasing numbers of overlapping and connected Internet devices, digital networks, and linked communities, a COAP-engaged user and artist contribute in the conceptual direction of a work and have the power to share in its production. Additionally, online collaborative creativity calls into question considerations at the heart of current art production, including issues of originality and reproduction, the relationship of artists to audience, and the nature of artistic authorship and collaboration. Instead of defining digital programmability and computation as the basis for new media, COAPs demonstrate how Internet artists examine the technological and social system created by the use of the Internet, uncovering the properties of that system as a basis for a new creative practice.

Whether COAPs fall into disciplines like performance, sculpture, installation, design, or multimedia, they may be better defined by their cross-disciplinary and hypermedia nature which is distinguished by multiple disciplines as well as being online. They may also be evaluated by their physicality as well as their digital time. For this reason, it is crucial that we not only look at the finished project but also scrutinize the technologies as well as the communities from which the works emerge. Furthermore, notions of participation, interconnection, and remediation are ever-present and become the means for investigation by COAP artists. Because COAPs develop as complex arrangements between art, technology, and society any new interpretations or evaluations need be equally multidisciplinary. The COAP artists are the new agents for viewing the use of technology by a society or group and will give us more clarity on current life as well as a glimpse into the future, recalling Marshall McLuhan's statement, "The serious artist is the only person able to encounter technology with impunity, just because he is an expert aware of the changes in sense perception."[1]

Light of Human Kindness

Self-proclaimed "kindness worker" Patience Salgado creates community by developing online shared collections that describe and heighten personal actions while striving to construct deep personal connections with others. This care to engage with a group is reflected on the streets of her hometown Richmond, Virginia, where starting a collection of online exchanges was equally as important as the physical connection made from meeting in person on the sidewalk. In fact, it seems not to matter whether the connections develop as a physical act or from an online post; Salgado strives to build better dialogue where she connects with people on a deep emotional level. For example, in her project called *The Mother of All Journals*, Salgado sought to create a community journal where she would collect comments and emails from anonymous people in order to produce a public journal. This archive would then be left in a specific public location where others can sift through the stories written by others and even add their own.[2] While Salgado's journal project asked people to engage with her through emails and comments, another project asked her online readers to activate a community directly and report back to her. In *Love Wins*, Salgado wanted to have the community show support for the recent Supreme Court decision concerning marriage equality by fastening a red heart onto random doors, take a photograph of that installation, and then share the image with her using a hashtag (Figs. 12.1 and 12.2).

In all of Salgado's guerilla kindness projects, the artist attempts to connect people to each other through the creation of shared spaces online as well in community

[1] Marshall McLuhan, "The Medium is the Message," in *Understanding Media: The Extensions of Man* (New York: McGraw-Hill, 1964), 9.

[2] "The Mother of All Journals," on Patience Salgado's official website, accessed September 26, 2013, http://kindnessgirl.com/2013/05/31/the-mother-of-all-journals/#comments

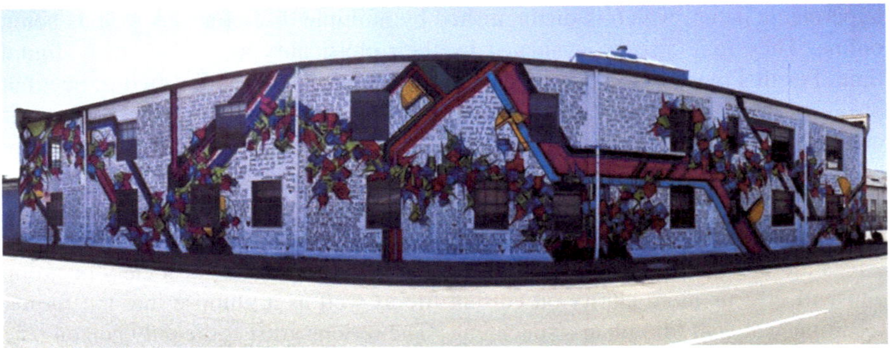

Fig. 12.1 "Finished Wall," *The Light of Human Kindness* RVA Street Art Festival September 11–15, 2013 (Courtesy Patience Salgado, Photo by Patience Salgado)

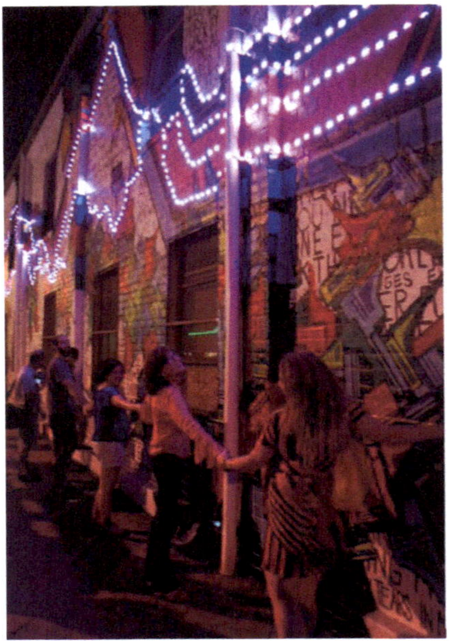

2 weeks ago

It's hard for most people to ask for help - even when they need it the most. Sometimes, it's even harder to offer help when you see someone in need - even a complete stranger. While I don't have much, as a single income family raising three kids with one in college, I try to help people whenever and however I can. Whether it's giving a colleague $5 for gas, so she can make it through to payday, providing shelter for a complete stranger who was broke down with kids and out of state, or gently offering a familiar contractor, who was over doing a small handyman project for a modest fee and expressed his own hardships about being evicted and not getting enough hours at work to take care of his young family, if I can buy him groceries and then delivering them to his house, every gesture - regardless of how small - helps.

I refer to it as a hand up - not a hand out. Everyone has something they can give - even love.

3 weeks ago

Blessed

I felt at my lowest when I was laid off from a temporary job and unemployment was exhausted after three months. Did I mention I had been laid off prior to that for two years not able to find full time employment. I had NO income and plenty of bills. Everything was behind an about to be disconnected. No agencies would assist me because I had no children, buying my home and two vehicles both which is over 13 years old. I felt as if I was drowning. I wanted to kill myself.
I knew it was going to be okay when my daughter who has four children, husband and bills of her own told me that it was going to be okay that I have always help her now she would help me. Then the one person I didn't expect to help me, my ex., helps me out financially and cuts my grass for me at no charge. He had hurt me beyond repair so I thought and I closed all doors on him when he decided to marry the person he cheated on me with. He has been there for me during this difficult time. Blessings are coming from a few people now. I still pray daily tofind a job in my field.
When nothing works out, I try to remember that GOD did not bring me this far to pass me by. He provided with a wonderful daughter and great friends and a sister that has been my rock.
When I think of where I am today, I realize that I not where I should be but I am not where I used be because I am truly loved by many, I feel lighter and more full of hope.

Fig. 12.2 (*Left*) "Hands in Front of Wall," *The Light of Human Kindness* RVA Street Art Festival September 11–15, 2013 (Courtesy Patience Salgado, Photo by Phil Riggan). (*Right*) Segment from online entries *The Light of Human Kindness* RVA Street Art Festival September 11–15, 2013

environments. Intriguingly, it is through these spaces that Salgado wishes to create her own space, a space where she can be in control of the content. Her created and managed spaces are filled with deep emotional content and connections. Salgado hungers for personal connection with her collaborators, which is why she has taken on the topic of kindness. For Salgado, her choice of kindness comes from her understanding that the act of kindness gets at a deeper understanding of humanity

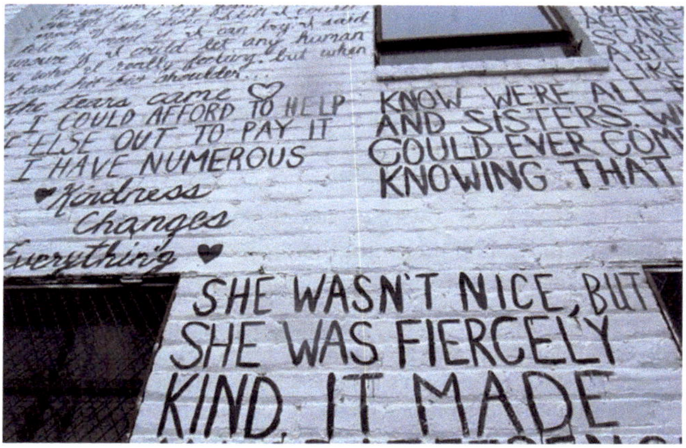

Fig. 12.3 "Fierce and Kind," *The Light of Human Kindness* RVA Street Art Festival September 11–15, 2013 (Courtesy Patience Salgado, Photo by John Murden)

than does being nice. Being kind instead of being nice reflects a specific act of personal engagement because kindness comes as an innate and more thoughtful action. She explains:

> Nice is societal and polite and makes us model citizens, which is great. We need that. But, kindness, is like empathy and compassion and makes us part of that human family. For me to take that jump and say if we are going to move that way of thinking, screw being model citizens, this thing is a little bit more on the edge, we will have to be more honest and direct about our dark and how are we not connecting in the world. The one thing that I heard over and over again when I asked people about light versus dark was they would often say that they were alone. There are too many people in the world who think they are along in their darkness. If we are all feeling that way and we are living our lives side by side what is it about that and how can kindness and how can we use art and technology. Technology and art stand on their own just fine but what would happen if we bring those things together (Fig. 12.3).

Located on the wall of the abandoned 80-foot-long transit authority administration building, *The Light of Human Kindness* existed as a place for community engagement as well as a multimedia online collaboration. Salgado took excerpts from the stories she collected online, oftentimes a one- or two-sentence selection summarizing the user's contribution, and created a mural out of them by enlisting street artist Hamilton Glass, known by his street name Ham. Salgado covered the building with hand-painted texts, which were then incorporated by Ham into the mural.

To engage the physical wall even further and to create an interaction where participants could participate with the work in the public city space, Salgado collaborated with The Martin Agency's Brian Mount and Jeff McDonald to create a new light installation using a connection from her website. The collaboration with Mount, McDonald, and others from Martin (a Richmond-based advertising firm) allowed the website's visitors who sent Salgado a story describing an act of kindness to activate a newly created LED light installation that was included into Ham's mural. The advertising agency also created an electrical current that could be initiated when visitors to the wall held hands. With all these components, the wall could be activated in a number of ways depending on how one wished to access the installation. Salgado's connection between the physical urban environment and the online posts allowed the project's participants to view their stories as a catalyst used to define a community in several ways. Once the online posts found their way into the urban setting and placed on this wall, the project became a work for the people. She describes this transition as follows:

> I guess the one thing I learned in this piece was that everyone had a place on that wall. You just had to figure out what your place was. If for you it was technology that was awesome. If for you it was the art that drew you in. Or if for you it was the fact that you shared your story of you shared something vulnerable or dark about your life. Everybody had an entry point and each of these three elements also had this base of pushing us back this togetherness. The surprise in all of it, when we asked people to share their stories in light and dark, I had no idea people would be so raw or so real. It was almost like they were waiting for a chance to say how they felt and to say what they were holding. So when those stories flooded in and we put them on the wall Hamilton turned to me and said, 'Patience, this is not a mural anymore, this is a movement. When the people themselves felt ownership over the wall it became the people's wall. People were drawn to it because they could find a part of themselves on that wall'.[3]

Community collaboration and partnership have become more of a focus for Salgado. Her performance, community activism, and archival work allow participants to feel that they have a say in how the work progresses. This is a welcome part of the creative process for Salgado. By eventually releasing any control or ownership over the work, her collaborators/participants assume more responsibility and bond with her audience. By releasing control, she is forming deeper human interaction and connections (Fig. 12.4). She states:

> I think when you crowd-source or when you invite a community to do something with you in a certain sense you no longer have control. That is an expression of our shared humanity. You have to honor that and then you're in service to honoring that art and their willingness to share. So the priority sort of shift, now it is not about us and we, it is not about I am going to take your content and make something. Now, it is about we doing something. I thought that this was going to be an expression but it turned into a movement together. Now we are acting on it. It didn't work unless the community did it. I think it is more meaningful when it belongs to everybody. Now we are crossing over to social impact.[4]

Not all of Salgado's installation came easy to her or her audience. Because the work existed in three components as a blog, a mural, and physical light installation, participants had to be able to access all parts of the work virtually and physically to get the full effect. Salgado recalls that the blog community was the easiest of the three and that setting up the site and sharing the project with others using

---

[3]Salgado, Patience, Interview by the author. Personal interview. Richmond, VA September 28, 2013.

[4]Salgado, Patience, Interview by the author. Personal interview. Richmond, VA February 21, 2014.

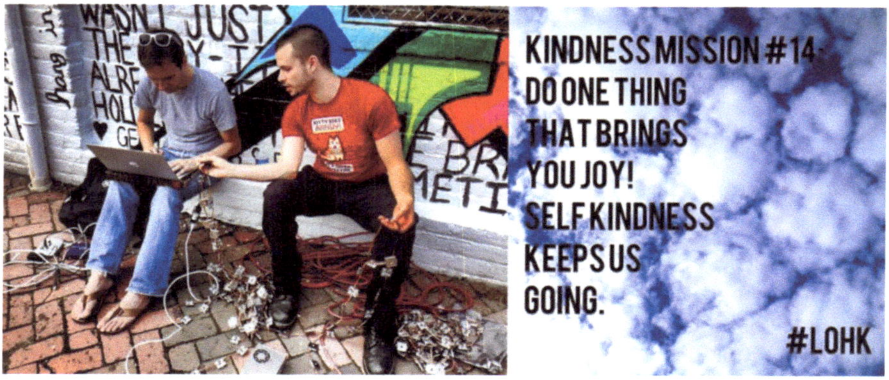

Fig. 12.4 (*Left*) Martin Agency Technologists installing LED lights for *The Light of Human Kindness* RVA Street Art Festival September 11, 2013 (Courtesy Patience Salgado). (*Right*) Online image from *The Light of Human Kindness* campaign RVA Street Art Festival September 11, 2013 (Courtesy Patience Salgado)

technologies most had become accustomed to allow for better access. It was the success of her online blog community that gave her audience a way into the project. In fact, the project may have not developed as well as it did if she had relied solely on physical connections that came later. It was the blog that allowed her first participants a way into the work, paving a way for the mural and light installation. Salgado explains:

> Once the stories started to be shared and people could read on social media that was when it really started. We actually had more connection and attention on the front end than when we did the community project when people came to help paint the stories on the wall. People were already in by that point and they were more excited because they were on the front end.[5]

For Salgado, the obstacle that proved to be the most frustrating was convincing people to engage with the light installation and map online. In this part of the project, users and participants could select a specific light on the wall while they were online and once they had uploaded text to the website. But a problem developed that the project's parts remained in several spaces, virtually and physically. Because the light show appeared in the side of the building in the middle of the city, those who walked by the installation did not have direct access to the blog or website. On the other hand, it was difficult to see the urban setting and meet others who were engaging with the wall while being online. Furthermore, to really activate the light installation and change its programming, a group of participants are needed to be on the site at one time and work together in order to activate the conductive paint and copper strips that triggered the light show. In this case, having multiple engagement paths

[5]Salgado, Patience, Interview by the author. Personal interview. Richmond, VA September 28, 2013.

and ways to participate divided the community into those online and those standing on the street. This is something that Salgado believes is an interesting part of future work but needs more work.

Coming Soon

Like the success seen in creating a blog and movement around a cause, the potential to engage a participatory online work of art while on the street may help promote in building the identity of a community or reconstructing the presence of a location. For the Bronx River Art Center exhibition *Virtual/Monumental*, Stephanie Rothenberg, known as REV, presented a virtual account around the physical changes planned for a community by attaching digital plans and announcements made viewable on mobile phones and tablets in a work titled *Coming Soon* (Fig. 12.5). The online and offline exhibition of new artwork took place between April and September 2012 and was an effort by the Bronx River Art Center to engage more participants within their surrounding community. Not only did the museum want to exhibit new work, it also wanted to engage with its community more, to have the community become of a participant in the institution's exhibition. As explained on their website, "Through the use of art and technology, we hope to allow citizens to play an active role in foreseeing their community's future." In

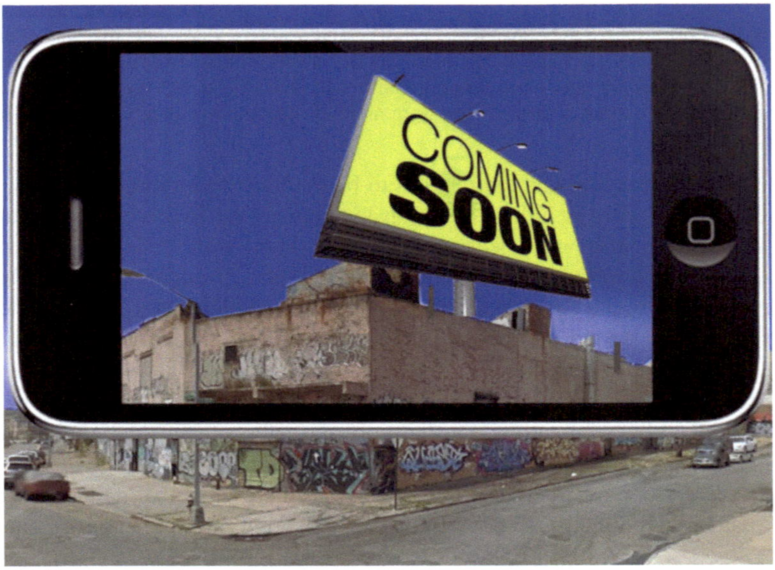

Fig. 12.5 PR Image, *Coming Soon*, The Bronx River Art Center *Virtual/Monumental*. Artist Stephanie Rothenberg. Fall 2012 (Image courtesy of Stephanie Rothenberg)

the case of *Virtual/Monumental*, the Bronx River Art Center hoped to build better representations of its purpose and place by mimicking the way its patrons and visitors engaged with their community.

What was distinctive to *Virtual/Monumental* is the consideration by an art and community institution to engage and build more community around a growing notion that today's technology will give new visitors another way to interpret the collection and new works of art. The Bronx River Art Center designed and envisioned *Virtual/Monumental* in order to engage the viewing public by asking them to continue to participate with the exhibition as they would with the landscape outside the museum's walls. According to the exhibition website, this shows integrated Internet technology, mobile devices, and physical landscape at the same time: "The first installment of *Virtual/Monumental* weaves a visual narrative around the evolving resources and untapped histories of the center, community and river, to overlay the historic past with future designs through Augmented Reality and QR codes, to be experienced in the present, on iPads and mobile devices."[6]

Rothenberg's *Coming Soon* seeks to invite participants out into the landscape of public spaces while employing mobile technologies. This multi-space project appears on mobile and Internet devices as a set of digital billboards that users may acquire by logging into their handheld digital technologies while standing in the urban environment of Bronx, New York. REV's billboards and digital images ask the viewer to imagine what their community will soon look like as they browse through plans collected from upcoming building developments. While standing on the corner of the street, a participant can pull up building proposals in order to see the next transformations of the urban space. The main component in this installation took place in a large housing complex that will take 10 years and $350 million to complete. Visitors could stand on a specific section of the street and see a proposed building's architectural renderings via their mobile devices and converse with others who may be interested in the neighbor's future. What REV hoped for in *Coming Soon* was a public discussion and shared knowledge about the future developments that were planned for this neighborhood (Fig. 12.6).

While REV's project took place on the streets, its content relied entirely on the ability of the user to connect to the information and online archive. Whereas *Coming Soon* spurs interesting considerations between virtual and urban living, there was a problem with access to the online information. According to REV, the people who participated with the work were instructed that the collection of online documents and plans existed because of the connection to the museum and exhibition. The general public, on the other hand, would have not known that the installation existed without direction or invitation from the museum or artist. To make the work successful, REV's participants had to both gain knowledge of the work and have access to the plans through their devices while standing in the specific

[6]"Virtual/Monumental," *Bronx River Art Center*, art exhibition in Bronx, New York, exhibition dates April 19, 2012 to September 30, 2012, accessed October 12, 2013, http://bronxriverart.org/events-archive.cfm?recordID=57

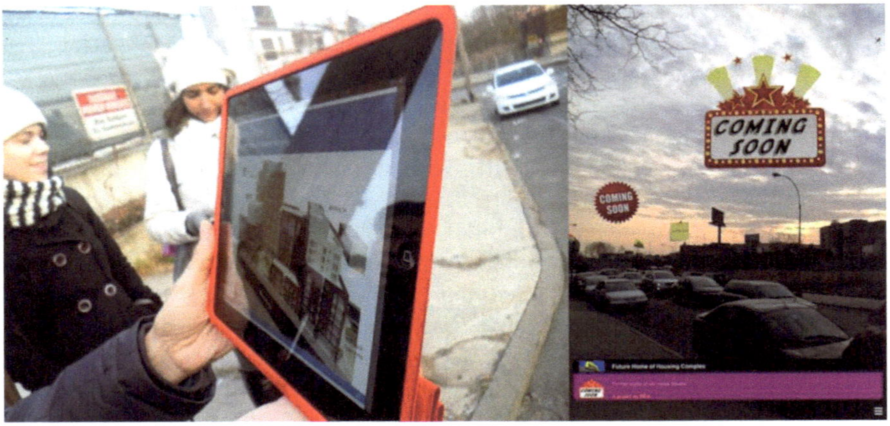

Fig. 12.6 (*Left*) Members of Parsons' Sheila D. Johnson workshop "Art, Environment, Action" viewing the building design from the street, *Coming Soon*, The Bronx River Art Center *Virtual/Monumental*. Artist Stephanie Rothenberg. Fall 2012 (Image courtesy of the artist). (*Right*) Virtual/Monumental opening and public tour, *Coming Soon*, The Bronx River Art Center *Virtual/Monumental*. Artist Stephanie Rothenberg. Fall 2012 (Image courtesy of Stephanie Rothenberg)

location. REV explains, "Participation was mostly enabled through the context of how it was promoted. For example, an AR [Augmented Reality] piece is basically invisible in the sense that people only know about if it's promoted unlike a physical sculpture that you see in a public square. So depending on how it was utilized or promoted resulted in the outcome of who participated."[7] Those without the ability to access the technology and did not have knowledge of the installation at the given moment would have been left out of the experience. While the work shows a serious step for art museums and galleries to consider technology as a way into the new viewing and learning experience, we need to ponder how a group uses technology as much as we do what technology can accomplish. *Virtual/Monumental* is a new direction into an inquiry between art, technology, and community, and I hope more cultural and community institutions follow their lead.

Energy of the Nation

Sometimes, when a COAP grows large enough, it may have the potential to produce an effective representation of a changing consciousness around technology as well as to ascertain and record a global point of view and sentiment. The success of recording large-scale impressions was evident in Sosolimited's *Energy of the*

[7] Stephanie Rothenberg, e-mail message to author, October 7, 2013.

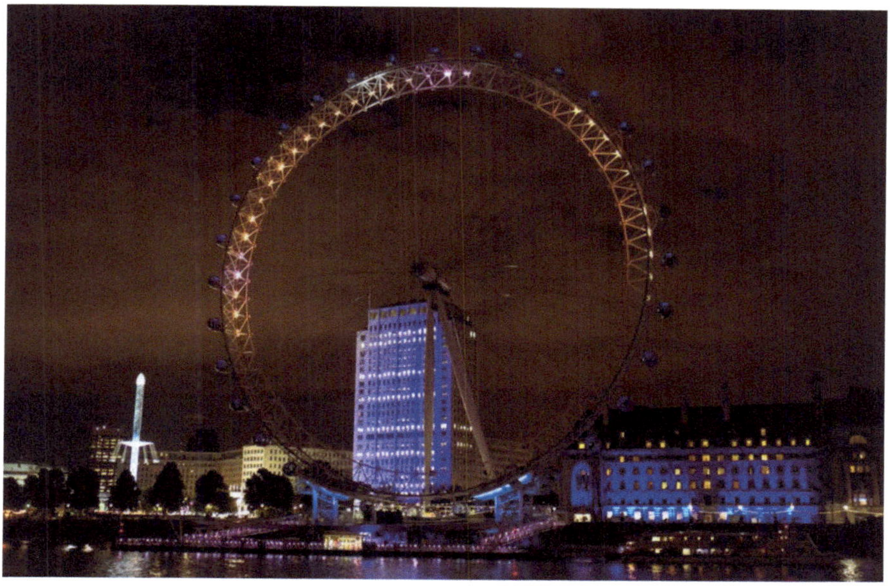

Fig. 12.7 *Energy of a Nation*, Installed on the London Eye during the 2012 London Olympics *Sosolimited* (2008 Copyright Sosolimited. Courtesy Sosolimited)

Nation. In this participatory work, visitors to the city of London were asked to post emotional responses they had to the 2012 Olympics while using Twitter. Once participants have tagged their emotional feeling of the event using hashtags and uploaded these sentiments into their social media profiles, software developed by the collaborative art group Sosolimited would find the reference and convert the collected data into the world's first social media light installation. The installation took place on the London Eye as a light display. The purpose of the work was to access shared information through the use of social media technology in order to ascertain the general sentiment of the experience in London and the Olympics (Fig. 12.7).

Sosolimited's artists and contributors – Justin Manor, John Rothenberg, Eric Gunther, Lauren McCarthy, Sam Kronick, and Wade Aaron – utilize online user processes, programs, and information from a range of sources and present them as visual documents, interactions, installations, and real-time performances. In *Energy of the Nation*, Sosolimited collected shared online data and projected the collected data as a light installation and in a graph diagram alongside the installation. This diagram included the live tweets and overall feeling of the day as well as emotional highlights that were created when the artist group analyzed the collected data for keywords (Fig. 12.8). Sosolimited's co-founder Justin Manor explains, "Our software will find that tweet and find the positive and negative words and whether you used emoticons, or how long your sentences were, and whether there was exclamation points or question marks and we score that with a number of points

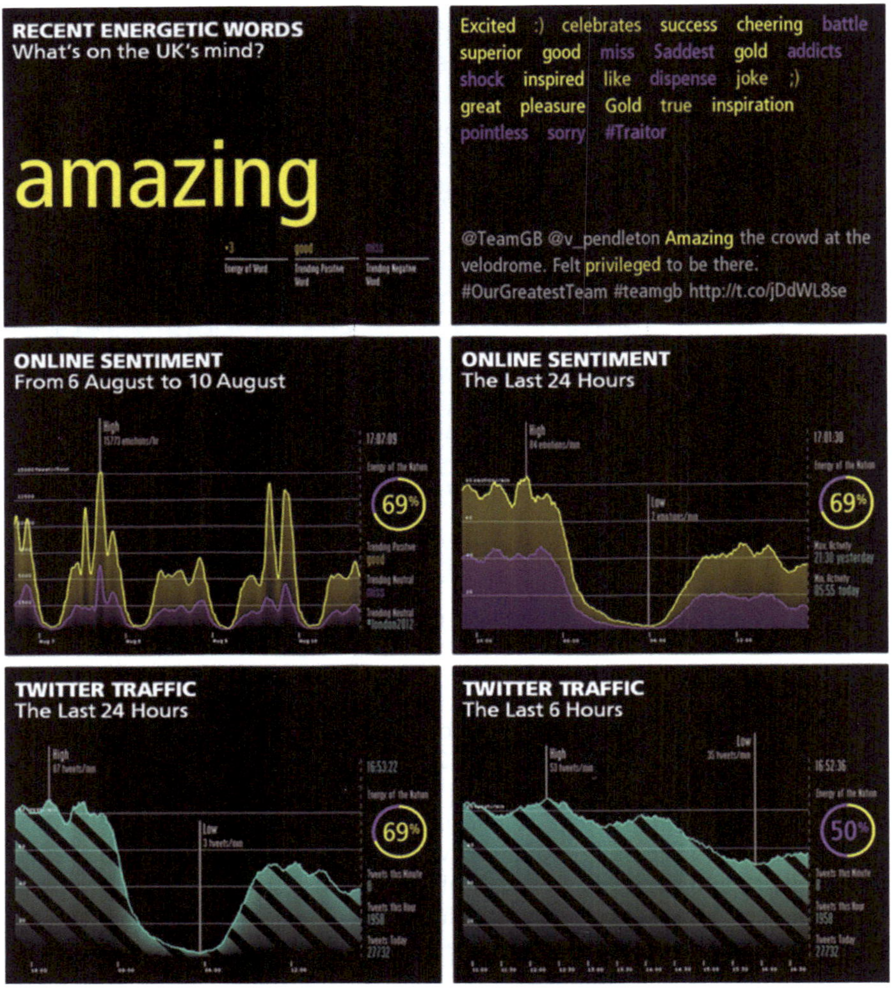

Fig. 12.8 "Energy of a Nation," Installed on the London Eye During the 2012 London Olympics *Sosolimited* (2012 Copyright Sosolimited. Courtesy Sosolimited)

for positive and negative. We tally those at the end of every day and turn that into a visual score, kind of a concert of lights that happen."[8]

[8]"EDF Energy Presents Energy of the Nation," YouTube video, 1:29, video documentation of public artwork installed for the 2012 London Olympics, posted by "EDF Energy" July 19, 2012, http://www.youtube.com/watch?v=5LYwVs7qwZY

Baltimore Slumlord Watch

For some artists and activists, online engagement is a chance for a community to address some hard issues that may not be possible within the politics of a city or town. The Baltimore Slum Watch project and the extending artist project, called Wall Hunters, are such examples.

Carol Ott's *Baltimore Slumlord Watch* started in January of 2009 as an online community activist blog in which members of the community could give the public notice of abandoned/derelict houses within the city of Baltimore, Maryland. According to the website, the project was conceived and developed by an anonymous city resident "who was tired of watching out-of-town 'investors' and others destroy neighborhoods as a result of their negligence."[9] The blog format gives readers the ability to post addresses of houses they feel needs attention. According to Baltimore's *City Paper*, the blog's owner goes as far as, "tracing problem property owners' political contributions, suspected tax cheating, and action (or lack of action) from City Hall."[10]

Each post on the blog directs the reader to a property's address, sometimes including several properties per entry, if they are grouped together in the urban environment. Each post has the address of the property, a photograph of its current state, and the contact information for the owners who have left it to deteriorate. The blog post even goes as far as including the contact information of the city council and state electives. In the January 22, 2014, post "A Trio of Blight: 4712, 4717, and 4716 Park Heights Avenue," Ott includes the owners name of the three residences, two of which are linked to the mayor and city council as well as the name of an agent for a connected trust, city council member of the 6th district, and contact links for Maryland Senator Lisa Gladden and State Delegates Carter, Oaks, and Rosenberg.[11] At the bottom of the post is a photograph of the three boarded-up Park Heights Avenue residences.

Community activism does not stop for this particular project with the online post generated by the public that sends Ott the property's address. In fact, this is just the start of a larger conversation and includes mural artists going to the property where they create new works of art on its walls in hopes to enliven the area and engage those who still live in the neighborhood.

Because of their current nature as ghostly structures with absent owners, these derelict buildings have two options: they can continue to fall apart and be an eyesore for the neighborhood or become material for public action and art. The second

[9]Ott, Carol. *Baltimore Slumlord Watch*, (blog). http://slumlordwatch.wordpress.com/about-us/. accessed February 5, 2014.

[10]Best Citizen Journalism, "Baltimore Slumlord Watch blog" *City Paper*. September 16, 2009. http://www2.citypaper.com/bob/story.asp?id=18704. Accessed February 5, 2014.

[11]Ott, Carol, "A Trio of Blight: 4712, 4717, and 4716 Park Heights Avenue," Baltimore Slumlord Watch (Blog), January 22, 2014. http://slumlordwatch.wordpress.com/2014/01/22/a-trio-of-blight-4712-4714-and-4716-park-heights-avenue/. Accessed February 5, 2014.

option has the potential to strengthen a community, which is why artists are working with the remaining residents in order to remedy the existing blight. The properties listed on Ott's blog become canvases for a collaborative group of street artists who seek to bring further attention to the building as an abandoned structure and, at the same time, create new public art installations.

Wall Hunter's *Slumlord Project* was founded in response to Ott's blog and connects artist collaboration with community action. The two projects now work together in order to build community and draw attention to specific neighborhoods in Baltimore area that are at risk of falling apart. Not only does the Wall Wallhunter mural project cultivate a new street art installation, but it also helps to construct new identities of a life that continues to grow even when homes are falling down. In much of the Wall Hunter mural work, artists use imagery and language from the streets that surround their creation – for example, the street artist Lunar New Year painted the portrait of Shawniece Smith, a resident who lived in the building next to the mural work.

Wall Hunter mural artists incorporate digital icons called QR codes into the images they leave on the abandoned structures. These codes link the building back to Ott's blog and give anyone with the mobile devices the ability to gain information from Ott's investigation while standing on the street. By adding a hyperlink away from the viewed work, the public may learn more about their community and participate in its life by contacting anyone on the list. The QR codes give a level of power and action back into the community because they form a direct connection between the spaces with the information, and position those in the community with more access to possible solutions, and direct correspondence to owners who may have been unseen before. According to Wall Hunter's website:

> By incorporating QR codes next to the art pieces that link to specific Slumlord Watch website posts, the project will educate community members about the housing and safety code violations on the property and the owners responsible for the property's decline. Additional to the QR codes, on the doors of the properties we will have text detailing the specific property violations. Communications strategies will be employed to maximize public exposure, result in ripple effect publicity for the project, and begin public dialog about vacants, their owners, and the role of street art in community building and as a form of beautification and political expression.[12]

Shawniece Smith notified Ott's blog looking for help and informed it that she had tried to contact the owners of 539 concerning its physical state but had no luck getting anything addressed with the property. She listed her worries with the building connected to her home as follows, "The property had just about everything imaginable wrong with it from leaking water (flowing, really) to a collapsed roof, from rodents to a nasty termite infestation that was making its way through the support beams into Shawnee's home. She'd made countless calls to the City and to SS3. Nobody was listening to her concerns even after thousands and thousands of

[12]"About Wallhunters," On *Wallhunters.org* website, accessed February 5, 2014, http://www.wallhunters.org/#!about/c240r

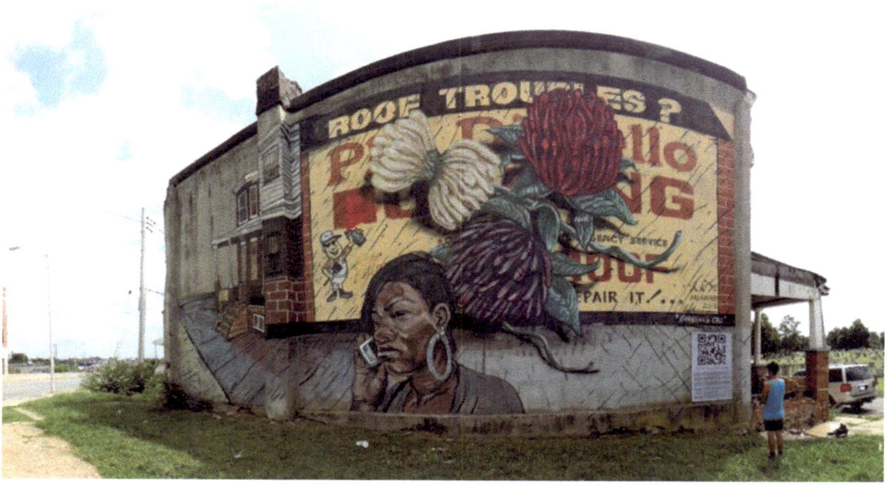

Fig. 12.9 539 N. Longwood Street. Permission for use granted by the artist Nether, taken on August 13, 2013 www.wallhunters.org (Downloaded from Flickr on February 24, 2014 https://www.flickr.com/photos/nether_street_art/9553757925/)

dollars in damage to her property."[13] Lunar New Year (LNY) "illegally" installed a mural with an adjoining QR code on the side of the abandoned row house and on top of a previous mural advertisement for a roofing business (Fig. 12.9). The mural depicts Ms. Smith frustrated and on her cell phone, as she attempts to get answers regarding the vacant building.

Google Maps Road Trip

Not only are artists able to access and represent emotional states of mind and communal archives developed by an online group, they may also recreate new experiences by simply setting up unique scenarios that are constructed purely with Internet materials. Take, for example, a 2009 project by artists Peter Baldes and Marc Horowitz that challenges notion of what is lived experience online. For their collaborative virtual performance called *Google Maps Road Trip* (Fig. 12.10), Baldes and Horowitz joined a video web stream to a community chat room in order to document the artists traveling across the country exclusively using Google

[13]Nether, "Nether on Wall Hunters' 'Slumlord Project'" *Vandalog* (Blog), September 6, 2013. Accessed February 5, 2014. http://blog.vandalog.com/2013/09/nether-on-wall-hunters-slumlord-project/

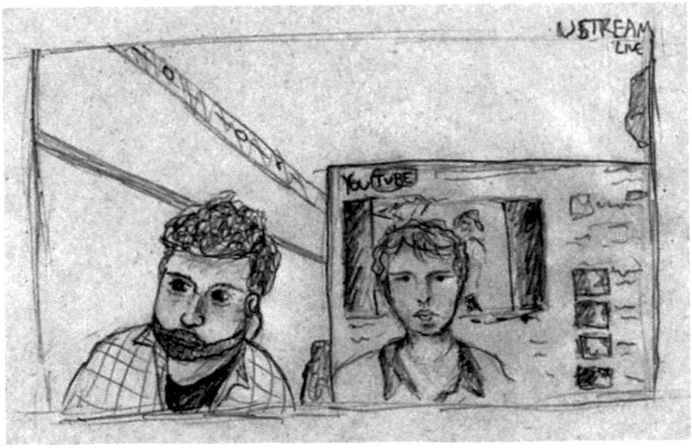

Fig. 12.10 Sketch of *Google Maps Road Trip*, Peter Baldes and Marc Horowitz *Google Maps Road Trip* August 10–18, 2009 (Drawing by Marc Horowitz, courtesy of Baldes and Horowitz)

Maps.[14] Over the course of 9 days, the two artists virtually "drove" from Los Angeles, California, where *Horowitz* lived, to Richmond, Virginia, where Baldes lived. They accomplished this by clicking through the street-view images on Google's online mapping system. Participants could join the trip in real time by logging into an online video service where Horowitz and Baldes presented their trip. Passengers could also join in by participating in a chat room set up by the artists.

Each morning, Baldes and Horowitz turned on their computer cameras and linked up to each other using web-streaming cameras and computer microphones. The artists would then open a chat group where they could connect to anyone wishing to ride with this that day (Fig. 12.11).

Every day, someone needed to "drive" the car. Driving meant that someone would be in charge of clicking through the street view in the Google mapping system. In much the same way passengers scroll through a radio station dial to locate the soundtrack for a car ride, virtual passengers on the *Google Maps Road Trip* searched the Internet for interesting things to see and do online as the trip progressed. The passengers looked for YouTube clips, collections of photos, information about various places, and roadside attractions or created digital mixed tapes by searching streaming radio stations and iTunes. When a passenger found information from the Internet about a specific US city or attraction, the driver would stop the virtual car to watch a video, research some fact of a place, or just scroll through the digital images of that location. Throughout the trip, Baldes or Horowitz decide on what town they would aim for in order to "stay the night," and the rest of the participants/passengers

[14]"Google Maps Road Trip," Ustream video and chat, Posted by Pete Baldes and Marc Horowitz, August 10–19, 2009, http://www.googlemapsroadtrip.com/

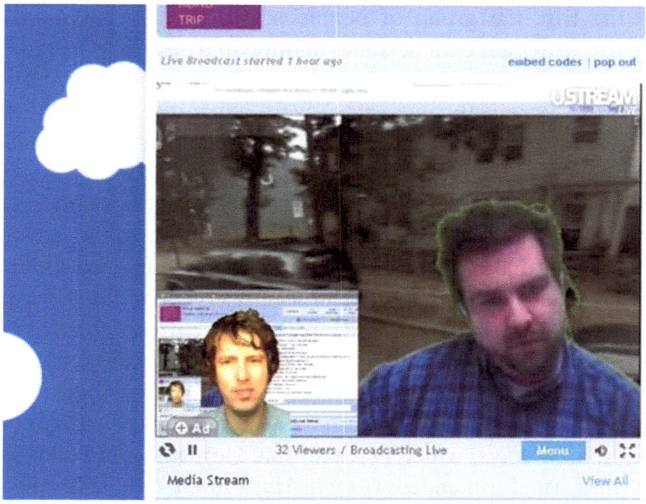

Fig. 12.11 Image from ustream.tv performance by Peter Baldes and Marc Horowitz *Google Maps Road Trip*, August 10–18, 2009 (Courtesy of Baldes and Horowitz)

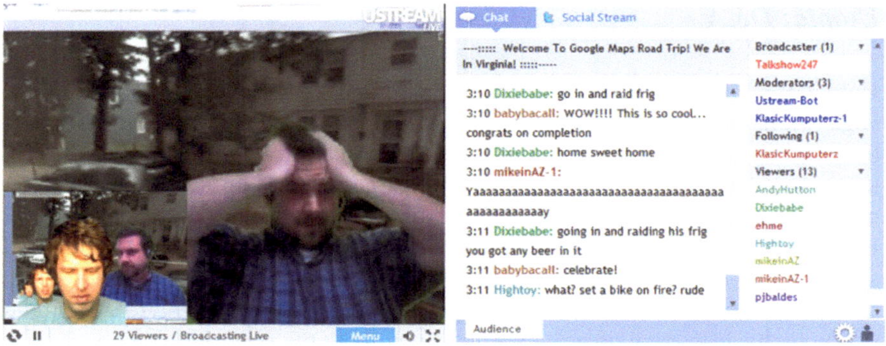

Fig. 12.12 Image from ustream.tv performance with chat comments, Peter Baldes and Marc Horowitz *Google Maps Road Trip* August 10–18, 2009 (Courtesy of Baldes and Horowitz)

would find out what to do between the starting point and ending points for the day. Sometimes we would end the day by either camping out inside the car or reserving a room in a hotel. In many instances, the artists, along with the help from participants, would decide on the town and then start calling hotels in that town to get room rates, accommodation information, and even types of breakfast plans.

Marc and Pete encouraged participation with virtual passengers while collectively accessing a wide range of digital data online through Internet queries, Ustream and Google Maps (Fig. 12.12). The constant tangents from the road traveled during the trip, endless stopping of the car, radio scanning, and odd conversations made to fill the time made the trip feel more real, more like a true cross-country expedition.

Throughout the trip, the group of people seated in the chat room collectively critiqued the trip with their own personal findings and desires to include experiences into the trip that they found from their own Internet searches. By the end of the trip, all of us, especially the artists, were exhausted, irritated, cramped, and feeling strangely claustrophobic. Because everyone had the power to access Internet searches, passengers in the project's virtual car shared in the progress of the trip.

By utilizing *Google Maps* "Street View" service along with communication technologies like chat rooms, digital networks, and radio, the artists revealed a new approach to art making. Horowitz and Baldes had created a new experience by collaboratively accessing what the Internet already housed in its vast networks of collections of data. According to the *New York Times* writer Matt Gross, the newness of *Google Maps Road Trip* was significant when it appeared in 2009. Gross writes, "To produce a real virtual vacation (and there is, I admit, something very wrong with the phrase 'real virtual vacation') you have to think bigger and combine existing Web services in new ways, which is what Marc Horowitz and Peter Baldes did."[15] By the time this virtual trip ended, *Google Maps Road Trip* had been on the radio, plastered all over the Internet and in print.

Utopia: Q&A

Unlike previous participatory works of art, COAPs possess the unique ability to engage a range of participants that are in different places and situations at a given time due to the command of the digital connections/networks that are enabled by interconnected technologies. But the ability to engage across technology is not new or distinct to the Internet. In 1971, artists Pontus Hulten, Fujiko Nakaya, and Billy Klüver connected telex machines in New York to Stockholm, Ahmedabad, and Tokyo.[16] Hulten, Nakaya, and Klüver asked their audience, who received the transmission of the broadcasted message in *Utopia: Q&A* (Fig. 12.13), to predict the future by answering prearranged questions about the year 1981.

Through the telex machine, the artists communicated with their audience, and the audience participated back with each other by replying to the statements from their collaborators. Even though the audience remained separated in different countries, the telex machine gave them the ability to conduct an exchange. The difference between *Utopia: Q&A* and our experiences with COAPs is that in order to participate with the telex machine you had to seek out the installation's location. A participant with *Utopia: Q&A* either happened upon the installation or had to go to the spot where the machine lived.

---

[15]Matt Gross, "Virtual Vacation," *The New York Times*, September 1, 2009, accessed October 30, 2013, http://frugaltraveler.blogs.nytimes.com/2009/09/01/virtual-vacations/

[16]"E.A.T. (Experiments in Art and Technology)," on *Compart: Database of Digital Art*, accessed June 3, 2013, http://dada.compart-bremen.de/Collective/13 (site discontinued).

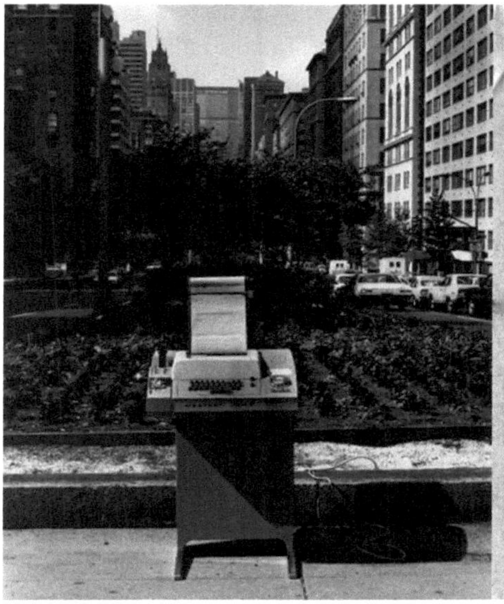

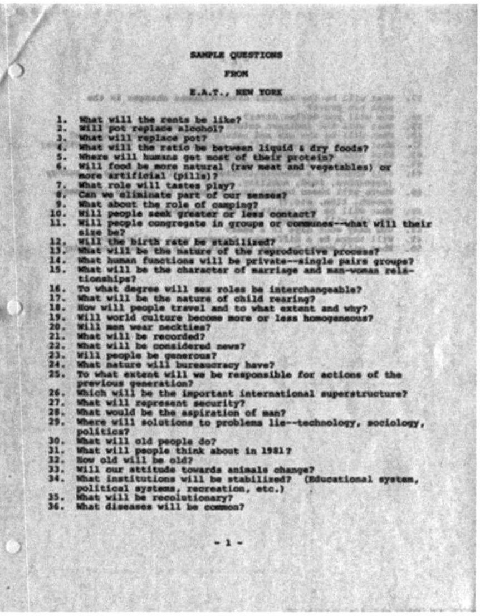

SAMPLE QUESTIONS

FROM

E.A.T., NEW YORK

1. What will the rents be like?
2. Will pot replace alcohol?
3. What will replace pot?
4. What will the ratio be between liquid & dry foods?
5. Where will humans get most of their protein?
6. Will food be more natural (raw meat and vegetables) or more artificial (pills)?
7. What role will tastes play?
8. Can we eliminate part of our senses?
9. What about the role of camping?
10. Will people seek greater or less contact?
11. Will people congregate in groups or communes--what will their size be?
12. Will the birth rate be stabilized?
13. What will be the nature of the reproductive process?
14. What human functions will be private--single pairs groups?
15. What will be the character of marriage and man-woman relationships?
16. To what degree will sex roles be interchangeable?
17. What will be the nature of child rearing?
18. How will people travel and to what extent and why?
19. Will world culture become more or less homogeneous?
20. Will men wear neckties?
21. What will be recorded?
22. What will be considered news?
23. Will people be generous?
24. What nature will bureaucracy have?
25. To what extent will we be responsible for actions of the previous generation?
26. Which will be the important international superstructure?
27. What will represent security?
28. What would be the aspiration of man?
29. Where will solutions to problems lie--technology, sociology, politics?
30. What will old people do?
31. What will people think about in 1981?
32. How old will be old?
33. Will our attitude towards animals change?
34. What institutions will be stabilized? (Educational system, political systems, recreation, etc.)
35. What will be recolutionary?
36. What diseases will be common?

- 1 -

Fig. 12.13 (*Left*) "Telex In the Street at New York City Terminal." *TELEX Q&A* 1971 (Photo by Julie Martin, image courtesy of Julie Martin). (*Right*) Questions asked of participants in Telex Q&A, *TELEX Q&A* 1971 (Photo by Julie Martin, image courtesy of Julie Martin)

Utopia: Q&A demonstrated that the general public could engage with art and others and, at the same time, remain in the urban environment. The level of access granted to the user of a technology and specified by artists produced a public connectivity that appeared, in part, as a work of performance art/sculpture. Furthermore, *Utopia: Q&A* indicated that distance could be disregarded and that connectivity could produce a level of freedom through the collapsing of space due to networked links. In this early work, participants could engage with each other and collaboratively create new art, just by accessing the technology's interconnected nature.

While the structure of COAPs extends from a long history of participatory art, they are new investigations into the varied relationships between users, participants, and digitally interconnected technologies. COAPs promote shared experiences even when members/participants may not physically be present at the time of the action and that they can participate virtually. This means that the technology is as much to account for the finished as the proposed action or the management of engagements. Additionally, the resulting artifact of the artist's work, whether it is a photograph, text, sound, or video of a performance, is placed back online as an archive of the experience. Therefore, artworks that involve community engagements may circulate back into the digital environment and become material for further experiences. This hyper-mediated transformation has the facility to exist as multiple experiences at a given moment and may be viewed through a wide range of connected devices, especially when users are activated within the community landscape and their mobile devises. COAPs exist as Internet data, formulated as a physical engagement within a community landscape and/or urban environment and finally transformed back into an online archive.

At present, the Internet is in everything we do. From driving our car to connecting with our friends, shared Internet technology has quickly reached into our daily actions whether we like that or not. Technology is part of our growing experience in life, even when we know that a technology may be too prominent. According to Arvind Rajagopal, "Technology is obtrusively present, in new and constantly changing ways. At the same time, it is everywhere and invisible . . . There is no "off" switch for technology, no place unaffected by it."[17] This overwhelming structure is symbolized by our highly connected Internet technologies, which is why we need to further investigate how new art projects relate, construct, and connect the world. Yet, the real point of technology is to help us, to aid us in our endeavors. What is most powerful about the technological moment and signified in community online art projects is that the collective body has the ability to create as a group because they utilize shared technologies.

References

About Wallhunters. Wallhunters. http://www.wallhunters.org/#!about/c240r. Accessed 5 Feb 2014

Best Citizen Journalism, Baltimore Slumlord Watch blog. City Paper, September 16, 2009. http://www2.citypaper.com/bob/story.asp?id=18704. Accessed 5 Feb 2014

E.A.T. (Experiments in Art and Technology), on Compart: Database of Digital Art. Accessed 3 June 2013. http://dada.compart-bremen.de/Collective/13 (site discontinued)

EDF Energy Presents Energy of the Nation. YouTube video, 1:29, video documentation of public artwork installed for the 2012 London Olympics, posted by "EDF Energy" July 19, 2012. http://www.youtube.com/watch?v=5LYwVs7qwZY

Google Maps Road Trip. Ustream video and chat. Posted by Pete Baldes and Marc Horrowitz, August 10, 2009 to August 19, 2009. http://www.googlemapsroadtrip.com/

Gross M (2009) Virtual vacations. The New York Times, September 1, 2009. http://frugaltraveler.blogs.nytimes.com/2009/09/01/virtual-vacations/

McLuhan M (1964) "The medium is the message". Understanding media: the extensions of man. McGraw-Hill, New York

Nether (2013) Nether on Wall Hunters' 'Slumlord Project'. Vandalog (Blog), September 6, 2013. http://blog.vandalog.com/2013/09/nether-on-wall-hunters-slumlord-project/. Accessed 5 Feb 2014

Ott C, Baltimore Slumlord Watch (blog). http://slumlordwatch.wordpress.com/about-us/. Accessed 5 Feb 2014

Ott C, A Trio of Blight: 4712, 4717, and 4716 Park Heights Avenue. Baltimore Slumlord Watch (Blog), January 22, 2014. http://slumlordwatch.wordpress.com/2014/01/22/a-trio-of-blight-4712-4714-and-4716-park-heights-avenue/. Accessed 5 Feb 2014

Rothenberg S, E-mail to author, October 7, 2013

Salgado P, Interview by the author. Personal interview. Richmond, September 28, 2013

Salgado P, Interview by the author. Personal interview. Richmond, February 21, 2014

Virtual/Monumental. Bronx River Art Center. April 19, 2012 to September 30, 2012, Bronx, New York. http://bronxriverart.org/events-archive.cfm?recordID=57

[17] Rajagopal. "Imperceptible Perceptions In Our Technological Modernity." P. 278.

Chapter 13
Digitized Street Art

Brian A. Brown

I think it takes much greater courage to create things to be gone, than to create things that will remain.

(Christo)

Abstract This chapter argues that "street art" is a mode of artistic expression reliant on the vagaries of the urban environment as its canvas and, as a result, is ultimately dependent on digital technologies to document, disseminate, and reproduce these inherently ephemeral artworks. Whether altered or destroyed by another artist or tagger, "buffed out" by overzealous municipal authorities, or simply decayed by the elements, street art is fundamentally ephemeral. It is this inherent ephemerality that requires the original piece be digitally documented and preserved. The digital camera and the Internet in particular, then, serve to preserve the work of street art that, in their absence, would otherwise be lost to time. By reference to firsthand field research undertaken in Detroit, Michigan, shortly after an "original" Banksy was relocated (and depending on one's perspective, destroyed or saved) by a local art gallery, this chapter concludes by exploring the idea that street artists working within the very physical and concrete confines of the urban city are better regarded as digital artists, albeit digital artists that go to great lengths in the preparation of their compositions.

Banksy in Detroit: An Introduction

In May of 2010, Banksy made his way to Detroit. Infamous yet anonymous,[1] Banksy is a street artist of world repute whose international fame is based largely

[1]Banksy's true identity remains a contentious topic of debate. Although never confirmed in an official capacity, the *Daily Mail* claims to have uncovered the "true" identity of the elusive street artist, believing him to be named Robin Gunningham (Tapper 2008).

B.A. Brown (✉)
Department of Communication Media & Film, University of Windsor, Windsor, ON, Canada
e-mail: bbrown@uwindsor.ca

F.T. Marchese (eds.), *Media Art and the Urban Environment*, Future City 5,
DOI 10.1007/978-3-319-15153-3_13

on the illegal application of aerosolized paint to walls, often with the assistance of a stencil. In the spring of 2010, he embarked on a "promotional" tour of the United States designed to create publicity for a documentary he directed called *Exit Through the Gift Shop* (2010). The film focused on the developmental history of "street art" and was eventually nominated for that year's Academy Awards for Best Documentary. In each city the film was to premiere in, Banksy would arrive a few days prior and create a handful of pieces of street art so as to promote the film in the hopes of generating media attention and, reciprocally, ticket sales at the box office. This promotional tour is what led him to the troubled city of Detroit in general and the dilapidated Packard Motor Car factory on the city's southeast side in particular.

Banksy created four individual pieces in Detroit, none of which survive to this day. While fragments of the piece he created at the Packard Plant remain, the work of art as the artist created it in situ was destroyed. The Packard Plant is an abandoned industrial manufacturing facility that once employed tens of thousands of individuals and whose three and a half million square feet now sit derelict among much of Detroit's notorious industrial refuse. The walls that remain have become a favorite location for local graffiti writers and street artists that see on (and within) them a place to practice and hone their skills while remaining relatively sheltered from the legal sanctions of the depleted police forces that struggle to maintain order in the increasingly troubled metropolis.

Located among hundreds of discarded tires, mounds of broken cinder block, crumbling walls, trash, and the valueless remnants of the industrial infrastructure left behind after years of scavenging, the piece created at the Packard Plant depicts a child holding a can of red paint and a paintbrush with a caption that reads "I remember when all this was trees." The piece itself is not particularly accomplished in terms of its artistic qualities and is rather crude in its message and design. As will become clear throughout what follows, however, this chapter is not interested in providing an aesthetic analysis of Banksy's works. It is, rather, much more interested in the assessment of the inherent ephemerality of street art and, reciprocally, in the examination of the pivotal place of the digital camera and the Internet in the preservation and dissemination of these fleeting works of art.

What follows below, then, answers the central research question that inspired the present chapter. If street art is a form of artistic production fundamentally linked to, and dependent on, the concrete and stone that constitute urban city space, then what role does the digital camera and the Internet play in its creation, preservation, and dissemination? Street art is an inherently ephemeral art form whose end products are destroyed (or at the very least altered to the point that they no longer bear any significant resemblance to that produced by the artist) due to their being created in the public, on the street, and thus free of any kind of physical protection. The five primary causes underlying street art's inherent ephemerality will be examined in much more detail below. For the time being, however, a number of qualifications need to be made so as to detail exactly what is being discussed and examined in the following pages.

The first qualification is better understood as an act of classification. Put simply, street art is not graffiti and graffiti is not street art.[2] The difference between the two eminently related, yet significantly different, art forms is located not so much in an historical/aesthetic turn, but much more so in an evolution of the art form over the years and in response to one of the primary causes for its ephemerality. In the late 1970s and early 1980s, graffiti was a burgeoning form of urban inscription[3] located primarily on the public transportation systems of New York City and Philadelphia (see Gastman and Neelon 2010). The municipal authorities of these cities regarded the rapidly proliferating tags and writings of some of their citizens on the subway cars, stations, and platforms as a threat to public order and as a visual blight that required rapid removal. Using a variety of tactics that are described in more detail below, the authorities in these municipalities waged aggressive and costly "wars on graffiti" that (in the space available at present to explain their outcomes) they effectively won.

Cleansing the subway systems of these illegal markings, however, had the unintended effect of driving them above ground, expanding the locus of this transgressive act beyond the confines of the subway system and thus converting all city streets into a makeshift and unsanctioned canvas for "writers." Throughout the years, the tags bearing the nickname of writers become increasingly complex. Developing from the rather crude markings of "Taki 183" (see *New York Times*, 1971), to the more accomplished work of a writer like Seen (active in New York in the early 1980s), to the so-called wild-style of graffiti, whose angular and geometric forms camouflage the name of the writer in complex arrangements of spray-painted confusion, graffiti cannot be regarded as a singular form of artistic inscription that obeys particular stylistic rules and/or conventions.

However, there are two central features that remain consistent throughout "graffiti" which assist in conceptually differentiating it from "street art." The first is the consistent use of aerosolized paint as its primary media. While some writers used broad-tipped felt markers to tag, the majority made use of spray paint to make their mark. The second consistent feature, and the one that most directly differentiates "graffiti" from "street art," is graffiti's all but ironclad dependence on typographic forms of urban inscription. That is, the repetitive yet always creative application of one's name, nickname, or moniker is the central feature that unites a wide range and amazing variety of graffiti writers throughout the years. As is to be expected, over the course of graffiti's evolution, the designs created by writers and taggers become increasingly complex. The development and profusion of "wild-style," a genre of

---

[2]Some authors argue that "street art" is better thought of as "post-graffiti" (Dickens 2008) due to its similar yet different mode of urban inscription. This chapter, however, employs "street art" as its primary signifier to describe in general terms the contemporary state of a subcultural artistic movement that creates art on the street.

[3]It should be noted, however, that there were stylistic progenitors to what is commonly known and referred to as "graffiti." The widespread diffusion of the "Kilroy Was Here" marking by American servicemen throughout World War II as well as the "Bozo Texino" marking (see Daniel 2005) used by so-called hobos in the mid-to-late nineteenth century are early iterations.

graffiti that remains as vital, interesting, and captivating today as it was in the past, was a signal that the process of applying one's name to a wall with spray paint had reached a pivotal moment of heightened abstraction very far from its modest origins.

By disguising one's name beneath the crisscrossing and seemingly incomprehensible markings characteristic of wild-style, graffiti's emphasis on legible typography begins to wane. If wild-style was the first indication that an easily and widely understandable tag was becoming less and less prominent, then street art takes its cue from wild-style by dispensing with graffiti's reliance on typography entirely. Street art, for the most part, abandons this reliance on typographic forms of inscription or at the very least begins to experiment with their combination alongside much more iconographic or pictographic forms. According to Luke Dickens, "the core component of graffiti writing, is increasingly being replaced by 'street logos'; a shift from typographic to iconographic forms of inscription" (2008, p. 474). While many of the arguments made below can be equally applied to the manifold styles emblematic of graffiti, for the sake of conceptual clarity and precision, this chapter makes specific and particular reference to street art and not to graffiti.

The second qualification requiring further explanation is the central difference between the sanctioned or legal form of street art and that is unsanctioned and illegal. The "art vs. crime" debates have dominated the scholarly discourse regarding graffiti and street art, and thus, they will not be rehashed here (see Halsey and Pendrick 2010; Halsey and Young 2006; McAuliffe and Iveson 2011; Young 2012). The central goal of this chapter is to better understand the influence of digital technologies on a form of artistic expression that is resolutely reliant on the physical qualities of the street. One of the many paradoxes that characterizes this influence, and one that escapes the boundaries of this chapter, is that the profusion of digital photographs of street art online has transformed the practice from one that is resolutely illegal in nature to one that has been embraced by certain jurisdictions due to its capacity to solicit tourism and tax revenue. In the absence of the digital camera and the Internet at large, this transformation would not have taken place.

In short, because of digital technologies, street art has become very popular and, hence, has the capacity to drive tourism and the municipal revenue streams associated with it. Street art festivals are held annually and around the globe in urban spaces as diverse as Perth, Australia (see FORM 2014); Hawaii, USA (see PowWowHawaii 2014); and New Delhi, India (see St. ART Delhi 2014), to name but a few of the more contemporary instances. These festivals designate particular walls throughout the city space as legal canvasses upon which some of the most talented artists are invited to create their work.

This kind of legal street art is not the focus of this chapter. Instead, the present chapter trains its critical lens on illegal forms of urban inscription created in the absence of municipal approval. While the evolution of the art form from one that catalyzed the aforementioned "Wars of Graffiti" in the late 1970s and early 1980s to one that is actively being promoted by municipal authorities for its capacity to generate tourist tax dollars is interesting in its own right, a detailed analysis of what can be roughly described as the "political economy of street art" is forestalled until a later time when it can be adequately assessed on its own terms.[4]

On first blush, the third and final qualification that requires clarification is one that is seemingly too obvious to mention. The present chapter considers only those instances of street art that are created and found in public space. While the "street" is obviously a fundamental component of street art, once again due to the popularizing impact of digital technologies on the art form, there has emerged a thriving market for street art prints and other forms of art created by "street artists" that are not be found in the street. Dickens' work focuses particularly on "the unique range of ways street artists are able to translate their edgy, exciting work on the street into commodity form" (2010, p. 63). Thus, consideration of this dimension of street art is not present in what follows. Therefore, unless otherwise noted, the following pages are concerned exclusively with illegal or unsanctioned "street art" that occurs in the public and on the street. However, they are also focused on an element of street art that has up until this point received scant attention in the scholarly literature.

The pivotal role occupied by the digital camera, the Internet in general, and blogs and social networks more specifically in the documentation, preservation, dissemination, and popularization of street art has never been given the critical attention it merits. As much as street artists rely on the brick, concrete, and stone that constitute the backdrop of the urban canvas, due to the inherent ephemerality of their works, they are also equally if not more reliant on the hardware and software of digital and networked modes of communication that serve to document, preserve, and archive these eminently fleeting works of art. That is, by extending the parameters of the central research question offered above, this chapter argues that as much as street artists can and should be considered as such, they should also (and at the same time) be considered digital artists, albeit digital artists that go to great lengths and put themselves at great risk in the preparation of their compositions.

Up until this point, the pivotal place of the digital camera and the Internet in the creation, preservation, and dissemination of one of the most "concrete" art forms to have ever influenced mainstream culture remains all but unexplored. This chapter, then, seeks to fill in this lacuna by providing a corrective reconsideration of what it means to create inherently ephemeral art that relies on the stochastic qualities of the urban environment with the aim of somehow preserving that which will undoubtedly disappear in the near term. In order to do so, it will make its way through three main sections. The first section will explore the underlying causes of street art's ephemerality. The second examines the importance and incorporation of very particular elements of quite specific locations into the creation of works of street art. When a handful of examples are considered, the locations chosen to create a piece of street art are anything but random and lead to the conclusion that the canvas upon which these works are placed constitutes an important ingredient in the overall work itself. When combined, then, the inherent ephemerality of the works of art as well as the importance of location in their creation leads to consideration of

[4]For an interesting assessment of legal graffiti walls, see Kramer (2010).

the third and concluding section of this chapter. Succinctly, street artists that make use of the concrete, brick, and stone characteristic of the urban environment can and should also be equally regarded as digital artists.

The Five Causes for Street Art's Inherent Ephemerality

In the late 1970s and early 1980s, a "war" was being waged on the streets and subways of New York City. In response to the rapid proliferation of graffiti "tags" on subway cars, New York's Metropolitan Transit Authority (MTA) launched an all-out offensive dedicated to eradicating the profusion of tags on its charges. As detailed in the documentary film *Style Wars* (Silver 1983), this offensive included a litany of measures meant to prevent and discourage the application of spray paint to the fleet of subway cars and stations that make up the system. Police officers, dogs, razor wire, criminal records, public advertising campaigns, and harsh chemical removers were all employed to either discourage or destroy the products of a then thriving subculture that made use of the urban environment as their primary canvas. For the most part, New York's "Graffiti Wars" were won by the MTA in that the subway system was eventually "cleaned up" with tags becoming less of an aesthetic nuisance, paving the way for the commercial appeals that now dominate.

Shortly thereafter, in 1984, photographers Martha Cooper and Henry Chalfant published *Subway Art* (1984). The book is one of the few documents of the era that serves to preserve the artifacts of this then fledgling subculture. Photographs of the tags taken by Cooper and Chalfant document the work of such "kings" of the subculture such as Dondi, Blade, Futura 2000, Skeme, Iz the Wiz, Seen, among others. In essence, this important moment in the history of unauthorized urban art would have been lost forever in the absence of Cooper's and Chalfant's cameras. The role of recording devices has from the very beginning been an important one in preserving the past and present of the art form. Graffiti and its artistic progenitor street art are, then, inherently ephemeral in that the artifacts created by the artists are destroyed very soon after they are created. They are for the following five reasons.

In a seminal treatise in the history of urban criminology, George Kelling and James Wilson argue their "broken windows" theory (1982). Schematically, Kelling and Wilson argue that if left unfixed, broken windows in a city building or neighborhood block of flats have the effect of signaling urban disorder and moral decay and that these signals will over time prompt other criminals to break even more windows and exacerbate the conditions that lead to the decomposition of the city's moral, law-abiding fabric. While the article itself makes scant reference to graffiti, the aesthetic disorder put on display by overt flouting of the terms of the social contract is easily applied to the art form. Much like the "wars on graffiti" of the 1970s and 1980s waged by the city of New York, its anti-graffiti units, and the MTA, most municipal councils around the world continue to aggressively eliminate any sign of aesthetic disorder that might result in further decomposition of the moral integrity of the urban fabric.

Known among writers and artists as "the buff," municipal authorities devote vast amounts of human and financial resources to the task of fixing these "broken windows" by painting over, pressure washing, or air-blasting the surfaces upon which the works are created. In London, England, for instance, "In 2005, the total cost of London graffiti was at least £23 m per annum. However, if damage to economic development and loss of capital value to people's homes was included, this figure rose to over £100 m" (Keep Britain Tidy N.D.). One of the key aspects of the process of fixing broken aesthetic windows that will be examined in more detail below is the fact that due to the vigilant efforts of most municipal councils, among other elements, street artists are fully aware of the fact that their work will disappear in the near future. They operate and go to incredible lengths in creating their art, under this very assumption.

In a particularly playful example that makes this fact explicit, a British street artist that goes by the name of Mobstr (the missing vowel an allusion, perhaps, to the photo-sharing social network Flickr) painted with the aid of stencils "Is This Shade of Grey Acceptable" on a wall in Newcastle, England. Very soon thereafter, as expected, the piece was painted over or "buffed out" by Newcastle's municipal authorities. The next day, Mobstr returned and painted in a different hue of gray: "Obviously not, HOW ABOUT THIS SHADE?" Once again, the piece was quickly buffed out. Mobstr responds the next day: "OR THIS ONE?" Again, the buff occurs. Finally, Mobstr's last salvo, painted in black this time, "I GIVE UP" (Mobstr 2010). Most municipal councils go to great lengths and devote vast sums of financial and human resources to eradicating street art from the environment under their charge.

The second central cause for graffiti and street art's ephemerality is private property owners. When municipal councils take too long to remove the offending tag or image, property owners will often step in and remove it themselves. Using many of the same methods used by the authorities, private property owners will paint over, wash off, or blast away the graffiti or street art from the building or structure in question. Often the kind of home one resides in has an influence on the kinds of graffiti or street art one encounters. In an apartment building with many individual units, the likelihood of a writer or tagger choosing its walls, instead of those of a single-family home, is greater. This predilection for surfaces that will be seen by a large number of individuals can be traced to graffiti's early history where the point of the art form was to get one's name in front of as many people as possible. Single-family homes, then, are not all that attractive to writers or artists because they provide very little exposure and will most definitely be buffed in the very near term. The particular facets and importance of some walls or surfaces to the exclusion of others is something that will be addressed in much greater detail below; for now, it suffices to acknowledge the fact that private property owners who see graffiti and street art on their property will remove it very quickly so as to maintain the aesthetic integrity and economic value of their home and/or the structures attached to it.

The third central cause that helps to explain the inherently ephemeral nature of street art is other artists, writers, or taggers. For street artists, the urban canvas is a contested domain where aesthetic battles between individual artists or affiliated groups of artists known as "crews" or "teams" are fought out on a nightly basis.

Street art has none of the protections afforded to other forms of socially sanctioned art. There is no lacquer or varnish applied to the finished canvas. There are no frames with coated glass that protect the works from ultraviolet rays. There are no velvet ropes cordoning off the work from the physical presence of others. There are no security guards keeping watch over the pieces themselves. There are none of the taken-for-granted assumptions of the gallery, where the works themselves are automatically granted cultural or economic significance and, hence, accorded their due respect, as a result of their presence within these socially sanctioned confines. Quite simply, if another artist wants to deface, cross out, or paint over the work of a rival artist, all he/she has to do is walk up to the piece and deface it by applying another layer of paint to the ever-changing palimpsest, that is, the city street.

One of the most infamous accounts of the battles waged by warring street artists is that to have taken place between King Robbo[5] and Banksy on the streets, canals, and alleyways of London, England (see: Preston 2011). In the early-to-mid 1980s, King Robbo was one of the founding figures of the then burgeoning graffiti scene in London. His prolific and accomplished pieces were at one time displayed in all of the usual places: underground trains, tube stations, and of course in various locations around the city. In a testament to the inherently ephemeral nature of graffiti and street art, the vast majority of his pieces have long since been destroyed. There was a single piece, however, whose faded outlines and all but indecipherable content survived in much degraded form.

Titled "Robbo Incorporated," the piece was placed underneath the British Transport Police Headquarters in Camden on a wall adjacent to a landing bordered by a canal accessible only by boat. The piece was largely regarded as the oldest remaining example of early graffiti in London and was ostensibly accorded a certain amount of respect for its place in the history of the subculture in this particular city. In reality, however, the piece had been tagged over numerous times with those tags being tagged over again by other writers and artists. To say that the original piece created by Robbo was degraded in some way, shape, or form would be to understate the amount of damage done to it over the years.

In 2009, Banksy committed the ultimate act of disrespect by creating a piece of his own over the ragged remnants of King Robbo's original. The fact that in 2009, the original piece created by Robbo bore very little resemblance to that of the original is often downplayed by those wishing to cast Banksy's act in an even more disparaging light. The act of painting over "Robbo Incorporated" brought King Robbo out of "retirement" and set off a turf war that reverberated around the alleyways of London for more than 2 years. Every time Banksy would put up a new piece in London, Team Robbo was there very soon after to destroy it. This high-profile example of the battles that ensue when one artist tags over the work of another serves to illustrate the ephemerality of the art form in that when it is left to wither on the street, one of the central causes that underlies its eventual demise is other artists or taggers.

[5]"King" is a term of respect bestowed upon those individuals who have proven themselves particularly adept at creating a large number of pieces that are artistically accomplished.

The fourth central cause underlying the inherent ephemerality of street art is, quite simply, the elements. As alluded to briefly above, works of art left on the street are offered none of the protections that other more traditional forms of art enjoy. Meticulously monitored humidity controls, UV protection, prohibitions against flash photography, and the like simply do not exist on the street. Once created, the artist abandons the work of street art, leaving it to wither on the very spot where it was originally created. Depending on the media chosen by the artist, the rate at which this decomposition will occur varies.

Wheat paste, one of the more common media used to create street art, is instructive in this regard. Depending on the artist, wheat paste is the combination of flour or some kind of starch and water applied to the surface of the wall with the aid of a brush or broom. Similar, yet far less resilient than wallpaper paste, wheat paste is a quick and relatively easy way of applying a creative work to the uneven surface of a city wall. It is, however, by no means archival. Shepard Fairey, the artist responsible for thrusting the "Obey" (Fairey N.D.) moniker into the mainstream, as well as the individual responsible for creating one of the most iconic images of Barack Obama in his historic ascent to the presidency of the United States,[6] has made a very successful career of pasting large-format photocopies on city walls with the assistance of wheat paste. Swoon (N.D.), one of the few female street artists to have gained some kind of public notoriety, also makes use of wheat paste in the application of her intricate and detailed works of art. Due to the media chosen by these artists, however, the half-life of their work is very brief.

The paper upon which their images are created is fragile and decomposes easily, and the adhesive, often applied with a broom, inconsistent in its coverage. Once again, there is no impermeable lacquer or varnish applied to the surface of the image in any way that might preserve it for the months or years to come. Depending upon the climate of the city in which the work of street art is created, the elements begin to deteriorate the artistic integrity of the work as soon as it is completed. The harsh summer sun, beating rain, penetrating frost, billowing snow, and gusting wind are inauspicious conditions for a work of art to survive unscathed. Whether made of paper and wheat paste, spray paint, or both, the elements will, over time, break down the work of street art to the point that the original bears very little resemblance to the piece created by the artist himself/herself.

The above four causes for street art's inherent ephemerality have been a consistent feature of the art form from the very beginning. Overzealous municipal councils seeking to assert their ideological and spatial authority over "their" city by buffing out any form of dissent that might provoke discomfort or anxiety, property owners seeking to maintain the aesthetic integrity of their dwellings, other taggers or writers seeking to make a name for themselves while others looking to save face, and the natural force of the elements have always influenced what survives of art

[6]In a telling indicator of the paradoxical nature of street art, a copy of the image of Obama placed on the street illegally by Fairey now resides in the Smithsonian National Portrait Gallery of the United States alongside the works of Gilbert Stuart, Edgar Degas, Irving Penn, and Paul Cézanne.

forms created in the public and on the street. The fifth underlying cause of street art's ephemerality is a relatively new addition to this list and is intimately, though perhaps paradoxically, tied to place of the digital camera in the preservation, dissemination, and, hence, popularization of the street art form. Due to the contemporary valence of this fifth cause for street art's ephemerality, it will be examined in more detail than the previous four.

In the contemporary art world, it is not uncommon for increasingly large and cumbersome walls made of concrete or cinder block to be placed under the gavel on the auction block. In May 2012, on the eve of Queen Elizabeth II's Diamond Jubilee celebrations, a piece by Banksy that depicts a toddler hunched over a sewing machine was created on the side of a "Poundland" store. After being placed behind protective Plexiglas so as to encourage tourism to this struggling area of London, the piece was removed by the legal owners of the wall. Using masonry saws, this entire section of the wall was removed and, depending on your perspective, either preserved, saved, or destroyed. As will be addressed in the next section of this chapter, the specific location that this piece (and many other pieces created by street artists) was placed is an incredibly important element in the overall meaning of the piece itself, and to extract it from the exact location within which it was created is to alter the work as the artist created it and intended it to be seen. What merits emphasis for the time being, however, is that the fifth underlying element for street art's inherent ephemerality is its monetary value on the open market or its cultural significance to institutions seeking to "preserve" these fleeting works of art.

In June 2013, the aforementioned piece created by Banksy on the exterior wall of a shop in London was sold to a private and anonymous collector for roughly $1.1 million (USD). Another piece created by the same artist, entitled "Kissing Coppers," first placed on the exterior wall of a pub in Brighton, England, in 2005, was sold in February 2014, reportedly for $575,000.00 (USD). In 2008, another of Banksy's pieces was sold via eBay to a private collector for $407,000 (USD) – plus the costs associated with extracting it from the wall. The seller indicated at that point in time that "this shouldn't cost more than 5,000 pounds" (Reyburn 2008). It merits mention that a very select number of street artists have the subcultural capital required to command this kind of attention and sums. For the most part, and, quite simply, unless your pseudonym is Banksy, street art does not attract that much attention from collectors or preservationists.

Returning to the beginning of this chapter, the piece created by Banksy at the Packard Plant in Detroit in May 2010 is interesting in this regard because it too was extracted from the wall upon which it was created, but this extraction neither was undertaken or accomplished by the legal owners of the wall nor was it initially extracted from its location with the intent of profiting from the sale of the piece. Instead this piece was "stolen," "saved," and/or "preserved" by a local not-for-profit art gallery in the city. Gallery 555 is a "nonprofit gallery and studio [whose] mission is to strengthen communities with the arts" (Gallery 555 N.D.). The struggling not-for-profit has moved locations on four different occasions in the past decade or so but seems to have found a relatively stable home in an abandoned Detroit Police Department precinct on the west side of the downtown core in what is commonly referred to as Mexicantown.

In May 2010, when news of Banksy's visit spread, individuals associated with the gallery packed up their tools and went to the Packard Plant. Armed with shovels, masonry saws, acetylene torches, and a backhoe, the volunteers at Gallery 555 set about trying to "save" the piece from its inevitable demise. According to Carl Goines, executive director and cofounder of the gallery,

> It's about preservation for us (. . .). We're watching this beautiful city crumble around us and we can't do anything to stop it. So with this fine-art piece – and it's not just everyday graffiti that you might whiz by – here was our opportunity to do something. It would have been destroyed if we didn't make the effort. (Stryker 2010)

The distinction made by Goines between the Banksy as a "fine art piece" and the "everyday graffiti" that one might "whiz by" is disingenuous. The walls of the Packard Plant are festooned with graffiti and street art that is, in some instances, much more artistically accomplished than the piece by Banksy. This is a derelict location where local writers and artists have been honing their skills for years on walls that are replete with exemplars of skill, passion, and commitment. To dismiss all of the other pieces within the factory so swiftly, at the same time as canonizing this relatively quotidian example of Banksy's work, is the first inkling that the stated intentions of the gallery are less than genuine.

This act of "preservation" sparked a fevered controversy within the subcultural confines of the street art world in general and that of Detroit more specifically. There were those that argued the piece should be left to wither and suffer its inevitable demise, those that praised the gallery for their efforts in preserving this important example of contemporary street art, those that questioned the actual motives of the gallery, and finally those that wanted to determine the true owner of the wall. A legal battle ensued with the gallery taking legal ownership of the wall in late June 2011 (Stryker 2011). On numerous occasions, Goines, along with other spokespersons for Gallery 555, claimed that their intention was not to sell the Banksy but to put it on display in their gallery so that the public could enjoy the piece. However, in March 2014, the gallery announced (Stryker 2014) that it was going to sell the piece so as to expand its capacity and ability to support the local arts community in Detroit.

This chapter is not interested in passing judgment on the actions of the gallery. Rather, it is focused on better understanding the underlying reasons for street art's inherent ephemerality and the role that the digital camera and the Internet play in its creation, preservation, popularization, and dissemination. In the last instance, the piece created by Banksy was removed from its original location by a group of individuals with the ostensible goal of saving it from its eventual destruction and putting it on public display for the citizens of Detroit to enjoy and contemplate. Thus, the fifth and final underlying cause of street art's inherent ephemerality is the cultural and/or economic value of the works themselves and the propensity of the owners (actual, assumed, or otherwise) of these works to extricate them from their very material foundations so as to sell them on the open market.

When considered in sum, the five elements detailed above constitute the underlying causes for street art's inherent ephemerality. The fact that these works of art are often made on or in stone, therefore, should not be equated with their durability or their permanence. Looked at from these five different perspectives, street art is one of the few "plastic arts" that is planned, designed, and created in full knowledge that the end product will disappear or be destroyed over time.

This inherent ephemerality is, of course, where the digital camera and the Internet become incredibly important implements in the street artist's quiver. As much as these artists are reliant on the physical and material qualities of cinder block, concrete, wood, and steel to exercise their creative vision, they are equally reliant on the immaterial, virtual, digitized, and distributed hardware and software characteristic of the contemporary era, to document that which will disappear forever. What merits emphasis at this point, however, is the street artist's appreciation of the eminently ephemeral nature of his/her work. When an artist makes the decision to create works of art on the street and in the public, he/she is aware of the fact that once completed, the work will eventually deteriorate for one of the five reasons listed above. If there were any inclination or desire on the part of the artist to create art that would exist beyond the immediate future, then they would not have created it on (and left it in) the middle of the city street. Street artists operate, therefore, in full knowledge of the fact that for one of the five reasons enumerated above their work will be destroyed.

However, under no illusions regarding the durability or permanence of their work, before turning their backs on their pieces and walking (or running away), street artists accomplish one last act of artistic creation. They step back from their piece, frame it within the viewfinder or screen of their camera, and capture a picture. This final act of artistic inspiration has garnered very little critical attention in any of the scholarly texts dedicated to furthering our understanding of this subcultural form.[7] Understandably, the emphasis is traditionally placed on the works themselves. Their political message; the skill, daring, and courage required to create them; the work's (il)legality; and the color, scale, and scope of the piece have all historically dominated the discourse on street art. However, the digital camera is not a passive recording device that serves only to document the works of art themselves. Instead, it has become an ever more important tool in the street artist's field bag.

In some cases, as a result of the digital camera, the audience presumed to be on the other end of its lens is in fact the intended audience for the work. This final act of preservation, then, not only documents that which will disappear, but it also alters

[7]For a rare exception, see the work of Gregory Snyder (2006). Snyder argues that the process of capturing photographs and publishing them in physical form, primarily via underground "zines," was an important element in writers honing their skills and being inspired. He argues, "Photographs made ephemeral graffiti pieces permanent, allowing writers to view the work of others without attachment to a specific place or time. The inclusion of these 'flicks' in magazines created a space where graffiti pieces from all over the world could come together to be judged, critiqued, and offered as instruction" (Snyder 2006, p. 93).

the intended audience of the work and by doing so transforms the street artist into a digital artist. The important place of the digital camera in the process and practice of creating street art requires further unpacking.

Location, Location, Location

By framing a piece of graffiti or street art with the lens of a camera, the camera cannot help but capture fragments of the broader contexts[8] within which the piece was created. The urban canvas, of course, bears very little resemblance to the traditional canvases of more traditional artists. It is anything but blank or empty. Its surfaces uneven, cracked, covered in dirt, and full of "content" before the artist arrives to create his/her work. These blemishes or found details have become pivotal to the work of many street artists in that they are ever more woven into the works of art themselves. The idiosyncrasies of particular walls, the very specific location of these same walls, and the appreciation and importance of this broader context to the works themselves have therefore come to play a much more important role in street art than has been previously acknowledged.

In October of 2013, for instance, Banksy embarked on an ambitious project in the streets of New York City. "Better Out Than In" was a relatively unique artist's residency in that the artist was not invited to, or stationed "in," a particular gallery but rather took it upon himself to create his work "out" on the streets of the city. The plan was to create a single piece of street art every single day throughout the month of October and was accomplished, except for a single day where a message to the artist's Instagram account used to document and authenticate all of his offerings during this period announced: "Today's art has been cancelled due to police activity" (Banksy 2013). The fact that Banksy made use of a digital camera and a social network as the media through which he captured, documented, authenticated, and communicated his work is, in and of itself, a telling indicator of the importance of the digital camera to contemporary street art but will become much more so when the details of a select number of pieces are considered.

While a comprehensive list of all of Banksy's works created throughout his residency is beyond the scope of this chapter, there are two pieces in particular that merit further examination.[9] The first is the image used to announce and promote the

[8]This was one of the central differences between the work of early graffiti photographers Chalfant and Cooper. Whereas Chalfant focused his lens as tightly as he could on the tag itself, Cooper framed her subject much more expansively so as to include the broader cityscape in the frame of the photograph.

[9]Though plentiful and easily retrieved online, none of the images referenced herein are included in this chapter. This was an intentional decision on the part of the author in that it further reinforces the archival and documentary dimensions of the digital camera and Internet and underscores one of the central arguments made throughout. It merits mention, however, that Banksy has a rather ambivalent relationship with copyright protecting his right to his creative work in some instances,

"Better Out Than In" residency in the streets of New York. Resembling a traditional flyer announcing a prototypical gallery show, though never printed and only ever distributed online, the image depicts the stenciled figure of a boy, roughly 10 to 12 years old, hunched at the waist, holding a can of spray paint at his side, and in the act of vomiting. However, in an element of Banksy's work that is becoming ever more prominent, rather than the vomit being rendered in spray paint, the artist chose to weave very distinct elements of the urban fabric into the work of art. Spewing from the mouth of the little boy is a cluster of wild flowers and foliage growing out of the cracks in the wall's surface that mimics the spread of the would-be vomit. Portions of the "canvas" upon which this piece was created are elemental ingredients of the work, and any attempt to remove it from this location would also destroy it. The various surfaces upon which street art is created are therefore anything but neutral and influence the creation of the artwork in a direct manner.

The second is commonly known as "Hammer Boy" and once again depicts a small boy, but this time, he is holding a carnivalesque hammer that is about to be brought down on an actual fire hydrant connected to an actual fire alarm. Once again, the boy is the only element of this piece created by the artist. The hydrant, alarm, and conduit connecting the two were there long before the artist arrived and will surely be there long after the work is destroyed. In this instance too, Banksy incorporates elements of the urban fabric into the composition of his piece in such a way that its extraction would destroy the piece as the artist created it. Therefore, both of these pieces and many others made by the artist (including to a lesser extent that made at the Packard Plant in Detroit, Michigan) are noteworthy not because they are particularly accomplished, intricate, or provocative in their message or detail but because of their incorporation of found elements of the city fabric into their compositions and, as a result, the emphasis placed on very particular locations, the digital camera, and the Internet in their creation.

Put simply, Banksy is well aware of the fact that much of his work has been either destroyed by councils, property owners, and other taggers or extracted from its original location and put up for sale via private auction houses. By incorporating elements of the urban fabric into his works, he is making the task of extracting these works from their locations much more complicated than it would be otherwise.

In the first example, a stencil of a boy hunched over at the waist, holding a can of spray paint, and in the act of vomiting was used as the promotional image to announce the artist's residency. This piece is particularly noteworthy because of the importance of the urban fabric in its composition. The image of the boy vomiting could have been created solely with the use of spray paint and stencils. However, the artist regards the cityscape as an important ingredient in the overall piece and

then claiming in one of his books: "Trademarks, intellectual property rights and copyright law mean advertisers can say what they like wherever they like with total impunity. Fuck that. Any advert in a public space that gives you no choice whether you see it or not is yours. It's yours to take, re-arrange and re-use. You can do whatever you like with it. Asking for permission is like asking to keep a rock someone just threw at your head" (2005, p. 160).

one that is vital to its appreciation and meaning. If the piece were buffed out and destroyed by municipal councils or private property owners, then a photograph of its creation remains as evidence. If another tagger destroys it, then once again, the photograph remains. Without question, as the seasons shift, the flowers which are a core component of the work will wither and die, thus rendering the meaning of the piece incomprehensible over time. Finally, if the property owner were to extract the piece from the wall, the flowers would die. This and in addition to the fact that if it were placed behind Plexiglas so as to preserve it, the flowers, pressed between the wall and the Plexiglas, would undoubtedly die. When this example is considered indicative of a particular strain of street art, the awareness of the artist regarding the inherently ephemeral qualities of his/her work is thrust to the fore, so too, therefore, is the importance of the digital camera and the argument being made throughout this chapter. Due to the fact that the artwork will eventually wither and fade for one of the abovementioned reasons and the fact that the piece loses all meaning if moved from the very specific location within which it was created, the street artist is better thought of as a digital artist even if he/she does not regard himself/herself as such.

This piece in particular was not created for the passerby who might happen upon it while browsing the shops in a retail district but for an audience that might happen upon it while browsing around online. It is this shift in the intended audience of so-called street art that signals the evolution of this subcultural form from one exclusively concerned with "getting up" (see Castleman 1982) on the street to one much more enamored with intentionally creating much more durable works of art for a much larger audiences online. The archival qualities of the online environment compensate for the inherent ephemerality of the physical works. This compensatory function has had a lasting influence on the physical works themselves in that artists weave seemingly random elements of the urban fabric into their creations in such a way that privileges the virtual audience over the physical. The digital camera and the Internet are not simply documentary tools used as an afterthought to preserve that which will eventually disappear but influence the process of creation in the first place. A close reading of our second exemplar will help to make this point more forcefully.

"Hammer Boy" is a simple stencil that with adequate preparation could have been painted on the street in very little time. The crux of the piece is much more dependent on the found elements of the urban environment for its semiotic force. In the absence of the hydrant and alarm, the piece itself would be nonsensical. By incorporating these found elements of the city space into the piece itself, Banksy is guarding against the commodification of his work via its extraction and sale on the open market. Cutting this piece from the wall and removing it from its original context would destroy the piece entirely. Alternatively, the stencil absent the hydrant and alarm would make little sense to anyone who sees it in the "white cube" (Austin 2010) of a gallery. Similar to the boy vomiting flowers referenced above, the exact location at which this piece was created is pivotal to the work itself. Once again, it is at this moment that the role of the digital camera and the Internet, not only in the preservation of these ephemeral pieces as an afterthought but also as the intended medium via which the artist communicates with his/her audience, comes to bear on the discussion.

These pieces, as well as the one created in the Packard Plant in Detroit and many others, are inextricably linked to the exact locations they were created and photographed in. To remove them from these locations is to change the work of art from something created by the artist to something else entirely. The social and political relationships that undergird these alterations are incredibly interesting in their own right,[10] but a critical examination of the political economy of street art will have to be tabled for another time and place. What merits emphasis at this juncture is threefold. The importance of location; the incorporation of infrastructure, found detritus, and/or details of the wall itself into the work of art; and the inherent ephemerality of the works force a reappraisal of the characteristic undervaluation of the digital camera and the Internet in the street artist's aesthetic arsenal.

The Street Artist as Digital Artist: A Conclusion and Provocation

It is not only the idiosyncrasies of particular walls that come to influence the kinds of art created on and for the street but also those stochastic elements of the urban fabric that sit alongside, up against, within, or adjacent to them that also interweave the importance of the digital camera into this subcultural art form. Once again, Banksy's work is informative in this regard. If we return to Detroit for a moment and consider the act of applying a stencil of a young boy to a wall in a derelict industrial facility and then captioning that image with the sentence "I remember when all *this* was trees" (emphasis added), the central argument made by the present chapter becomes clear.

It is this "this" in Banksy's caption that emphasizes the importance of this very particular location to the meaning of the piece and the intention of the artist that created it. When extracted or removed from *this* exact place for whatever reason, the artwork itself may continue to be interesting or noteworthy because of its providence and/or history, but it should no longer be thought of as a work by Banksy. The work as the artist envisioned and created it is fundamentally dependent on the exact location within which it was placed. To change that location is to change the work of art into something other than that created by the artist himself/herself. It is, therefore, to destroy the work of art in the misguided hope of preserving it. Whatever the piece becomes once extracted from the location that serves as one of its primary referents and elements of composition, it is no longer a Banksy. Similar to the destructive

[10] See Luke Dickens' article (2008) on the "journey" of the Peckham Rock for an excellent take on how the manifold interests involved in street art fundamentally alter the work of the artist himself/herself.

process of carving out a small corner of a painter's canvas with a scalpel, to carve out this section of wall from the derelict location that is as important to the semiotic integrity of the work as the spray paint itself is to destroy the piece the artist created.

Once again, the urban fabric is not a blank canvas upon which street artists exercise their creative vision, but one that is always already loaded with content and meaning. This content is being used not only as inspiration for the creation of particular pieces but also is being interlaced into the works themselves. Therefore, to remove this small section of the street artist's "canvas" is to destroy it in such a way that it becomes something other than that created by the artist. In other words, to extract the "street" from the work of "street art" is to transform it into something other than that created by the artist. In this way, then, the object that more faithfully represents the intentions and/or vision of the artist is not the physical work itself, removed from its original context or extracted from the broader urban canvas that is elemental to its composition, but the digital representation thereof that captures this contextual urban canvas and the intentions of the artist more faithfully than the tattered remnants housed in a gallery. It merits reemphasis that artists are aware that their work will be destroyed. Therefore, much better at depicting the creative intentions or artistic vision of the street artist is the digital photograph of the work as he/she envisioned it within the location it was created. This leads to the conclusion that as much as these individuals can and should be regarded as street artists, they can and should also be regarded as digital artists, albeit digital artists that go to great lengths in the preparation of their compositions.

References

Austin J (2010) More to see than a canvas in a white cube: for an art in the streets. City 14(1–2):33–47

Banksy (2005) Wall and piece. Random House, London

Banksy (Director) (2010) Exit through the gift shop [Motion Picture]. UK

Banksy (2013, October 23) banksyny. Retrieved April 24, 2014, from Instagram. http://instagram.com/p/f0Jfw1K-5K/#

Castleman C (1982) Getting up. The MIT Press, Cambridge, MA

Cooper M, Chalfant H (1984) Subway art. Thames & Hudson, London

Daniel B (Director) (2005) Who is Bozo Texino? The secret history of Hobo Graffiti [Motion Picture]. USA

Dickens L (2008) Placing post-graffiti: the journey of the Peckham Rock. Cult Geogr 15:471–496

Dickens L (2010) Pictures on walls? Producing, pricing and collecting the street art screen print. City 14(1–2):63–81

Fairey S (N.D.) About. Retrieved April 24, 2014, from ObeyGiant. http://www.obeygiant.com/about

FORM (2014) FORM. Retrieved April 24, 2014, from Public. http://www.form.net.au/project/public/

Gallery 555 (N.D.) Mission. Retrieved April 24, 2014, from Gallery 555. http://www.555arts.org/mission/

Gastman R, Neelon C (2010) The history of American graffiti. Harper Collins, New York

Halsey M, Pendrick B (2010) The game of fame: mural, graffiti, erasure. City 14(1–2):82–98

Halsey M, Young A (2006) 'Our desires are ungovernable': writing graffiti in urban space. Theor Criminol 10(3):275–306
Keep Britain Tidy (N.D.) Facts and figures – British Transport Police – Statistical Bulletin 2008/2009. Retrieved April 24, 2014, from Keep Britain Tidy. http://kb.keepbritaintidy.org/graffitiv2/FactsandFigures/headline.pdf
Kelling G, Wilson J (1982, March) Broken windows: the police and neighborhood safety. The Atlantic, pp 29–38
Kramer R (2010) Painting with permission: legal graffiti in New York City. Ethnography 11(2):235–253
McAuliffe C, Iveson K (2011) Art and crime (and other things besides . . .): conceptualizing graffiti in the city. Geogr Compass 5(3):128–143
Mobstr (2010) Playing with the buff. Retrieved April 24, 2014, from Mobstr.org. http://mobstr.org/image%20pages/progressions/playing%20with%20the%20buff.html
New York Times (1971, July 21) 'Taki 183' Spawns Pen Pals. New York Times, p 37
PowWowHawaii (2014) About. Retrieved April 24, 2014, from PowWowHawaii. http://powwowhawaii.com/
Preston J (Director) (2011) Graffiti wars [Motion Picture]. UK
Reyburn S (2008, January 14) Banksy Mural on London Wall Sells for $407,000 in Ebay Auction. Retrieved April 24, 2014, from Bloomberg. http://www.bloomberg.com/apps/news?pid=newsarchive&sid=alxBQ4i1IUFk&refer=muse
Silver T (Director) (1983) Style wars [Motion Picture]. USA
Snyder G (2006) Graffiti media and the perpetuation of an illegal subculture. Crime Media Cult 2(1):93–101
St. ART Delhi (2014, January) About. Retrieved April 24, 2014, from St. ART Delhi. http://st-artdelhi.org/#!/about
Stryker M (2010, May 15) Graffiti artist Banksy leaves mark on Detroit and ignites firestorm. Retrieved April 2014, 2014, from Detroit Free Press. http://www.freep.com/article/20100515/ENT05/100514077/Graffiti-artist-Banksy-leaves-mark-Detroit-ignites-firestorm
Stryker M (2011, September 10) 555 Gallery gets OK to display Banksy mural. Retrieved April 24, 2014, from Detroit Free Press. http://www.freep.com/article/20110910/ENT05/103020001/555-Gallery-gets-OK-display-Banksy-mural
Stryker M (2014, March 14) Gallery's plan to sell Banksy's Packard Plant mural sparks uproar. Retrieved April 24, 2014, from Detroit Free Press. http://www.freep.com/article/20140314/ENT05/303140017/banksy-555-gallery-detroit-packard
Swoon (N.D.) SwoonStudio. Retrieved April 24, 2014, from Facebook. https://www.facebook.com/SwoonStudio
Tapper J (2008, July 12) Banksy uncovered: the nice middle-class boy who 'became the graffiti guerrilla'. Retrieved April 24, 2014, from MailOnSunday. http://www.mailonsunday.co.uk/news/article-1034613/Banksy-uncovered-The-nice-middle-class-boy-graffiti-guerrilla.html
Young A (2012) Criminal images: the affective judgment of graffiti and street art. Crime Media Cult 8(3):297–314

Chapter 14
Solar Cycle 24: 15 Nightly Projections

Malin Abrahamsson

Abstract "Solar Cycle 24" is a public art installation with a focus on placemaking and digital aesthetics in the urban environment. Presented in a "nonart" venue, the temporary project coincided with the expected peak of an actively occurring astronomical event, solar cycle #24. This astronomical event is a recurring phenomenon of solar storms and the resulting spectacular activity of aurora borealis, or northern lights, which ebb and flow on 11-year cycles with high visibility around the North and South Poles. The installation "Solar Cycle 24" was an investigation of digital animation as public art with a focus on placemaking in the urban environment. Coproduced by the New York-based arts organization chashama, "Solar Cycle 24" was publicly visible between dusk and dawn between February 23 and March 10, 2013.

Introduction

At no previous point in history has technological innovation been so rapid and seductive. "Solar Cycle 24" is a public art installation with a focus on placemaking and digital aesthetics in the urban environment. Presented in a "nonart" venue, the temporary project coincided with the expected peak of an actively occurring astronomical event, solar cycle #24. This astronomical event is a recurring phenomenon of solar storms and the resulting spectacular activity of aurora borealis, or northern lights, which ebb and flow on 11-year cycles with high visibility around the North and South Poles. The installation "Solar Cycle 24" was an investigation of animation of digital animation as public art with a focus on placemaking in the urban environment. Coproduced by the New York-based arts organization chashama, "Solar Cycle 24" was publicly visible between dusk and dawn between February 23 and March 10, 2013, and its near-complete integration of our liveshas forced

M. Abrahamsson (✉)
Independent Artist, www.malinabrahamsson.com

F.T. Marchese (eds.), *Media Art and the Urban Environment*, Future City 5,
DOI 10.1007/978-3-319-15153-3_14

us to collectively rethink several fundamental aspects of existence. The definitions of concepts such as time, place, and space have been expanded in both practical-existential and philosophical terms and have forever changed the way we look at the world. Today, digital media is a much expected and entertaining part of the public urban environment, primarily in the form of large-scale advertising and live broadcasts. And in the art world, countless artists are trying to make sense of the global digital impact of technology and grasp the magnitude of its effect on life and art. Expanding beyond my background in painting and public art, "Solar Cycle 24" became a personal exploration of animation as public art, driven by a curiosity about what role digital art currently plays in public space. By and large a personal statement, this essay is a descriptive summary of the making of the project that attempts to explain the context in which it was experienced and some of the reactions it catalyzed.

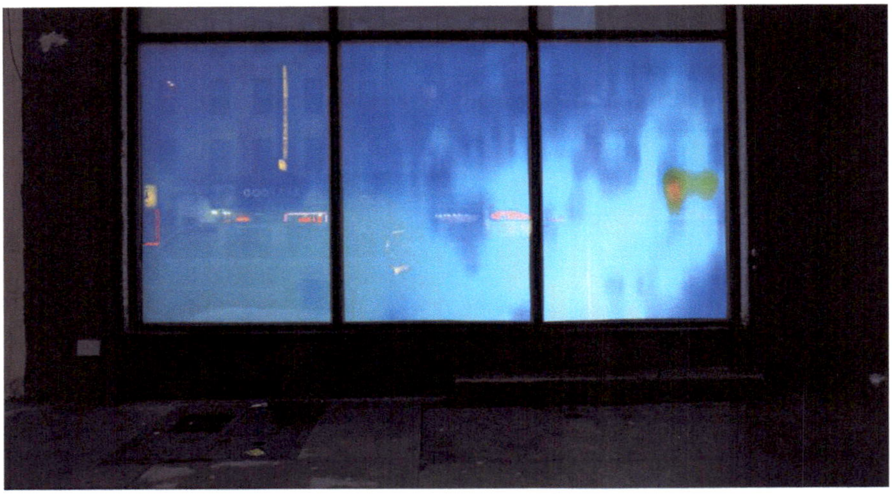

Installation view of "Solar Cycle 24" at dusk

The Solar Cycle

A cosmic, magnetic phenomenon, northern lights or aurora borealis are triggered by the ions in the solar wind captured by the Earth's magnetosphere and travel to the poles. These are recurring events that run on consecutive and overlapping 11-year cycles with high visibility at the peak of each cycle. It is difficult to perfectly predict when and where northern lights will occur, and although it does happen, this phenomenon is rarely seen on latitudes that are too far from the polarcircles or above very large, well-lit cities. In the fall of 2012, sightings of northern lights were increasing in northern Scandinavia and Siberia, and according to reports by NASA[1], solar cycle number 24 was slowly nearing peak mode. Often compared to a "curtain of electricity," the northern lights are an eye-catching, multi-chromatic phenomenon that becomes apparent in the dark sky after sunset. Their visibility ranges from a second or two to several hours, depending on the atmospheric conditions. On a clear night, the lights may perform a dance across the sky that oscillates between washed-out blues and greens, deep purples, and vibrant reds, while continually transforming in shape, size, saturation, and speed.

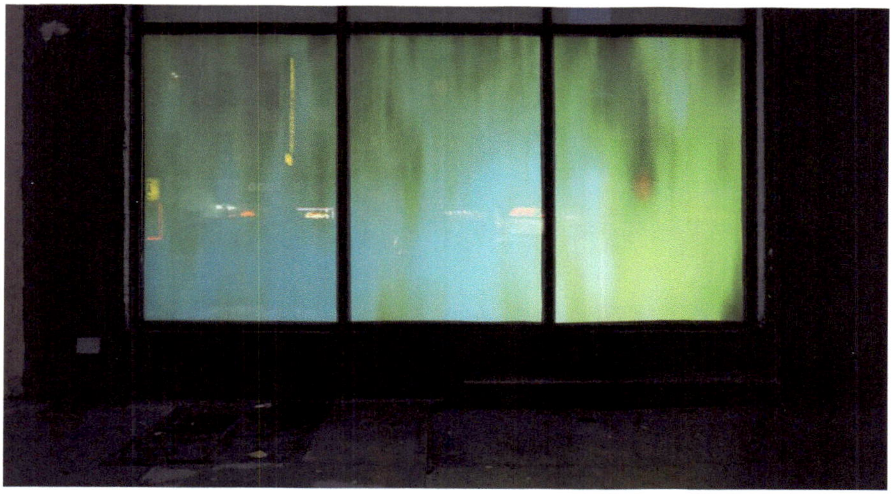

Installation view of "Solar Cycle 24" at dusk

The main component of the project "Solar Cycle 24" was 15 nightly public projections of an abstract and soundless animation – intended as a digital, hyper-chromatic version of northern lights. chashama,[2] a New York-based, nonprofit arts organization, provided logistical and material support to realize this project. With a mission tc connect art and real estate by transforming underutilized properties into work and presentation spaces for artists, chashama provided 24-h access to a vacant commercial space. Called "chashama 1351," the location is a street-level storefront that faces the sidewalk on the east side of Amsterdam Avenue in upper Manhattan, just north of 125th Street. This part of the city is a densely populated, urban area that has left the vertical frenetic light show of Times Squarebehind for

[1]NASA: http://www.nasa.gov

[2]Chashama: ht:p://www.chashama.org/

a somewhat more horizontal and slower light experience. The brightly lit, big-box stores that dominate the commercial area around 125th Street are less prominent on Amsterdam Avenue and on the few blocks closer to Hudson River. Primarily residential and working class, the area is generally dominated by smaller, local establishments and sprinkled with occasional industrial buildings as well as blacked-out, empty storefronts.

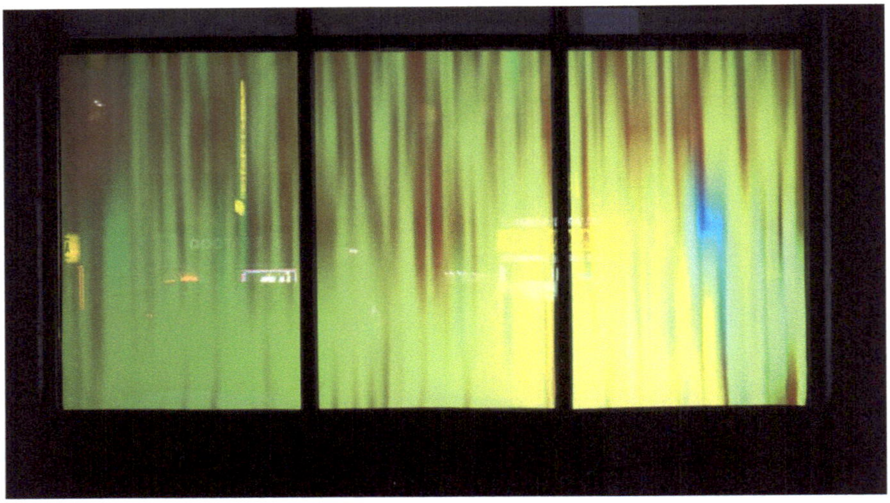

Installation view of "Solar Cycle 24" at night

Moving Light

Within the empty space, a projector was set up to beam onto thin sheets of Mylar[3] that were taped up on the inside of the storefront windows. The semitranslucent material created a temporary projection screen, which made the animation perfectly visible from the outside. On the street, passersby could see the animation slowly unfold: it changes in color from deep purple and dark blue to vibrant green; pulsating spots in bright orange briefly appear, and a for a few seconds, canary yellow dominates, with thin sections of intensely red light. The secure location allowed for unsupervised all-night screenings. The 12:30-min long animation would gradually appear every evening at sunset and then continue to loop back and forththroughout

[3]Brand name for stretched polyester film. Its properties make it desirable for commercial applications such as electric insulator and gas/odor barrier.

the night. Its intense colors gently embraced pedestrians on the sidewalk, and its deliberate movements were reflected in puddles on the ground and the windows of cars and buses in the street. Since daylight effectively made it impossible to see the projection, it was primarily visible in the dark before and after regular business hours. While this cycle of visibility was a direct result of equipment and technical setup, it conveniently mimicked the natural appearance of northern lights. While perfectly visible throughout the night, it would slowly begin to fade away at the first light of dawn.

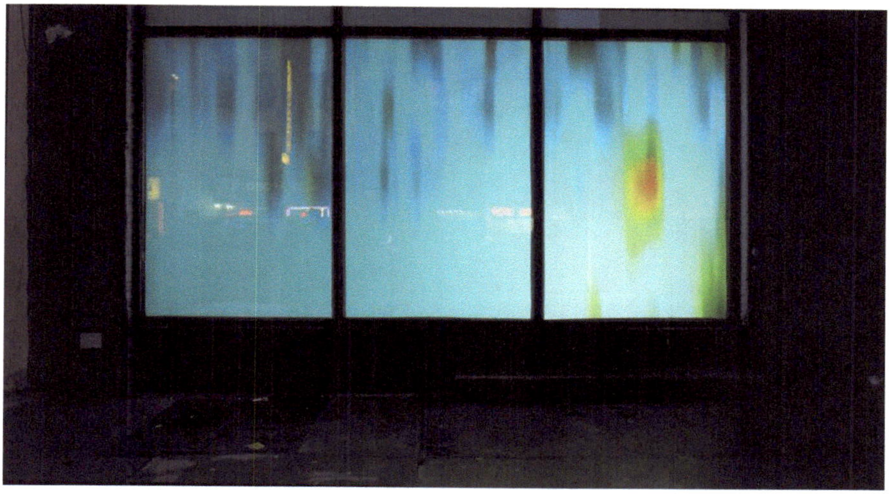

Installation view of "Solar Cycle 24" at dawn

The rear-projection technique made it possible for viewers on the sidewalk to walk up to the 10-ft-tall by 20-ft-wide animation without casting a shadow onto the moving image: a proximity that is not primarily associated with art. By effectively eliminating all distance between spectators and the projected image, there was an atmosphere of quiet intimacy surrounding the installation despite its urban, public location. Slow-moving and abstract, "Solar Cycle 24" looks nothing like the fast-blinking and multicolored electric signs that competed for attention on the neighboring businesses, and their glaring reflections in the glass in front of the projected image served as a visual reminder of the striking differences. Unlike the multinational fast food chain restaurant, glitzy pawnshop, brightly lit liquor store, small local Mexican eatery, and busy hairdresser across the street, "Solar Cycle 24" has no distinct message. It includes no text, and while the installation took up three full storefront windows, it was only visible after dark.

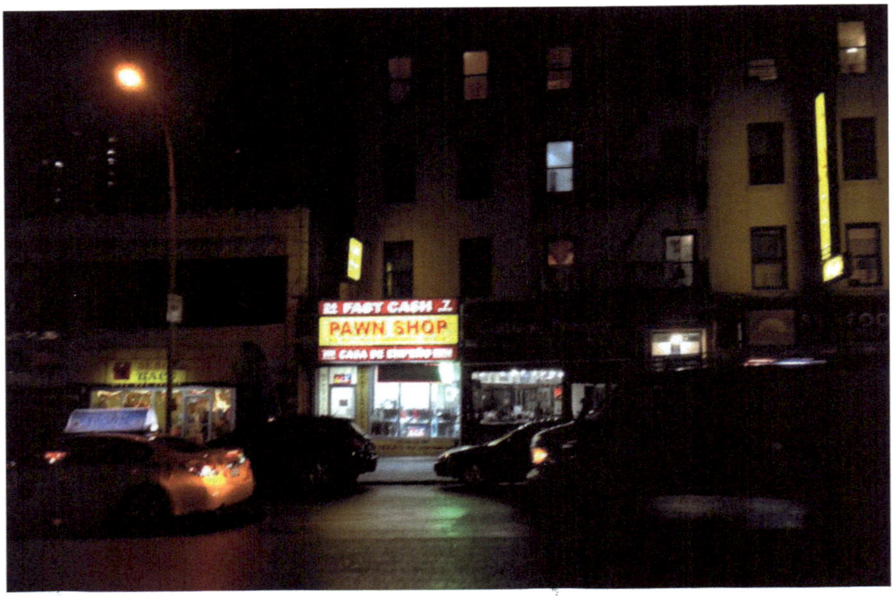

View of businesses directly across from the installation on Amsterdam Avenue

The project's relatively brief appearance was based on the natural visibility of northern lights and was specifically scheduled to coincide with the currently peaking solar cycle. The conceptual connection between the elusive northern lights and the constant urban throb of New York City was palpable to those who had previously experienced aurora borealis and were familiar with its electrifying quality. For others, the intention to channel poetic stirrings of the universe down to a New York City sidewalk may not have been immediately evident. Since the context and location of the installation did not provide an obvious purpose or message, any meaning had to be individually discovered by the viewer. "Location is not a set of coordinates," writes Emma Ota (2008), "it is not something static and easily measurable, it is not a case of physical geography but is a state which exists through the complex interplay of history, culture, socio-politics, economics and technologies. Location is a multifaceted context, a situation and state of being and is not necessarily linked to the ground beneath our feet." By installing the project in this empty storefront, "Solar Cycle 24" successfully transformed the venue into a destination. The silent, polychromatic illumination of the previously vacant location made it clear that it meant *something*. Pedestrians on the double-wide sidewalk often slowed down as they passed by. Some would come to a complete stop in order to look, while others hesitated briefly before stepping closer. An older gentleman, resting on his cane for a few minutes, asked if he was looking at a church.

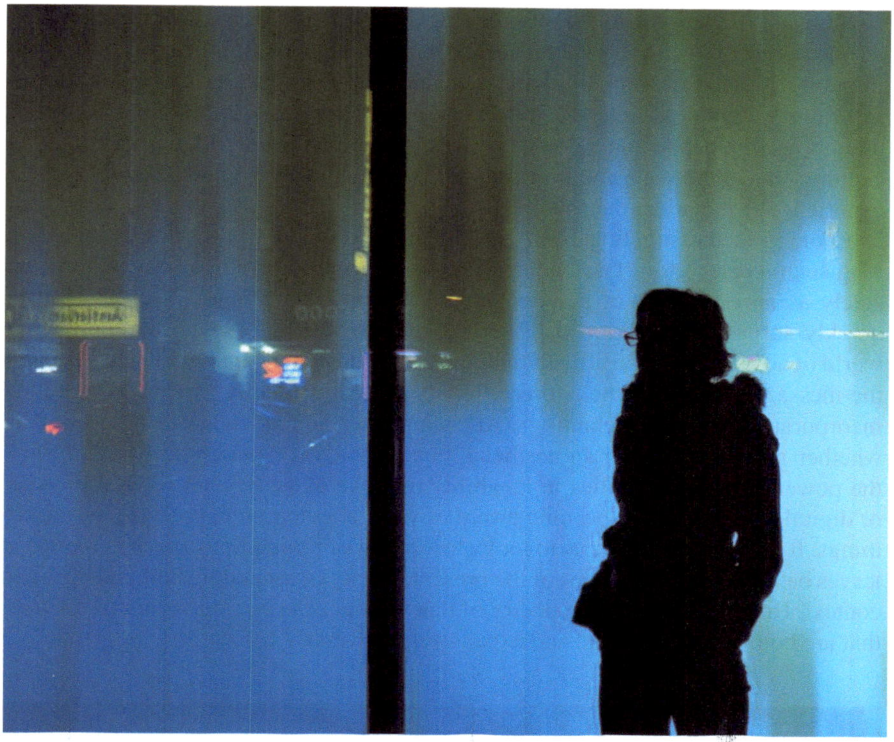

"Solar Cycle 24," close up (Photo: Isaiah Davis, 2013)

Afterglow

"Solar Cycle 24" was a relatively low-tech exploration of digital animation as public art. Aesthetically, the project resembled a large-scale, slow-moving painting, and it relied equally on color, concept, and code in its attempt to offer a poetic experience of space, place, and pace. The temporary presentation in a "nonart" venue emphasized a focus on placemaking in the urban environment and successfully transformed an empty storefront in upper Manhattan into a destination for digital art. With a solid foundation in science, the time- and light-sensitive installation provided a conceptual link to the electromagnetic stirrings of the universe that emphasized not only the viewer's specific location on our planet but by extension also Earth's position in the solar system.

"Solar Cycle 24" was created within the small but quickly growing niche of public digital art: an artistic category that often triggers fantastical ideas of special effects, interactive software, and expectations of displays on spectacular scale. Our communal fascination with large screens and massive electronic advertising reveals an uncritical embrace of digital media and how it is still considered a relative novelty in public space. This general global technological optimism has played a big part in bringing digital art into the public, urban environment: although at this point ubiquitous in most people's everyday lives, new digital technology continues to be exciting on a grand scale and is quickly and systematically immersed into the visual experience of the urban metropolis.

Looking at the situation from an art critical standpoint, one can wonder how long such a celebratory fascination with digital media is going to last. Art created solely as an exhibition of new technology tends to briefly fascinate but is often quickly surpassed by the next technological innovation. Outside the commercial world of advertising, it is difficult to imagine a future where the medium has become the message and where the appreciation of art will primarily rest on its ability to incorporate innovative technology. Art is a deeply human endeavor after all. And whether approached from an aesthetic, conceptual, or experiential point of view, the power inherent to art lies in its ability to make us go beyond its own physical or digital limitations and become aware of what is going on beyond the surface of things. It is when art invites us to look closely that our mind finds space to roam and leaves behind the practicalities of the present, briefly transcending both medium and context. However fleeting, the poetry of that moment can be transformative in ways that are both deeply personal and expansively universal.

Installation view of "Solar Cycle 24"

Reference

Ota E (2008) Inaccurate coordinates. In: Proceedings of ISEA2008: the 14th international symposium of electronic art. http://isea-web.org/archives/pdfs/proceedings/ISEA2008.pdf. Accessed 21 Nov 2013

Index

F.T. Marchese (eds.), *Media Art and the Urban Environment*, Future City 5,
DOI 10.1007/978-3-319-15153-3

MIX
Papier aus verantwortungsvollen Quellen
Paper from responsible sources
FSC® C105338

If you have any concerns about our products,
you can contact us on
ProductSafety@springernature.com

In case Publisher is established outside the EU,
the EU authorized representative is:
Springer Nature Customer Service Center GmbH
Europaplatz 3, 69115 Heidelberg, Germany

Printed by Libri Plureos GmbH
in Hamburg, Germany